Variorum Revised Editions and Reprints:

JOSEPH R. STRAYER
The Royal Domain in the Baillage of Rouen

JOHN WILLIS CLARK
The Care of Books. An Essay on the Development of Libraries and their
Fittings, from the Earliest Times to the End of the 18th Century
Cambridge 1902 definitive edition

In the Collected Studies Series:

WALTER ULLMANN
The Church and the Law in the Earlier Middle Ages

WALTER ULLMANN
The Papacy and Political Ideas in the Middle Ages

WALTER ULLMANN
Scholarship and Politics in the Middle Ages

BERNHARD BLUMENKRANZ
Juifs et Chrétiens — Patristique et Moyen-Age

PAUL MEYVAERT
Benedict, Gregory, Bede and Others

EDMOND-RENE LABANDE
Spiritualité et vie littéraire de l'Occident. Xe-XIVe s.

GILES CONSTABLE
Religious Life and Thought (11th-12th Centuries)

MICHEL MOLLAT
Etudes sur l'économie et la société de l'Occident médiéval (XIIe-XIVe s.)

DAVID HERLIHY
The Social History of Italy and Western Europe, 700-1500

PAUL J. ALEXANDER
Religious and Political History and Thought in the Byzantine Empire

HENRY MONNIER
Etudes de droit byzantin

MILTON V. ANASTOS
Studies in Byzantine Intellectual History

Studies in Medieval Physics
and Mathematics

Professor Marshall Clagett

Marshall Clagett

Studies in Medieval Physics
and Mathematics

VARIORUM REPRINTS
London 1979

British Library CIP data   Clagett, Marshall
　　　　　　　　　　　　　　Studies in medieval physics and mathematics.
　　　　　　　　　　　　　　— (Collected studies series; CS103).
　　　　　　　　　　　　　　1. Mathematics — History
　　　　　　　　　　　　　　2. Physics — History　　3. Science, Medieval
　　　　　　　　　　　　　　I. Title　　II. Series
　　　　　　　　　　　　　　510'.9'02　　　　　　QA23

　　　　　　　　　　　　　　ISBN 0-86078-048-1

Published in Great Britain by　　Variorum Reprints
　　　　　　　　　　　　　　　　21a Pembridge Mews London W11 3EQ

Printed in Great Britain by　　Kingprint Ltd
　　　　　　　　　　　　　　　Richmond Surrey TW9 4PD

　　　　　　　　　　　　　VARIORUM REPRINT CS103

CONTENTS

This volume contains a total of 366 pages

PREFACE

The articles I have chosen for inclusion in this volume represent the two main interests of a research career that extends over forty years: medieval physics and medieval mathematics. They also illustrate my concern with the transmission of scientific learning from Greek antiquity to the Latin West. And finally they demonstrate my preoccupation with publishing texts that remain obscured in manuscript.

My initial interest in medieval physics was kindled by a seminar in medieval learning conducted at Columbia University by my master Lynn Thorndike in the spring of 1939, when I decided to investigate the unpublished works of the Italian physician and natural philosopher Giovanni Marliani (d. 1483). That investigation culminated in a dissertation that was published at New York in 1941 under the title: *Giovanni Marliani and Late Medieval Physics*. After an absence of five years from the academic scene during World War II, I renewed my interest in medieval physics. Paper No. I ("Some General Aspects of Medieval Physics") attempted a broad survey of the limits and concerns of physics in the Middle Ages. I would no doubt present a considerably different survey were I to attempt it today. Still that early article unwittingly presaged my future research interests. At that time I was engaged in a detailed examination of a long and complex work, the *Liber calculationum* of Richard Swineshead (fl. ca. 1340–55), an extraordinarily subtle schoolman of Merton College, Oxford, who applied quantitative procedures to some of the principal problems of natural philosophy. The first fruit of that investigation was Paper No. III ("Richard Swineshead and Late Medieval Physics"). It treats of the first tract of the *Liber calculationum* and stresses Swineshead's conceptions of quantitative functions and infinitesimal analysis. Swineshead's work has drawn the attention of a number of scholars and the reader will find an excellent treatment of it with copious references to recent literature in the article on Swineshead by John E. Murdoch and Edith D. Sylla in the *Dictionary of Scientific Biography*, Vol. XIII, New York, 1976, pp.184–213. My own interest in medieval mechanics continued with the publications of three books: (1) *The Medieval Science of Weights (Scientia de ponderibus)*, ed. with Ernest A. Moody, Madison, 1952; repr. 1960; (2) *The Science of Mechanics in the Middle Ages*, Madison, 1959; repr. 1961; new Italian ed., Milan, 1972; (3) *Nicole Oresme and the Medieval Geometry of Qualities and Motions*, Madison, 1968. Paper No. II ("Some Novel Trends in the Science of the Fourteenth Century") is a summary article that reflects the research of these volumes and more specifically of the research of the last two volumes. In it stress is placed upon the various efforts to quantify physics in the Late Middle Ages. Paper No. VI ("Francesco of Ferrara's *Questio de proportionibus motuum*") presents the complete edition of an Italian schoolman's paraphrase and commentary on Thomas of Bradwardine's *Tractatus de proportionibus*. It thus gives the full text of a work that had received only passing attention in my *The Science of Mechanics*. Similarly Paper No. IV ("The Pre-Galilean Configuration Doctrine") reflects the subject of my volume on Oresme by presenting the text of an anonymous treatise on the configuration doctrine that was composed in the middle of the fourteenth century. No. V ("The Use of Points in Medieval Natural Philosophy") is a short paper that attempts to throw light on the ontology of the geometrical concepts that were employed by Oresme in pursuing his efforts at quantification, while Paper No. VII ("Leonardo da Vinci: Mechanics") examines in detail the ties that bind Leonardo's notebooks to medieval mechanics and at the same time stresses those areas of mechanics in which Leonardo produced novel conceptions. I hardly need say that a great deal of research has been done on medieval mechanics in the last fifteen years or so, and the reader may find appropriate references to the literature in various articles of the *Dictionary of Scientific Biography* (e.g., in the articles on Blasius of Parma, Thomas of Bradwardine, Jean Buridan, John Dumbleton, Gerard of Brussels, Jordanus de Nemore, Nicole Oresme, and other schoolmen of the thirteenth and fourteenth centuries). Indeed I first thought I would include my long article on Oresme from the *Dictionary*, and perhaps a similar article ("Nicole Oresme and Medieval Scientific Thought") from the *Proceedings of the American Philosophical Society*, Vol. 108, 1964, pp.298–309. But I decided against their inclusion since the essential points made in those articles were also made in Paper No. II. Still the reader may consult these articles for bibliographical purposes. The reader will also find useful the bibliographical essay appended by Edward Grant to his *Physical Science in the Middle Ages*, New York, 1971, the notes and introductions to the documents in Grant's splendid *A Source Book in Medieval Science*, Cambridge, Mass., 1974, and the notes cited in the appropriate articles in *Science in the Middle Ages*, ed. D. C. Lindberg, Chicago, 1978. Students of Oresme will also wish to consult the various texts of his works that were begun in my seminar in medieval mechanics. Two of these have been published, namely those of Edward Grant: *Nicole Oresme: "De proportionibus proportionum"* and *"Ad pauca respicientes"*, Madison, 1966, and *Nicole Oresme and the Kinematics of Circular Motion: Tractatus de commensurabilitate vel incommensurabilitate motuum celi*, Madison, 1971. Three others remain as doctoral dissertations: Claudia Kren, "The *Questiones super de celo* of Nicole Oresme", University of Wisconsin, 1965; Garret Droppers, "The *Questiones de spera* of Nicole Oresme", University of Wisconsin, 1966; Bert Hansen, "Nicole Oresme and the Marvels of Nature: A Critical Edition of his *Quodlibeta* with English Translation", Princeton University, 1973 (to be published soon by Brill). Further literature on Oresme will be found in the soon-to-be-completed dissertation of Peter Marshall under the direction of Professor James John at Cornell University: "Nicole Oresme, *Questiones de anima*". See also the edition prepared by Albert J. Menut and Alexander J. Denomy with English translation by Menut: Nicole Oresme, *Le Livre du ciel et du monde*, Madison, 1968. Finally on the subject of medieval mechanics and its literature, I should cite the exhaustive treatment of medieval statics found in Joseph E. Brown's dissertation, "The *Scientia de ponderibus* in the Later Middle Ages", University of Wisconsin, 1967, and Brown's recent more general article appearing in Lindberg's *Science in the Middle*

Ages noted above.

The second area of research embraced in the papers of this volume, namely medieval mathematics, caught my attention later than the first, but since the publication of Paper No. VIII ("The Medieval Latin Translations from the Arabic of the *Elements* of Euclid") in 1953 my interest in medieval mathematics has continued to develop. In Paper No. VIII I attempted to order the complex manuscript materials that bore on the translations of the *Elements.* While the classification of translations established in this article has in general persisted in the investigations of this subject, John Murdoch has brilliantly extended it in two articles: "The Medieval Euclid: Salient Aspects of the Translations of the *Elements* by Adelard of Bath and Campanus of Novara", XIIe Congrès International d'Histoire des Sciences, Colloques, in *Revue de synthèse*, Vol. 89 (1968), pp.67—94, and "Euclid: The Transmission of the Elements", *Dictionary of Scientific Biography*, Vol. IV, New York, 1971, pp.437—59. The reader will also wish to examine the edition of H. L. L. Busard: *The Translation of the Elements of Euclid from the Arabic into Latin by Hermann of Carinthia (?) [Books I—VII]*, Leiden, 1968, *Books VI—XII*, Amsterdam, 1977. Also pertinent to this subject are two editions begun in my seminar: Sister Mary St. Martin Van Ryzin, O.S.F., "The Arabic-Latin Tradition of Euclid's *Elements* in the Twelfth Century", University of Wisconsin, 1960, and Thomas J. Cunningham, "Book V of Euclid's *Elements* in the Twelfth Century: The Arabic-Latin Traditions", University of Wisconsin, 1972.

Papers Nos. IX—XII of this volume are all related to the study of Archimedes in the Middle Ages. The ninth and tenth present and discuss two small texts which are not by Archimedes but fit into a treatment of his Latin tradition. Paper No. XI ("Archimedes and Scholastic Geometry") examines the nature of scholastic additions made to the Archimedean texts and particularly to the many Latin versions of his *De mensura circuli.* Paper No. XII ("Archimedes in the Late Middle Ages") is a summary article of the Latin Archimedean traditions. The reader will find the subjects embraced by these last articles treated in great detail in my *Archimedes in the Middle Ages*, Vol. I, Madison,

1964; Vol.ş 2–3, Philadelphia, 1976–78, Vol. 4 (in press).

The final paper (No. XIII: "The Works of Francesco Maurolico") falls chronologically outside of the limits of the volume, since it is concerned with the establishment of the canon of the works of the great sixteenth-century mathematician of Messina. But I have included it because of the great interest shown by Maurolico in medieval mathematics and astronomy and because it complements the long treatment of Maurolico in Vol. 3 of my *Archimedes*. I further hope by its inclusion to interest scholars in the analysis and publication of the many works of Maurolico that appear only in manuscript. Such I believe would be a rewarding venture.

MARSHALL CLAGETT

The Institute for Advanced Study
Princeton, N.J.
March, 1979

MEDIEVAL PHYSICS

I

Some General Aspects of Physics in the Middle Ages*

THOSE working in the field of medieval science use the term "physics" in at least two different senses. It is used in its modern meaning to cover the activity of the medieval schoolmen in those special branches which are considered a part of physics today: heat, magnetism, optics, mechanics, etc. It is also used in a medieval sense as the science of nature, or natural philosophy, which by the thirteenth century is equivalent to its Peripatetic definition, namely, "the study of the material world in so far as it is carried in the stream of change, *motus*." [2]

---

* ABBREVIATIONS OF WORKS FRE-QUENTLY CITED IN FOOTNOTES

ACLC = Al-Fārābī, *Kitāb iḥṣā al-ʿulūm*, Arabic edition from Escorial manuscript by Angel Gonzalez Palencia, *Alfarabi Catalogo de las ciencias*, Madrid, 1932. Includes also two medieval Latin and one modern Spanish translation.

AOS = Al-Fārābī, *De ortu scientiarum*, edited by C. Baeumker in *Beiträge zur Geschichte der Philosophie des Mittelalters. Texte und Untersuchungen*, Band 19, Heft 3, Münster, 1916.

CLMP = M. Clagett, *Giovanni Marliani and Late Medieval Physics*, New York, 1941.

DHP = P. Duhem, "Physics-History of" in *Catholic Encyclopedia*.

DOS = P. Duhem, *Les origines de la statique*, vol. 1, Paris, 1905.

DSM = P. Duhem, *Le système du monde*, vols. 1–5, Paris, 1913–17.

GDP = Domingo Gundisalvo, *De divisione philosophie*, edited by L. Baur in *Beiträge zur Geschichte der Philosophie des Mittelalters. Texte und Untersuchungen*, Band 4, Heft 2–3, Münster, 1903. This includes a long essay on medieval classification theory by Baur.

GGSM = M. Grabmann, *Geschichte der scholastische Methode*, vol. 2, Freiberg, 1911.

HSVD = Hugo de Sancto Victore, *Didascalicon de studio legendi*, edited by C. H. Buttimer, Washington, D.C., 1939.

LAL = G. Lacombe, *Aristoteles Latinus*, Rome, 1939.

SIHS = G. Sarton, *Introduction to the History of Science*, vol. 1–2, Baltimore, 1927–1931.

THME = L. Thorndike, *History of Magic and Experimental Science*, vols. 1–6, New York, 1923–1941.

---

[1] The nucleus of this paper was contained in an address delivered to the Medieval Club of New York at Columbia University, April 1947.

[2] M. de Wulf, *Philosophy and Civilization in the Middle Ages*, Princeton, 1922, p. 91.

Clearer understanding of the general province of physics in the middle ages can be obtained by examining those works current in the twelfth and thirteenth centuries which treat of the classification of the sciences. By the twelfth century important modifications in, and extensions of, the seven liberal arts were taking place in the direction of increasing complexity and specialization. These changes were reflected in the divisions of philosophy followed in treatises of schoolmen of the twelfth century. The pedagogical and classificatory handbook of Hugh of Saint Victor (*d.* 1141) entitled *Didascalicon de studio legendi* was the most important and influential of those treatises using exclusively Latin sources. In his division of philosophy Hugh follows a modified Boethian-Aristotelian tradition, dividing philosophy into theory, practice, mechanics, and logic.[3] In discussing the sciences included in these main divisions Hugh gives attention to the pseudo-historical origin of each of the sciences. Particularly interesting is his belief that before the sciences became formal arts with rules and precepts, they had developed in a rudimentary and customary way by usage: [4]

Earlier there were common discourses and literature, but the doctrine of discourses and literature had not yet been made into an art. Thus far no precepts of speaking correctly or disputing had been given, for all of the sciences were in the state of use rather than in that of art. But then men thought that "use" could be converted to "art" and that what had been vague and unrestrained before could be restricted by certain rules and precepts. And so they began, as it has been said, to reduce to an art custom which had grown up partly by accident and in part naturally.

Of the four main divisions of knowledge we are concerned here with theory, which is subdivided into theology, mathematics, and physics. As was customary in these treatises, mathematics included the *quadrivium* of arithmetic, music, astronomy, and geometry. In discussing physics, Hugh relies on Boethius, but he does not bring into prominence the Aristotelian point of view of physics as the study of change and motion exemplified in matter: [5]

Physics investigates the causes of things in their effects and their effects from their causes.

Whence come tremblings of the earth, the force to make deep seas swell,
The powers of herbs, the passions and angers of wild animals,
All kinds of shrubs, stones, and also reptiles.[6]

Physis is translated "nature" (*natura*). Whence

Boethius has named natural physics in the upper division of theoretical philosophy. It is also called "physiology," i.e., the discourse treating of natural things, because the latter investigates the same cause. Sometimes physics is sweepingly accepted as of an importance equal to theoretical philosophy. According to this viewpoint certain philosophers divide philosophy into three parts, i.e., physics, ethics, and logic. Mechanics is not contained in this division. . . .'[7]

In another chapter,[8] where Hugh treats of the distinctive properties or provinces of each art, the distinctive property of physics is revealed as "to attend in an unmixed manner to things which are actually mixed, for the actually existing bodies of the world are not pure bodies but are rather compounded out of the acts of pure bodies." And Hugh adds that "it ought not be passed over that physics alone properly treats of *things*. All the others (e.g., mathematics and metaphysics) treat of the *intellectual comprehensions* of things. Logic and mathematics are prior in the order of learning to physics." Thus, Hugh explains, "anyone ought to be informed with regard to them before he undertakes physical speculation." Hugh concludes this discussion with high praise for "reason" vis à vis "experience." "It was necessary that logic and mathematics, therefore, put their consideration not in the actual state of things, where experience (*experimentum*) [9] is deceitful, but in reason alone, where unshaken truth

[3] *HSVD*, Bk 3, Chap. 1, p. 48.
[4] *HSVD*, Bk 1, Chap. 11, p. 21.
[5] *HSVD*, Bk 2, Chap. 16, pp. 34–35.
[6] The first line is from Vergil, *Georgics*, 2, 479. The source of the other two lines is unknown to me.
[7] Reference here is to the Platonic type of classification where physics is considered one of the principal divisions of philosophy.
[8] *HSVD*, Bk 2, Chap. 17, p. 36.
[9] In much of medieval literature *experimentum* is used interchangeably with *experientia*. *Experimentum* in the early works rarely seems to mean "controlled testing" as it does today and

dwells. Then with reason provided for, they could descend to the experience of things. . . ."

Less important than, but similar to, Hugh's classificatory remarks are the anonymous passages from Bamberger and Munich manuscripts of the twelfth century published by Martin Grabmann.[10] In one of them we are told on the authority of Boethius that study should proceed from the practical to the speculative. In the course of this ascent we arrive at speculative philosophy. There we first study mathematics and then pass up to the study of nature, "for physics is comprised in the nature of things." Of some interest also are two schematic diagrams of the classification of the sciences that appear in Munich manuscripts.[11] Both follow the so-called Platonic tradition of Cassiodorus and Saint Isidore of Seville, dividing philosophy into physics, ethics, and logic. And both follow Isidore in subordinating the *quadrivium* and thus mathematics to physics. One adds physiology to physics; while the other adds mechanics and medicine. Lest there be confusion in the term "mechanics," it should be noted that it is used in other places much like our own expression, the "mechanical" arts. The medieval treatises speak of it as the "adulterine" science, the science of ministering those things which are necessary because of the weakness of the body. Hugh says that: [12]

Mechanics contains seven sciences: cloth manufacture, armament making, navigation, agriculture, hunting, medicine, and the theatrical science. Of these, three pertain to the extrinsic clothing of nature by which nature protects itself from discomforts and four to the intrinsic things by which nature sustains itself with nourishment and healing. This division is indeed similar to the trivium and quadrivium. . . .

Pursuing the Latin tradition one step further, we can see in the commentary of Radulfus de Longo Campo to the *Anticlaudianus* of Alanus de Insulis in 1216 a further development of the province of physics.[13] He first notes that physics "treats of the invisible causes and (their relationship to) the nature of things." Later he says that physics concerns "the nature of all things which are sublunar and even of some things which are above the moon." Piecing together his various comments, we find that physics includes three subdivisions: elementary, terrestrial, and celestial. The elementary and celestial receive no further division, but under terrestrial physics is included medicine which has two subclassifications, inferior theory and inferior practice. This subclassification of particular sciences into theoretical and practical was common enough, particularly in the Arabic tradition, and it is clear that this author was influenced by that tradition.

To this point we have examined the place of physics in the classifications of the science which developed out of the liberal arts and the cathedral schools, i.e., those largely influenced by Latin authors: Augustine, Cassiodorus, Boethius, Isidore, *et al.* We must now turn our attention to another more important current of classificatory writings, which flows from scattered Greek headwaters through broad Islamic beds to twelfth century Europe. One of the most influential of the Islamic works in this latter current was the work of al-Fārābī (*ca.* 870–*ca.* 950 A.D.) entitled *Enumeration of the Sciences*.[14] It was twice translated during the course of the twelfth century, the second time in masterful fashion by Gerard of Cremona, the most distinguished of the translators at Toledo in the second half of the twelfth century. Having first treated of language and logic at some length, al-Fārābī then takes up mathematics, which he states has seven principal parts, to each of which he devotes a chapter. It is of interest to the subject of this paper to note that mathematics embraces among its parts

at least did occasionally in the thirteenth and fourteenth centuries.
[10] *GGSM*, pp. 31–48.
[11] *GGSM*, p. 44.
[12] *HSVD*, Bk 2, Chap. 20, pp. 38–39.
[13] *GGSM*, pp. 48–49.
[14] My translations are from the Arabic text in *ACLC*; occasionally I have followed the Cairo MS where the reading seemed better.

the science of aspects (including perspective and the science of mirrors), the science of weights, and a science of mechanics (or more properly, a science of mathematical devices). A translation of the principal parts of the chapter devoted to optics (i.e., aspects) will throw light on the reasons for its inclusion in mathematics: [15]

The science of aspects treats of the same things as geometry: figures, magnitudes, positions, orders, equality, and inequality, and similar things. But it does so not according as they are in lines, surfaces, and bodies in an abstract way; while geometry does just that.[16] Hence, geometry is the more general.

It is necessary to single out the science of aspects, even if it embraces in toto the same things of which geometry treats, because many of the things which geometry necessarily studies, in that they have some condition of shape, position, order, etc., are changed when they are observed into something which is the opposite of what they really are. For instance, there are things which are in truth squares but, when looked at from afar, appear to be circular, and things which are concurrent and equal that appear to be unequal. . . . Furthermore some coplaner objects appear to be lower, and some higher. . . . Thus this science distinguishes between that which appears to the sight different than it really is and that which appears as it really is.

This art makes it possible for one to know the measurement of that which is far distant . . . for example, the height of tall trees and walls, the width of valleys and rivers, the height of mountains and the depth of valleys and rivers . . . then the distances of the celestial bodies and their measurements. . . .

Everything which is observed or seen is seen only through a ray which pierces the air and every transparent body, going from a point where it touches our sight up to a point where it falls on the object seen.[17] The rays which penetrate transparent bodies up to the object seen are direct, reflected, reverse (converse) or broken. Direct rays are those which upon leaving the sight are extended in the straight line of the sight until they arrive at the end (the object) and so are cut off. Reflected rays are those which upon extending their penetration from the sight are met in their way, before they can pass on to the end, by a mirror which deflects them from their straight line passage; and so they are reflected obliquely on one side of the mirror. . . . Reverse rays are those which return from the mirror in the same path over which they had originally travelled. (They travel this reverse route) until they fall on the body of the observer from whose sight they had gone out. . . . Broken rays are those which return from the mirror toward the observer from whose sight they had originally gone out, but they are extended obliquely to one side of him, and so they fall on something else which is behind him, on his right, or his left, or above him. . . .

The media between the sight, the thing seen, and the mirror are, in short, transparent bodies such as the air or water or some body composed artificially of glass or something of this kind. Some of the mirrors which reflect rays and prevent them from following their course are made artificially of iron or some other material. Some are hard wood which has been moistened. Water and other bodies also (act like a mirror). . . .

Thus the science of aspects inquires into everything which is seen or observed by means of these four rays in each one of the mirrors, and it investigates everything which pertains to the object viewed. It is divided into two parts. The first of these investigates what is seen by direct rays; while the second inquires into what is observed with non-direct rays. The latter is particularly ascribed to the science of mirrors.

The science of weights, included in present-day statics, was considered as mathematics because it was the study of the *basic principles* of the balance, lever, and other lifting machines, that clearly involved geometrical principles such as geometrical symmetry, similar triangles, etc. Al-Fārābī does not give much detail to this science, devoting only one short paragraph to it: [18]

The science of weights is concerned with matters of weights in two fashions: (1) according as it investigates weights from the point of view of their being measured or something being measured with them, i.e., the investigation of the fundamental principles of the discourse on the balance; or (2) according as it investigates weights which are moved or something being moved with them, i.e., the investigation of the basic principles of instruments with which heavy things are lifted and upon which they are carried from place to place.

---

[15] *ACLC*, Arabic text, p. 36.

[16] This is a confused passage. Palencia prefers the Cairo reading (*ACLC*, Arabic text, p. 71) in his translation. I have followed the Escorial MS (*ACLC*, Arabic text, p. 36) as making better sense.

[17] This is the theory of light that Greek tradition associates with the Pythagoreans.

[18] *ACLC*, Arabic text, p. 43.

Some General Aspects of Physics in the Middle Ages

From mathematics al-Fārābī passes in his *Enumeration of the Sciences* to physics. Here, the dependence on Aristotle is predominant. He shows an acceptance of the Aristotelian concepts of matter and form, and the parts of physics for al-Fārābī are the Aristotelian and pseudo-Aristotelian physical works current in his day. He begins by noting that "natural science (physics) investigates natural bodies and the accidents which are in these bodies (i.e. the sensible attributes of the bodies)." [19] After some preliminary definitions and examples, he states the concepts of matter and form thus: [20]

The being and accidents of any natural body whatsoever depend on two things: (1) that thing resident in it which serves the same purpose as the sharpness of the sword, i.e., the "form" of that natural body; and (2) that thing resident in it which is like the iron of the sword, i.e., the "matter" of the natural body, the substratum. It is like the carrier of the form (of artificial bodies) except that the form and the matter of the sword, the bed, the clothing, and the remaining artificial bodies are verified by the sight and the senses. . . . The forms of qualities and the matters of natural bodies are not sensibles. We are certain of their existence only by syllogism and apodictic demonstration.

Al-Fārābī also alludes to and accepts the Aristotelian doctrine of the four causes: material, formal, efficient, and final: [21]

The science of physics makes known natural bodies in that it establishes everything in them which is "sensible" and everything in them which is "not sensible." It teaches with regard to every natural body, its matter, form, efficient cause, and the final cause for the sake of which that body exists. It does the same thing with regard to the accidents of the body. . . . This science, then, gives the principles of natural bodies and the principles of their accidents.

Physics, then, teaches of these four causes in bodies. These causes were commonly called in the middle ages the "principles" of bodies.

The section on physics in the *Enumeration of the Sciences* is concluded by a discussion of the eight major parts of physics, each part being contained in a work or a pseudo-work of Aristotle: [22]

The physical science is divided into eight major parts: (1) the first treats of that which is common to all natural bodies — simple bodies as well as those compounded of elements. All of this part is included in the book entitled *The Course of Physics* (i.e., the *De naturali auditu* or *Physics* of Aristotle).

(2) This part is occupied with whether simple bodies exist, and if so, what kinds of bodies they are, and how many of them there are. It is, then, the study of the world, what it is, what kind and how many are its prime parts and whether they are three or five in number. It is also the study of the heavens and their distinction from the rest of the world, establishing that the matter of the heavens is only one. All of this is treated in the first book of the *Book of the Heavens and Earth* (of Aristotle). It then examines the elements of the compound bodies, whether they exist in the simples whose existence has been demonstrated, or are distinct bodies derived from them. . . .

(3) This part is concerned in general with the generation (i.e. the coming into being) of natural bodies and their corruption (i.e., their passing away), as well as with the things peculiar to them. It studies how the elements are engendered, how they are corrupted, and how compound bodies are afterwards engendered from them. And it is concerned with the principles of all that which is contained in the *Book of Generation and Corruption* (of Aristotle).

(4) The fourth part treats of the principles of the accidents of elements and the peculiar effects of those elements, (considered) individually and (thus) excluding those bodies which are compounded of them. This matter is treated in the first three books of *The Superior Impressions* (i.e., the *Meteorology* of Aristotle).

(5) The fifth part is occupied with the study of bodies compounded of elements, as follows: Some of these bodies are of similar parts; others are of dissimilar parts. Those of similar parts are also of two classes. There are some whose parts are composed of dissimilar parts, like meat and bone. The others are those which have no part but to serve as a basis for a natural body of dissimilar parts, for example, salt, gold, silver. . . . All of this figures in the fourth book of *The Superior Impressions*.

(6) This part is contained in the *Book on Minerals*.[23] It treats of bodies compounded of similar parts which are not composed of dissimilar

---

[19] *ACLC*, Arabic text, p. 45.
[20] *ACLC*, Arabic text, p. 46.

[21] *ACLC*, Arabic text, p. 48.
[22] *ACLC*, Arabic text, pp. 48–50.

34

parts. These are mineral bodies, stones, and the various species of minerals and stones. It considers that which is peculiar to each of their species.

(7) The seventh part is contained in the *Book on Plants*.[23] It is concerned with those things in which the species of plants share and with those things which are peculiar to each of their species. This part is the first of two parts of the study of bodies compounded of dissimilar parts.

(8) The eighth part is contained in the *Book of Animals* and in the *Book of the Soul* (both of Aristotle). It studies that which the diverse species of animals share and that which is peculiar to each of their species. It is the second part of the study of those things compounded of dissimilar parts. . . .

The name of al-Fārābī is also associated with a short work entitled *De ortu scientiarum* which is extant in Latin of the twelfth century only, the Arabic original, if any, not being known. It treats from a logical rather than a historical point the rise or origin of the sciences. Physics is presented as the science of change in substance. The author expresses this idea by saying that "because substance sometimes takes on a natural color and sometimes pales, sometimes is lengthened and at times shortened, sometimes is increased and sometimes diminished, sometimes generated and sometimes corrupted, sometimes becomes infirm and at times well, therefore it was necessary to have a science which would demonstrate this whole matter, evidently one by which we would arrive at knowledge of how any kind of permutation takes place. . . ."[24] Passing on to the parts of physics, the author tells us that "according to what older savants have said" they are eight: the science of judgments, the science of medicine, the science of nigromancy according to physics, the science of images, the science of agriculture, the science of navigation, the science of alchemy, and the science of mirrors.[25] This is certainly a motley collection of sciences and pseudo-sciences that hardly seems to coincide with viewpoints expressed in the mature *Enumeration of the Sciences*. We might well cast doubts on its attribution to al-Fārābī.[26] The author himself seems impressed with the range of subjects included in physics, for he suggests separating out the science of judgments (astrology) and medicine.[27]

The principal heir of the Arabic tradition of classification was Domingo Gundisalvo, the Toledan translator and philosopher, who about 1150 composed the most important classificatory work of the twelfth century, *De divisione philosophie*.[28] While Gundisalvo is deeply influenced by al-Fārābī, he shows considerable catholicity of taste and some originality. The Latin authors are often quoted, with perhaps Isidore being the most frequently cited. This treatise of Gundisalvo is more highly organized than any of its predecessors. He maintains the traditional Aristotelean divisions of theoretical and practical philosophy with the division of theoretical philosophy into physics, mathematics, and metaphysics.[29] Then for each science Domingo includes its definition, genus, material, species, parts, utility, aim, instrument, who uses it, its etymology, and in what order it should be studied. At the outset he tells us that "natural science is the science considering only things unabstracted and with motion." We are told that its "genus" is that it is the first part of philosophy. It is first with respect to us, "for we apprehend matter simultaneously with form by the senses earlier than we apprehend form without matter by the intellect." The material with which natural science deals, i.e., its "matter," is body. But it does not treat body according to its being (*esse*) and it does not seek the answer to the question of what its substance is,

[23] The works *On Minerals* and *On Plants* are Pseudo-Aristotelian. The former is an Avicenna mélange; the latter is probably the work of Nicholas of Damascus. (See *LAL*, pp. 91–92).

[24] *AOS*, p. 20.

[25] *AOS*, p. 20.

[26] Cf. H. G. Farmer, *Al-Fārābī's Arabic-Latin Writings on Music*, Glasgow, 1934, pp. 51–52.

[27] *AOS*, p. 20.

[28] Edition in *GDP*.

[29] *AOS*, p. 19; succeeding quotations are consecutive from pp. 19–28.

I

nor what is composed out of the two principles of matter and form. The author indicates that there are other sciences which take up bodies from these points of view. Physics considers bodies only "according as they are subjected to motion or change," a true Aristotelian point of view, and one which will dominate the schools of the thirteenth and fourteenth centuries.

In delineating further the province of natural science, Domingo establishes the distinction between "universal" sciences and "particular" sciences. "Those are spoken of as universal which contain many other sciences under them." Physics, then, is universal since it contains eight sciences under it. These eight sciences are the same as those listed in the small treatise *On the Rise of the Sciences*, the contents of which have been analysed above. One present-day historian of medieval philosophy (Wulf) sees great importance in the distinction found between universal and particular sciences in the thirteenth century, the latter being based on the collection of sensory data and supplying the materials for the more general objectives of the former.[30] There is, however, much evidence to show that many of the so-called special or particular sciences were studied *after* the general sciences, and there is little indication that the conclusions of the general sciences were strictly based on those of the various particular sciences.[31] While there was much important experimental work in the particular sciences of magnetism, optics, statics done in the thirteenth century, as I shall indicate briefly below, it does not appear to have altered the conclusions of physics as a universal science, the point of view of which in the thirteenth century was almost completely Aristotelian. This is less true of the fourteenth century where a combination of speculation and observation produced serious breaches in the Aristotelian generalizations, particularly with regard to dynamics (see section III below).

With regard to the "parts" of physics Domingo quotes almost verbatim the section from al-Fārābī's *Enumeration of the Sciences*, associating the parts of physics with the works of Aristotle. He is also dependent on al-Fārābī for his discussion of the four causes as principles (*principia*) of bodies: "And indeed the matters and forms of bodies, and their agents and ends through which they exist are called the principles of bodies, and if they relate to the accidents of bodies, they are called the principles of the accidents which are in the bodies."

Treating the remaining aspects of physics which he had promised to discuss in the introduction, the Spanish philosopher tells us that the "aim" of natural science is the "cognition of natural bodies." "The 'instrument' of this science is dialectical syllogism." So far as the "artificer" of this science is concerned, he is "the natural philosopher who proceeding rationally from causes to effects and from effects to causes seeks out principles." Finally, we are told that "it ought to be read and learned after logic."

As in the treatise of al-Fārābī, optics, weights, and mathematical devices are included among mathematics. In fact, the discussion of these sciences is taken in toto from the abbreviated translation of al-Fārābī's *Enumeration*, known to the Latins under the title *De scientiis*.

᭑᭥

The two main currents of thought on the division of knowledge, Latin and Islamic, fuse in the thirteenth century works, particularly in that of Robert Kilwardby, *De ortu et divisione philosophie*. Kilwardby was an English Dominican who taught at Paris and Oxford, became Archbishop of Canterbury, and died in 1279. After immediately distinguishing speculative and practical philosophy, he divides the speculative into natural science (physics), mathematics, and metaphysics.[32] He notes the pre-

---

[30] M. de Wulf, *History of Medieval Philosophy*, vol. 1 (1925), p. 271. Cf. his *Philosophy and Civilization in Middle Ages*, pp. 85–91.

[31] The University curricula do not bear Wulf out. See, for example, L. Thorndike, *University Records and Life in the Middle Ages*, New York,

1944, pp. 64–65, 279–282.

[32] I have used what appears to be a summary of part of Kilwardby's work in a manuscript owned by the New York Academy of Medicine (no. 6 in the de Ricci *Census*); also summaries in *GDP*, pp. 368–380 and *THME*, 2, pp. 81–82.

occupation of physics with movable and material things, and he makes the divisions of physics follow the Aristotelian and Pseudo-Aristotelian works.[33] He is concerned, as Domingo had been to some extent before him, with establishing the relation of the mathematical disciplines to physics. He emphasizes that while physics treats continuous quantities in their actual increase, decrease, and division in bodies, mathematics treats them only as abstract quantities, a point that both al-Fārābī and Domingo Gundisalvo had raised before him. He suggests that one can raise questions as to whether astronomy and perspective (optics) belong to natural science or geometry.[34] A similar realization of the possible dual classification of the science of weights is made in a thirteenth century commentary on weights contained in a manuscript of the Bibliothèque Nationale: [35] "The science of weights is subordinated to both geometry and natural philosophy. It is then necessary in this science that certain propositions receive a geometric proof, others a philosophic proof."

Briefer treatments of the division of the sciences are found in many other thirteenth and fourteenth century authors, including Roger Bacon, Thomas Aquinas, Egidius Romanus, etc.[36] They follow the general lines laid down in the treatises we have already discussed. Bacon's treatment is interesting from the point of view of his distinction between what he calls "the things common to the natural sciences" and the special sciences.[37] The latter include perspective, astronomy (both judiciary and operative), the science of weights treating heavy and light bodies, alchemy, agriculture, medicine, and experimental science.[38] Much has been written on Bacon's advocation of experimentation,[39] although how much he himself practiced the experimental technique is questionable, and certainly the results obtained were not striking in their originality. His definition of the experimental science is of interest, a science "which induces full certitude out of the perfection of experience." [40]

II

Enough has been said of the place of physics in medieval thought. Before passing on to a review of some of the physical ideas, we can note briefly some of the principal works of Antiquity and Islam that passed over to the West in the twelfth and thirteenth centuries and served as the basis, or at least the point of departure, for much of the physical work of the high and late middle ages.

Most important for the study of physics as a science of motion are the following:

(1) Aristotle, *Physics* in four translations, the earliest about 1150 from the Greek.[41]

(2) Aristotle, *De caelo*, three translations, the earliest by Gerard of Cremona, in the second half of the twelfth century.[42]

(3) The commentaries of Averroes on the first works; translated after 1200, perhaps both by Micheal Scot.[43]

(4) The encyclopedic commentary of Avicenna entitled *Kitāb al-shifā'* (*Book of the Healing*, etc.), at least part of the physical sec-

tion of which was translated by John of Seville (?) and Domingo Gundisalvo about or before 1150.[44]

(5) Al-Fārābī, *Distinctio super librum de naturali auditu*, translated by Gerard of Cremona.[45]

(6) Alexander of Aphrodisias, *De Motu et tempore*, translated by Gerard of Cremona.[46]

(7) Simplicius, a commentary on the *De caelo* of Aristotle, translated by William of Moerbeke in the thirteenth century.[47]

There were also a number of works that were basic for the special sciences that we regard as a part of physics. A few of the most important are the following:

---

[33] New York Academy MS no. 6, f. 105 r; GDP, p. 370.

[34] GDP, pp. 370–371, 380.

[35] Cited in DOS, p. 131.

[36] GDP, pp. 376–397.

[37] R. Bacon, *Communia naturalium*, edited by R. Steele, in *Opera hactenus inedita*, fasc. 2, p. 3.

[38] *Ibid.*, p. 5.

[39] See particularly R. Carton, *L'expérience physique chez Roger Bacon etc.*, Paris, 1924 and

THME, 2, chap. 61, and their copious citations.

[40] Bacon, *op. cit.* in note 37, p. 9.

[41] LAL, p. 51.

[42] LAL, p. 53.

[43] LAL, pp. 104–105.

[44] SIHS, 2, pp. 171, 173.

[45] SIHS, 2, p. 340.

[46] SIHS, 2, p. 340.

[47] LAL, p. 98.

(1) The *Optics* ascribed to Ptolemy and translated by Eugene the Amīr in Sicily about 1154.[48]

(2) Diocles, *De Speculis comburentibus.*[49]

(3) The *Optics of Alhazen*, translated by Gerard of Cremona.[50]

(4) Al-Kindi, *De Aspectibus; de umbris et de diversitate aspectium.*[51]

(5) Aristotle, *Meteorology*, three books translated by Gerard of Cremona, the fourth book by William of Moerbeke; and other translations.[52]

(6) Alexander of Aphrodisias, commentaries on the *Meteorology.*[53]

(7) Archimedes, *Floating Bodies*, translated by William of Moerbeke in 1269.[54]

(8) Hero, *Catoptrics*, attributed to Ptolemy, and translated by William of Moerbeke.[55]

(9) Thābit ibn Qurra, *Liber charastonis*, (Book of the Roman Balance), translated by Gerard of Cremona.[56]

(10) Several fragments on weights attributed to Euclid.[57]

(11) The miscellaneous *Problemata* attributed to Aristotle, translated by Bartholomew of Messina in the thirteenth century and also earlier by an unknown translator.[58]

I have left out of this listing any reference to the numerous astronomical works that influenced physical ideas.

III

While the schoolmen of the thirteenth century were absorbing Aristotelian physics with only occasional departures therefrom (particularly in the sphere of cosmological thought where Ptolemaic ideas were beginning to emerge victorious in a struggle with Aristotelian theories),[59] there were significant developments in those special sciences which we note as a part of physics: optics, statics, and magnetism. One of the factors in this development is the growing use of the experimental technique. Although Bacon, Witelo, and Peckham (all of the thirteenth century) went little beyond Alkindi's and Alhazen's treatises in optics, it is worthy to note their repetition of earlier experiments and the conducting of some new ones, e.g., some new values for angles of refraction. Particularly promising was the theory of the rainbow posited by the Dominican, Dietrich of Freiberg (before 1311).[60] Dietrich by a number of experiments with glass balls filled with water to produce spectra similar to rainbows supported the theory that the rays making the bow visible are reflected on the inside of spherical drops of water, and "he traced with great accuracy the course of the rays which produce the (primary and secondary) rainbows respectively."[61] It is Dietrich, who, while respecting the authority of Aristotle, maintained that "according to that same philosopher (Aristotle) one ought never renounce what has been made manifest according to the senses."[62] The progress of this theory of the rainbow can be traced from the time of Dietrich to that of Descartes.[63]

The principal contributions to medieval statics are found in a series of manuscripts *On weights* that bear the name of Jordanus de Nemore, who may be identical with the Dominican master general, Jordanus Saxo, who succeeded to the generalship in 1222.[64] The eminent physicist Pierre Duhem has attributed this series to multiple authorship.[65] Later writers tend to ascribe all of the works on weights bearing his name to Jordanus himself, thus playing down what seemed to Duhem as internal contradictions within the treatises.[66] From this confusion of material, however, emerge several original propositions of interest to the historian of physics: (1) a proof of the law of the equilibrium of the lever by use of what is at least the germ of the principle of virtual

[48] *SIHS, 2*, p. 346.
[49] *SIHS, 2*, p. 342.
[50] *SIHS, 2*, p. 23.
[51] *SIHS, 2*, p. 342.
[52] *LAL*, pp. 56–57.
[53] *LAL*, p. 96.
[54] *SIHS, 2*, p. 764.
[55] *SIHS, 2*, p. 830.
[56] *DOS*, pp. 79–93.
[57] *DOS*, pp. 62–79.
[58] *LAL*, pp. 86–87.
[59] This struggle is outlined in detail in *DSM*, vol. 3.

[60] *De iride et radialibus impressionibus*, edited by J. Würschmidt in *Beiträge zur Geschichte der Philosophie des Mittelalters*, vol. 12 (1914), Heft 5 and 6, parts 2 and 3 particularly.
[61] *DHP*, p. 50.
[62] *De iride, edit. cit.*, p. 61.
[63] *DHP*, p. 50.
[64] *DOS*, pp. 98–108.
[65] *DOS*, p. 107; also Duhem's *Études sur Léonard da Vinci, 1*, p. 316.
[66] See particularly the citations and statements of B. Ginzburg, "Duhem and Jordanus Nemorarius" in *Isis, 25* (1936), pp. 345–346.

38

displacements,[67] as even Duhem's most determined critic admits;[68] (2) a demonstration of the equilibrium of the bent lever, again with the principle of virtual displacements at least implicit;[69] and (3) the concept of gravity according to position (*gravitas secundum situm*), i.e., the component of weight in the line of the trajectory or apparent weight on inclined planes. This concept is not stated with mathematical precision and is erroneously used in certain lever problems, but it is employed correctly in formulating (4) the law of the equilibrium of weights on inclined planes; and the proof given is very neat.[70]

So far as the study of magnetism is concerned we see its beginnings in the letter of Peter the Stranger of Maricourt *On the Magnet*. Written in 1269 by a man who was praised by Bacon as the only Latin to understand the experimental science,[71] it is a model of the observational and experimental technique in physics. It has no concern with supposed magical properties of magnets but simply describes their physical properties.[72] It shows how to determine the poles of lodestones and describes the action of the poles on each other, the effect of lodestones on iron, the fact that on breaking the magnet the two pieces form new magnets, etc. He also details the construction of compasses. It is not excelled until the publication of Gilbert's work on the magnet in the sixteenth century.

While the thirteenth century showed some excellent advances in the special sciences of physics and in the use of experiment, the fourteenth is fertile in the "general" science of physics, i.e., dynamics, the science of motion. This does not mean of course that development in the special sciences came to a halt. I need only mention the work of Blasius of Parma *De ponderibus*, which aside from including the earlier material on statics also has one book on hydrostatics.[73] The development in dynamics takes the form of modifications, corrections, and criticism of earlier Peripatetic ideas. It forms a background for the accelerated interest and activity in mechanics in the sixteenth century. Highlights in this development appear to me to be the following:

(1) Out of theological discussions of the thirteenth century as to the possibility through God's omnipotence of the earth's having motion, came discussions in the fourteenth century of the real rather than hypothetical possibility of the earth's motion. While Jean Buridan (one of the earliest and principal exponents of the new dynamics at the University of Paris, variously mentioned there from 1329–1358)[74] advanced, only to reject, the possibility of the diurnal rotation of the earth,[75] Nicholas Oresme, his disciple, later in the century may well have adopted such a movement.[76]

(2) At almost the same time a doctrine which posited that all bodies tend to unite their centers of gravity with the center of the earth was taken up by the Parisian doctors of the fourteenth century, Jean Buridan and Albert of Saxony. Jean and Albert added to it the belief that geological processes were constantly changing the center of gravity of the earth. Since the center of gravity of the earth after these geological changes would not then be at the center of gravity of the universe, the earth would be continually shifting its position so that these centers of gravity would coincide. Albert would explain the precession of the equinoxes by this shifting of the earth.[77]

---

[67] *DOS*, pp. 121–123.
[68] Ginzburg, *Op. cit.*, p. 353. He complains, however, the principle "is found surrounded with obscurities." This is fully admitted. One cannot expect later precision in thirteenth century material.
[69] *DOS*, pp. 141–144.
[70] *DOS*, pp. 144–147.
[71] R. Bacon, *Opus tertium*, edited by Brewer, London, 1859, chap. 13, p. 43.
[72] See *SIHS*, 2, p. 1032 for various editions and translations.

[73] *DOS*, pp. 147–155.
[74] *DSM*, 4, pp. 124–127.
[75] Jean Buridan, *Questiones de caelo et mundo*, edited by E. A. Moody, New York, 1942, Bk. 2, questio 22, pp. 226–233.
[76] Nicole Oresme, *Le livre du ciel et du monde*, edited by A. D. Menut and A. J. Denomy in *Medieval Studies*, vol. 4, (1942), Bk 2, chap. 25, 137d–144c, pp. 270–279.
[77] *DHP*, p. 51. Cf. Buridan, *Op. cit.*, pp. 231–232.

(3) The possibility of the existence of a plurality of worlds like our own was discussed at the end of the thirteenth and the beginning of the fourteenth centuries.[78] Strict Peripatetics thought such a plurality to be impossible on the grounds that everything tended to seek its natural place, the heavy bodies down and the light bodies up. Hence, any plurality of worlds must necessarily coalesce. Buridan mentions the theory that each of the several worlds would arrange itself according to the center of gravity of its own world, the heavy things toward the center, the light things further out. This theory Oresme adopted. The development of this doctrine could displace the theory of natural place by one which assigned position according to relative density. The doctrine of relative density was by no means a new concept with Galileo or his immediate predecessors since it had already appeared in the writings of a Jewish philosopher of the fourteenth century, Crescas (1340–1410), who was subject to Islamic influence.[79]

(4) At the beginning of the fourteenth century, almost concurrently at Oxford and Paris, there arose the quantitative analysis of forms and the application of this kind of analysis to local motion. This may well be the most important of the developments we are summarizing since it resulted by the middle of the century in a quantitative description of accelerated motion and the basic kinematic theorem for the expression of that motion, namely, that with respect to space described in a given time a uniformly accelerated motion can be represented by a uniform motion equivalent to its mean speed.[80] This theorem received an arithmetical or formally logical proof by the Englishmen: Richard Swineshead (before 1350), John of Dumbleton (fl. at Oxford between 1331–1349), and William Heytesbury (mentioned variously at Oxford from 1330 to 1371).[81] Oresme at Paris demonstrated this theorem graphically with singular neatness.[82] The Italians of the fifteenth century leaned more heavily on the English logicians, but they were not unaware of the coordinate system of Oresme.[83] It is this theorem with its geometric proof that Galileo applies to the free fall of bodies to derive his version of the law of free fall.[84]

(5) Attention was also given in the fourteenth century to the fact that falling bodies undergo some sort of acceleration. Buridan seems to imply that they have a uniform acceleration with respect to time.[85] Albert of Saxony mentions two possibilities: that the velocity of fall is proportional to the time of fall or to the distance of fall.[86] It was extremely important to decide this question, for once it was decided that the fall was uniformly accelerated with respect to time, then, by the application of the kinematic theorem described in the preceding paragraph, the law of free fall would follow. Albert made no such application of the theorem however. Nor was Galileo the first to do so, but rather a little-known Spaniard of the sixteenth century, Domingo Soto. Soto was trained at Paris and in 1545, while discussing the kinematic theorem, cited a falling body as an example of uniformly accelerated motion.[87] There is no evidence that he supported this declaration with any experimental efforts such as the use of an inclined plane by Galileo.[88] It should be added in passing that neither Stevin nor Galileo was the first to observe the incorrectness of the so-called Peripatetic law on motion: that velocity is proportional to the motive force (weight in the case of falling bodies) and inversely proportional to the resistance (the medium),

[78] *DHP*, p. 51.

[79] H. A. Wolfson, *Crescas' Critique of Aristotle*, Cambridge, U.S.A., 1929, pp. 238–239. The doctrine is implicit in ideas of Plato and Democritus, as well as in certain Islamic authors. Cf. S. Pines, "Quelques tendances antipéripatéticiennes de la pensée scientifique islamique" in *Thales*, vol. 4 (1940), pp. 216–217.

[80] See *CLMP*, chap. 5 for various proofs of this theorem.

[81] *CLMP*, pp. 40, 102, 105 (note 8), 119–123.

[82] *CLMP*, p. 120 (note 25).

[83] *CLMP*, pp. 120–121 (note 25).

[84] Galileo Galilei, *Two New Sciences*, translated by Henry Crew and Alfonso de Salvio, New York, 1933, pp. 173–174.

[85] See the quotation from Buridan translated in the body of the article above note 96.

[86] *DHP*, p. 51.

[87] *DHP*, p. 56; P. Duhem, *Études sur Léonard da Vinci*, 3, Paris, 1913, pp. 555–562.

[88] Galileo, *Op. cit., edit. cit.*, pp. 178–179.

40

by resorting to the experiment of dropping weights. John Philoponus in his sixth century commentary on the *Physics* says: [89]

"According to Aristotle if the medium through which the motion takes place be the same, but the moving bodies differ in weight, their times must be proportional to their respective weights. . . . but that is wholly false, as can be shown by experience more clearly than by logical demonstration. For if you let two bodies of very different weights fall simultaneously from the same height, you will observe that the ratio of motion does not follow their proportional weights, but there will be only a slight difference in time, so that if their difference in weight be not very great, but one body were, say, twice as heavy as the other, the times will not perceptibly differ."

(6) One of the main points of criticism of Aristotelian dynamics current in the fourteenth century centered in Aristotle's explanation of projectile motion.[90] Aristotle described motion as arising from two forces, motive power and resistance, both of which must be continually present for motion to continue. There can be no movement in a void because there would be no resistance. Acceleration results when either the resistance is decreased or the motive force is increased. "Natural motion" occurs when bodies seek their natural places, the heavy bodies down and the light bodies up. "Violent motion" takes place when a motivating force moves a body contrary to its natural motion, i.e. away from its natural place. Ordinarily projectile motion is violent motion. In violent motion continued substantial contact must be maintained between the mover and the moving body. However, when a projectile is thrown the initial mover obviously does not continue to have direct contact with the moving body. Continued contact of the mover and moved, according to the Peripatetic position, is to be found in the action of the ambient air as an intermediate contacting agent, i.e., the motor imparts motive force to the air which then acts in some way as the continuator of the motion of the projectile. This explanation is set aside by numerous references to experience as early as the sixth century by John Philoponus, if indeed it had not already been abandoned earlier. He posits instead that an incorporeal kinetic power is impressed in the projectile by the projecting agent.[91] This impressed power continues the motion of the projectile after there is no longer contact with the projector until finally it is overcome by the tendency of the body toward its natural place. This theory is the germ of the *impetus* theory of the schoolmen of the fourteenth century. It passed to the Arabs (particularly Alpetragius and Avicenna) [92] and from them to the West, where it was criticised by Bacon, Aquinas, and others in the thirteenth century, and possibly accepted in a peculiar teleological form by Peter John Olivi [93] (*d.* 1298). It matured in the minds of the fourteenth century schoolmen, particularly with Jean Buridan, who emphasized that *impetus* varied directly with the quantity of first matter of the moving body and the velocity imparted to the body. He described it in short much as Newton described quantity of motion (momentum), namely the product of mass and velocity. In addition, Buridan suggested that impetus would last indefinitely if it were not diminished by resistance. A translation of the pertinent passages I believe will be of some interest (the italics are my own):

Therefore, it seems to me it should be said that a motor in moving a body impresses in it a certain impetus or a certain motive power capable of moving this body in the direction in which the mover moves it, either up or down or laterally or circularly. *By the same amount that the motor moves the same body more swiftly, by that same amount is more powerful the im-*

[89] J. Philoponus, *In physicorum libros quinque posteriores commentaria*, Prussian Academy edition of Vitteli, vol. 17, p. 683; translation of W. A. Heidel, *Heroic Age of Science*, Baltimore, 1933, p. 187. The translation is slightly abbreviated and would be more explicit if it read: ". . . their times must be [inversely] proportional to their respective weights."

[90] For the following narrative see particularly: Aristotle, *Physica*, 215a, 255b, 266b–267a; *De*

caelo, 288a, 300a, 301b.
[91] J. Philoponus, *Op. cit.*, edit. cit., p. 642.
[92] S. Pines, "Études sur Awḥad al-Zamân Abu'l Barakât al-Baghdâdi" in *Revue des études juives*, new series, 3 (1938), pp. 3–64, vol. 4 (1938), pp. 1–33.
[93] B. Jansen, "Olivi, der älteste scholastische Vertreter des heutigen Bewegungsbegriffs" in *Philosophisches Jahrbuch der Görresgesellschaft*, 33 (1920), pp. 137–152.

petus which is impressed in it. And it is from this impetus that the stone is moved after that which hurls it ceases to move; but on account of the resisting air and the gravity of the stone which inclines it contrary to the way which the impetus has power to move it, the impetus is continually weakened. Therefore, the movement of the stone will become continually slower. At length this impetus is so diminished or destroyed that the gravity of the stone prevails over it and moves the stone down to its natural place. . . . I can throw a stone farther than I can a feather, and iron or lead fitted to hand farther than a piece of wood of the same size. I say that the cause of this is that the reception of all forms and natural dispositions is in matter and by reason of matter. Therefore, the greater quantity of matter there is the more that body can receive of this impetus and the more intensely can it receive it. *Now in a dense and heavy body there is more of first*

matter *than in a rare and light one. Therefore, a dense and heavy body receives more of the impetus and does so more intensely. . . .* A feather, moreover, receives such a weak impetus that immediately it is destroyed by the resistance of the air. *And so likewise if one moves equally swiftly by hurling a light piece of wood and a heavy piece of iron of the same shape and volume, the iron will move farther because there is impressed in it a more intense impetus. . . .*[94]

Many posit that the projectile after leaving the projector is moved by an impetus given by the projector, and it is moved as long as the impetus remains stronger than the resistance, *and the impetus would last indefinitely if it were not diminished by a resisting contrary,* or by an inclination to a contrary motion; and in celestial bodies there is no resisting contrary. . . ."[95]

Buridan also applied the impetus theory to the explanation of the acceleration of falling bodies, rejecting the explanation that the proximity to the natural place of the body accounted for the acceleration. He says:[96]

This seems to me to be the reason why the natural motion of a heavy body downward is continually accelerated, for at first the gravity alone moved the body and, therefore, it moved more slowly. But in moving it impressed in

that body an impetus. This impetus indeed together with its gravity moves that body. Therefore, a faster movement takes place. By the amount that it is made more rapid, so is made the impetus more intense.

Duhem and Michalski have noticed many advocates of the impetus theory in the fourteenth, fifteenth, and sixteenth centuries.[97] I wish to note only three whose works they have not analysed, Roger Bacon [98] in the thirteenth, Gerard Oddo in the fourteenth, and Thomas Bricot in the fifteenth century. Bacon seems to have been the first of the important schoolmen in the West to have discussed the theory in its more primitive form as *virtus* theory. Like Aquinas he rejects it.[99] In the seventh book of his *Questions on the Physics* [100] he refutes the continuation of the virtue or power of the projector in the projectile because there would be no substantial conjunction of the mover and the moved, and such conjunction is necessary for projectile movement. With Aristotle he relies on the air to continue the motion.

That Gerard Oddo, Franciscan general, 1329–1342, accepted some such virtue theory we have on the testimony of Johannes Canonicus in his *Questions on the Physics*.[101] We are told that Gerard opposed Aristotle in discussing the possibility of motion in a void, declaring that there could be both natural and violent motion in *vacuo*. As to the question of where the source of motion of the projectile is, he answered that the projectile is moved "naturally" by the *virtus* of the projector. The

[94] Translated from Latin passage printed by E. J. Dijksterhuis, *Val en Worp etc.,* Groningen, 1924, p. 72.
[95] *Ibid.,* p. 73.
[96] *Ibid.,* p. 73.
[97] P. Duhem, *Études sur Léonard da Vinci 3,* Paris, 1913, *passim;* K. Michalski, "Les courants philosophiques à Oxford et à Paris pendant le XIVe siècle," in *Bulletin international de l'Académie polonaise des sciences et des lettres. Class de philologie. Classe d'histoire et de philosophie,* les années 1919, 1920. Cracovie, 1922–1924, pp. 59–88; "Le criticisme et le scepticisme dans la philosophie du XIVe siècle" *Ibid.,* l'année 1925, Cracovie, 1927, pp. 41–122; "La physique nou-

velle et les différents courants philosophiques au XIVe siècle," *Ibid.,* l'année 1927, Cracovie, 1928, pp. 93–164.
[98] I understand that Miss A. Maier in her *Die Impetustheorie der Scholastik,* Vienna, 1940 has made such an analysis.
[99] Thomas Aquinas, *In libros Aristotelis de caelo et mundo expositio,* Bk 3, chap. 2, lectio 7, Edition of the Dominicans by order of Leo XIII, vol. 3, Rome, 1886, p. 252.
[100] Edition of F. M. Delorme and R. Steele, Oxford, 1935, Bk 7, p. 338.
[101] Johannes Canonicus, *Super libros physicorum Aristotelis questiones,* Venice, 1481, Bk 3, quest. 5 (no pagination).

use of the term "naturally" seems contradictory in a discussion of violent motion. It is not explained further.

The final one of these three additional notes refers to Thomas Bricot, a fifteenth century Parisian advocate of the impetus theory, noted but not analysed by Michalski. Bricot describes the theory in his *Textus abbreviatus* of the *Physics*: [102] (a) Objections are raised to the Aristotelian theory of the role of the ambient air. He here repeats many of the same empirical objections raised by the fourteenth century schoolmen. He notes further that one can not move a candle flame by means of the air at a distance of twenty or thirty feet. Nor can the ambient theory account for the fact that you can not throw a bean as far as a ball or a stone of one pound. (b) The impetus he believes to be a secondary quality, a disposition (to motion), distinct from the moving body itself. Here he is following Buridan, Albert of Saxony, and others in rejecting a theory that seems to have grown up under the influence of William of Occam, the great English nominalist of the fourteenth century, who has said in one passage that "the moving thing (*movens*) in such a motion (i.e., projectile motion) after the separation of the moving body from the prime projector is the very thing moved according to itself (*ipsum motum secundum se*)." [103] (c) The impetus is also spoken of as an instrument which begins motion under the influence of a principal particular agent (the projector) but which continues it alone. Even though the continued motion of the projectile is from an intrinsic source (*principium instrinsecum*) when moved by an impetus, it does not follow that the motion is a natural one (ordinarily described as arising from an intrinsic source) because the source in projectile motion is violent and contrary to the nature of the projectile (which is to seek its natural place and fall to the earth). (d) He uses almost the same words as Buridan in explaining that the impetus is a motive power by which the projectile can be moved in the direction intended by the projector. (e) All impetus are essentially and specifically different, this difference arising from the diverse methods of impressing them and the diverse movements of the hand. He notes a distinction between the impetus moving a heavy body up and one moving the same body down. The former is dissipated by the gravity of the body, the latter assisted by it. Here we see him applying impetus to falling bodies as did his fourteenth century predecessors. All in all, although quite extended, Bricot's discussion is much inferior to those of Buridan and Oresme in the fourteenth century.

The evaluation of the dynamics of the fourteenth century and its importance for the development of modern mechanics in the sixteenth and seventeenth centuries is a subject of an increasingly large and contentious literature.[104] For my own part, I should like to summarize and review, in concluding this paper, what appear to be the most important dynamic and kinematic ideas of the fourteenth century which became a part of the stock of ideas of the sixteenth century. For the purpose of historical analysis these ideas can be divided into three groups: (A) Ideas incorporated directly into the newly developing mechanics with little change except in precision and terminology. (B) Ideas germinal to the new theories, but transformed through the application of a changed point of view or a new technique. (C) Ideas discarded and replaced in the formation of the new theories. The concluding summary follows:

---

[102] Thomas Bricot, *Textus abbreviatus Aristotelis supra octo libris physicorum*, etc., Paris, 1494, f. 101r, c. 2.
[103] William of Occam, *Questiones et earumdemque descisiones super quattuor libros sententiarum*, Lyon, 1495, quest. 26, part. m.
[104] See works cited in *CLMP*, p. 125 (note 1). In addition, we should note A. Koyré, *Etudes Galiléennes*, Paris, 1939 and A. Maier, *Die Impetustheorie der Scholastik*, Vienna, 1940.

A. CONCEPTS EMPLOYED DIRECTLY

1

Concept: Motion can be uniformly difform with respect to time (i.e., uniformly accelerated).

Comment: The kinematic study of accelerated motion begins at least in the first half of the fourteenth century at Paris and Oxford. The definition of uniformly accelerated motion given by Galileo is more precise and the terminology is considerably altered, but there can be no doubt he is the direct heir of the schoolmen in the study of accelerated motion.

2

Concept: With respect to the space transcribed in a given time the latitude of motion uniformly difform is equivalent to its mean degree.

Comment: We have already noted above that this is the basic kinematic formula which states that a body moving with a uniformly accelerated velocity will transcribe as much space in a given time as one moving with a uniform velocity equal to the mean of the final and initial velocities of the uniformly accelerated body. Used directly in the sixteenth century by Soto and in the seventeenth by Galileo, it was joined by both of them with the first of the possibilities presented in concept 3 below to obtain their versions of the law of free fall.

3

Concept: The velocity of a falling body is proportional either to the *time* of fall or the *distance* of fall.

Comment: These are the two possibilities presented by Albert of Saxony in the fourteenth century. Leonardo da Vinci decides correctly for the first of them, declaring the velocity to be proportional to the time of fall. Da Vinci also understood that this applied to fall on an inclined plane. Galileo repeats the two possibilities, demonstrating the correct one.

B. MODIFIED GERMINAL CONCEPTS

4

Concept: The continuation of projectile motion results from the impression in the projectile by the projector of an *inclinatio, virtus, impetus, redeur, forza,* or *impeto* (all of these words employed).

Comment: The impetus concept appears to have been employed in developing ideas of quantity of motion or momentum, however far away it was from the inertial concept. This is borne out by the concepts indicated immediately below.

5

Concept: The intensity of the impressed impetus is directly proportional to the quantity of first matter in the moving body and to the speed imparted to the body.

Comment: This concept emerges in the seventeenth century as the mathematical definition of the quantity of motion of Descartes and Newton, i.e., momentum, the product of mass and velocity.

6

Concept: If not diminished by the resistance of the air or the gravity of the body, this impetus imparted to the body will continue indefinitely.

Comment: In this statement of Buridan we find the closest approach of the schoolman of the fourteenth century to the idea of inertia. At that, William of Occam's identification of the source of movement of the body with the body itself appears to be closer to the Newtonian concept. If we substituted momentum for impetus in the sixth concept, it would be correct.

7

Concept: The continued swing of a pendulum as well as the phenomena observed in the impact and rebound of a ball are accounted for by the impetus theory.

Comment: Buridan's explanation of these phenomena by means of the impetus theory presents interesting similarities to their explanation in the seventeenth century by means of momentum.

8

Concept: A body appears to fall in a straight line in spite of the suggested diurnal rotation of the earth because it shares the horizontal motion of the rotating earth.

Comment: This idea was advanced by Nicholas Oresme to meet the criticism levelled at the theory of diurnal rotation of the earth by Ptolemy, namely, that a body thrown into the air would not fall to the earth at the point of projection but rather at some point behind it. He seems to imply that the rotating earth imparts an horizontal impetus to the falling body. Bruno in the sixteenth century makes this idea explicit. It is important in developing a directional analysis of impetus. From this concept it becomes increasingly evident that the gravity of the falling body is acting vertically at the same time that an impetus imparted to the body by the rotating earth propels the body horizontally, thus producing an apparent straight-line fall. Oresme had already analysed the motion of fall as a composite motion relative to something at rest with respect to the falling body and the rotating earth. A further significant step was taken by Benedetti in the sixteenth century toward a directional analysis of impetus when he proposed that a body whirled in a circle will, upon being set free, be impelled by the imparted impetus to move off in a straight line tangent to the circle at the point of release.

9

Concept: In the case of falling bodies a natural continual acceleration is produced because the gravity is continually acting on the body to impress in it more and more impetus. Thus the acceleration is produced by the gravity and the continually increasing impetus imparted to the body by the gravity. The velocity is proportional to the impressed impetus.

Comment: In this concept of Buridan we see the germinal idea of the later association of an acceleration with a continually acting force. It has implicit in it that gravity is continually introducing impetus so that the impetus is increasing proportionally to the time elapsed. The logical deduction from this concept is that the velocity too is increasing proportionally to the time, i.e., the first alternative of concept 3.

C. CONCEPTS LATER COMPLETELY DISCARDED

10

Concept: The proportion of velocities in motions follows the proportion of the motivating force to the resistance.

Comment: This is the so-called Peripatetic law of motion. It had great currency in the middle ages, both in the form here stated and in a somewhat more subtle form elaborated in the fourteenth century. It was probably suggested to Aristotle by the action of a balance. But when applied to the free fall of bodies, it is obviously contrary to experience, as John Philoponus in the sixth century pointed out. He noted that the time of fall of different weight bodies appeared to be about the same. The law was criticized severely in the fifteenth century by Giovanni Marliani and in the sixteenth by Benedetti. The work of Stevin and Galileo made final disposal of it.

11

Concept: After the projectile leaves the projector a period of initial acceleration occurs.

Comment: Although not held by Buridan or Oresme, this idea of Aristotle had considerable support in the fourteenth century. It remained to plague many of the sixteenth century authors. It is absent from Galileo's discussion of projectile motion.

University of Wisconsin

II

SOME NOVEL TRENDS IN THE SCIENCE OF THE FOURTEENTH CENTURY

t seems evident that the announced topics of the two papers that Mr. Drake and I have prepared represent the upper and lower limits of Renaissance science. In fact, while much of what I have to say has some bearing on science in the fifteenth and sixteenth centuries, not all of the trends here singled out as being novel in fourteenth-century speculation were actually of much substantial influence in the succeeding centuries, even though they do represent a changing tone by bringing cosmological theory closer to the rather more impersonal approach of terrestrial physics. Let me say further that what I have to say in the second half of the paper concentrates on efforts at quantification in the fourteenth century since out of these efforts arise some concepts that are rather more significant to the later Scientific Revolution.

Even the most cursory glance at the rich manuscript holdings in European libraries that bear on fourteenth-century science will convince the student of this period that, aside from the extraordinary increase in the volume of scientific material, there is a perfect explosion of philosophical speculation beyond the confines of Aristotelian philosophy. In part, this may be associated with the check put by the Condemnations of 1277 at Paris to considering the Aristotelian philosophy, with its unique cosmos, as the only possible natural explanation of the way things are. At least this is the theory held by Pierre Duhem in his celebrated studies completed some two generations ago.[1] And while this thesis has not been tested in any extensive way (and does not seem of great significance in some of the the cases discussed by Duhem) so that we cannot affirm it to be the principal cause of the varied speculation of the fourteenth century, it may have played some role in certain cosmological speculations like those connected with the possibility of the existence of a plurality of worlds,[2] certainly not a new topic but one handled with considerable skill and novelty by the schoolmen of

---

[1] P. Duhem, *Études sur Léonard de Vinci*, Vol. 3 (Paris, 1913; new printing, 1955), vii. See *Le Système du monde*, Vol. 6 (Paris, 1954), in general; see particularly p. 80: ". . . l'Université parisienne ne cessa de maintenir avec fermeté les condamnations portées par Étienne Tempier; en fait, le décret de 1277 resta, pendant toute la durée du xive siècle, une code révéré des maîtres parisiens; nous en aurons maintes fois la preuve."
[2] A. Koyré was highly skeptical of the significance of the Condemnations on the develop-

the fourteenth century. For example, let us look at the manner in which Nicole Oresme, whose ideas will provide us with some of our exemplary material, takes up this problem. Before turning to Oresme's bold speculations, let me remark on one general device that catches on in scientific speculation in the fourteenth century in an all-pervasive manner (in fact, enough so to constitute a trend in itself); this is the device of the *ymaginatio*, i.e., a kind of thought experiment or imaginative scheme constructed as if it existed in nature, with the purpose of illustrating basic theoretical ideas. Such schemes were usually not thought to be possible (and indeed I would say that many were clearly invented with the understanding that certain features of the real world were to be set aside). The main point to emphasize in connection with the *ymaginatio* is that it was the chief instrument of the speculative science of the fourteenth century, and this is particularly true of the schoolmen at the University of Paris. One such *ymaginatio* is the detailed picture of a possible plurality of worlds as visualized by Oresme. The account I wish to quote is from his *Livre du ciel et du monde*, written in 1377 at the end of his brilliant career:

All heavy things of this world tend to be conjoined in one mass (*masse*) such that the center of gravity (*centre de la pesanteur*) of this mass is in the center of this world, and the whole constitutes a single body in number. And consequently they all have one [natural] place according to number. And if a part of the [element] earth of the other world was in this world, it would tend towards the center of this world and be conjoined to its mass. . . . But it does not accordingly follow that the parts of the [element] earth or heavy things of the other world (if it exists) tend to the center of this world, for in their world they would make a mass which would be a single body according to number, and which would have a single place according to number, and which would be ordered according to high and low [in respect to its own center] just as is the mass of heavy things in this world. . . . I conclude then that God can and would be able by His omnipotence (*par toute sa puissance*) to make another world other than this one, or several of them either similar or dissimilar, and Aristotle offers no sufficient proof to the contrary. But as it was said before, in fact (*de fait*) there never was nor will there be any but a single corporeal world. . . .[3]

I perhaps should comment in passing that this passage has some importance in reflecting the growing interest in the use of the concept of

ment of scientific thought, see "Le Vide et l'espace infini au XIVᵉ siècle," in *Études d'histoire de la pensée philosophique* (Paris, 1961), pp. 33–41.

[3] M. Clagett, *The Science of Mechanics in the Middle Ages* (Madison, 1959), pp. 592–93. For the original text, see Nicole Oresme, *Le Livre du ciel et du monde*, Bk. I, Chap. 24 (ed. A. D. Menut and A. J. Denomy in *Mediaeval Studies*, Vol. 3 [1941], 243–44).

center of gravity in large bodies, an interest which in its mature development in the seventeenth century was to be of considerable importance to the marriage in astronomy of force physics and kinematics.

While the plurality of worlds was clearly rejected by Oresme, certainly that idea was to grow in the Renaissance, where, joined with Copernican astronomy and the substitution of Euclidian infinite space for the finite cosmos, it was to have some success with Bruno and others.

A less successful novel speculation concerned with the wider field of the supposed structure of this one world, but like it, a speculation that seemed to shake the Aristotelian dichotomy of two physics—a physics of the heavens and that of the earth—was that concerned with the celestial movers themselves. A standard view in the Middle Ages, and this was equally true of the philosophers of the fourteenth century, was that the movers of the heavens were the intelligences, moving the heavenly bodies as they willed. Hence, forces like these natural forces of the terrestrial realm were not involved. This was spelled out in great detail on several occasions by Oresme. For example, listen to him distinguish between natural and voluntary force in his *Questiones de spera:*

I posit three distinctions. . . . The first distinction is that certain force is natural and certain voluntary and free, so that I distinguish natural in opposition to voluntary. And they differ in two ways: (1) Natural force moves with some exertion or effort as some horse strives and exerts itself to draw a cart, but the force which is a pure, voluntary one does not move with some effort but by volition alone, so that these forces are of differing natures. (2) The second difference is that natural force is designatable by numbers and ratios. . . . But the force which is will is not designatable by numbers or ratios. Therefore, one such voluntary force ought not be described as twice or triple another. And accordingly we say that the first kind (i.e., natural force) is a certain quality which is receptive to increase and decrease. But the second (i.e., voluntary force) is pure substance itself which is not susceptible to comparison [by ratios]. The second distinction concerns resistance, which can be understood in two ways. The first and proper way is as a certain difficulty or quality contrary to some force. Spoken of in this first way, the velocity of motion increases or decreases according to its (i.e., the resistance's) decrease or increase. . . . In the second and improper way, resistance is nothing but a certain impossibility of moving faster or slower. And thus it can be said that a heaven resists being moved faster. The third distinction is that sometimes a certain effect or motion arises from a natural force, such is the motion of a stone downward, while sometimes a motion arises primarily from a force which is an intelligence . . . , such is the motion of a heaven. . . . [Hence,] I say that between an intelligence and a [celestial] orb there is no ratio, but only something analogous, in a way similar to that of things [here] below. For we see that a certain velocity arises from a definite ratio, and the cause is that every ratio, if it is rational, is signifiable by numbers. But an intelligence cannot be signified by numbers.[4]

The point to underline in this passage is that the action of intelligences is not quantifiable in terms of proportionality theorems and no force mechanics seems allowable. Hence, this speculative effort seems far removed from early modern mechanics. But there were speculations on the part of some natural philosophers that throw some doubt on the role of intelligences and seem to offer an opening wedge for force physics. One of the most interesting is that of Oresme's supposed master, Jean Buridan. He makes a suggestion that impetuses (like the impetuses found in this world that move projectiles after the initial force introducing the impetus is no longer in contact with the body) move the heavenly bodies. He was led to his suggestion by the idea that such an impetus tends to permanence, although in this terrestrial realm impetus is never permanent because of the continual presence of resistance and contrary force, which destroy the impetus. But if such impetuses were implanted in heavenly bodies where there is no resistance and no contrary force, then there is no reason why those bodies would not continue to move indefinitely with a uniform speed. Hence, since impetus was ranked by Buridan as a force, he is clearly, in this one instance, at least suggesting the possibility that the motion of heavenly bodies is explicable by forces akin to natural forces rather than by intelligences. Listen to Buridan:

. . . it does not appear necessary to posit intelligences of this kind [as celestial movers], because it could be said that when He created the world God moved each of the celestial orbs as He pleased, and in moving them impressed in them impetuses which moved them without His having to move them any more. . . . And these impetuses which He impressed in the celestial bodies were not decreased or corrupted afterwards because there was no inclination of the celestial bodies for other movements. Nor was there resistance which would be corruptive or repressive of that impetus.[5]

And in another place, Buridan specifically ties the kind of impetus present in the terrestrial realm to the suggested celestial impetuses:

Many posit that the projectile after leaving the projector is moved by an impetus given [it] by the projector, and it is moved so long as the impetus remains stronger

---

[4] I have published the Latin text of this passage in my "Nicole Oresme and Medieval Scientific Thought," *Proceedings of the American Philosophical Society*, Vol. 108 (1964), 300, n. 20.
[5] Clagett, *The Science of Mechanics*, pp 524–25. (I have here changed the translation slightly.)

than the resistance. And it would last indefinitely *(in infinitum)* if it were not diminished or corrupted by a resisting contrary or by a contrary motion; and in celestial motion there is no resisting contrary. . . .[6]

But Buridan's disciple, Oresme, in this case clearly would not go along with the master, primarily because he viewed impetus in a different way. For him it was expendable as it acted, and thus it could not be conceived of as tending toward permanence.[7] Furthermore, he seemed to connect impetus with acceleration. Hence, he considered it inapplicable to the uniform motion of the heavenly bodies. As we have already seen, he tended to accept the intelligences as movers. But he was never one for complete consistency; he liked to follow the argument where it led him, and so we see that he too made suggestions in the *Livre du ciel et du monde* that seemed to contradict or modify conclusions stated elsewhere in that and other works. Two of his suggestions are of particular interest. In the first he says:

And, according to truth, an intelligence is simply immobile, and it is not logical that it exists throughout the heaven which it moves nor that it exists in a part of such a heaven, it having been posited that the heavens are moved by intelligences. For, perhaps, when God created them (i.e., the heavens), He placed in them motive qualities and powers, just as He placed weight into terrestrial things, and He put in them resistances [to] counter the motive powers. And these powers and resistances are of a nature and a matter differing from any sensible thing or quality which exists here below. And these powers are so moderated, tempered and accorded to their opposing resistances that the movements take place without violence. And with violence excepted, it is somewhat like a man making a [mechanical] clock and letting it go and be moved by itself. And so God let the heavens be moved continually according to the ratios that the motive powers have to the resistances and according to the established order.[8]

It is clear that here Oresme's suggestion is completely contradictory to what he has previously said about the inapplicability of ratios and numbers to the heavenly movers. A further passage elaborates and repeats this suggestion of implanted powers and resistances (if not impetuses):

It is not impossible . . . that the heaven be moved by an inherent corporeal quality or power [and that it be moved] without violence and without work *(travail)*, for the resistance which is in the heavens inclines it to no other move-

[6] *Ibid.*, p. 524, n. 38.
[7] *Ibid.*, p. 552.
[8] *Le Livre du ciel et du monde*, Bk. II, Chap. 2 *(ed. cit. Mediaeval Studies*, Vol. 4 [1942], 170).

ment nor rest but only so that it does not move more rapidly [or, we could add, more slowly].[9]

In other words, Oresme has joined to his suggestion of implanted forces and resistances the idea of improper resistance, spelled out in our earlier quotation. While one might be inclined to see some inertial adumbration in Oresme's suggestion, I think it must be observed that it had very little, if any, direct influence on early modern authors so far as inertial ideas are concerned, although it certainly was one of several efforts to make celestial motion open to force analysis.

It can be pointed out further that Oresme seemed to feel no unease in transferring the basic proportionality theorem describing the relationship of velocity and force as present in natural motions to the heavens in a work entitled *Ratios of Ratios*[10] in spite of his oft-repeated assertion that the intelligences as celestial movers are not susceptible to treatment by ratios and numbers. If challenged, he would perhaps have answered that, although he believed intelligences to be movers, this is not certain and it might well be argued (as in fact he did in the previous quotations) that there are analogous (but essentially different) implanted forces and resistances in the celestial bodies, which forces and resistances might be related to the velocities of celestial motions by the same basic formula embracing movements arising from natural forces. At any rate, this tendency to take positions that seem contrary to most of the tenets of natural philosophy that he supports gives a strongly tentative and speculative tone to his natural philosophy that certainly appears to distinguish his approach from that of Galileo or Copernicus. In fact, the more one reads of his rather extensive bibliography, the surer he becomes that Oresme has been significantly influenced by the skeptical tendencies so evident in the philosophy of the first half of the century.[11] Further, in his *Quodlibeta* we hear him twice say that when it comes to natural knowledge (as distinguished from the true knowledge of faith), the only thing he knows is that he knows nothing.[12] It would appear to me, then, that such skeptical tendencies nurtured the speculative approach

[9] *Ibid.*, Chap. 3, p. 175.
[10] See the new edition by E. Grant: Nicole Oresme, *The "De proportionibus proportionum" and "Ad pauca respicientes"* (Madison, 1966).
[11] The classic work on the skepticism of the fourteenth century is that of K. Michalski, "Le Criticisme et le scepticisme dans le philosophie du XIVe siècle," *Bulletin international de L'Académie Polonaise des Sciences et des Lettres. Classe de philologie. Classe d'histoire et de philosophie*, L'Année 1925, Part I (1926), pp. 41–122; Part II (1927), pp. 192–242.
[12] *Quodlibeta*, MS Paris, BN lat. 15126, 98v: "Ideo quidem nichil scio nisi quia scio me nichil scire." See also 118v: "Et quamvis multis appareant faciles, mihi tamen difficiles videntur. Ideo nichil scio nisi quia me nichil scire scio."

that would just as well consider (and even support) one view as another. But lest I oversell this idea and leave you with the impression that Oresme everywhere introduces and supports opposing views of the basic problems of natural philosophy, may I say that, whether skeptical or not in some instances, Oresme was highly rational in his approach to natural phenomena and he often did take determinate positions. He did, as far as was possible for a believing Christian of the fourteenth century, seek only natural causes for the events of this world. In this connection, he was highly critical of the tendency on the part of many to seek supernatural causes for remarkable natural events. In the *Contra divinatores* which precedes the *Quodlibeta* we read: "From the said arguments it is evident that the diversity and plurality of effects here below arises by reason of matter and passive and immediate causes rather than from superior things."[13] And further, he says: "Every diversity here below can be saved by natural dispositions. Therefore, such particular [celestial] influences are posited in vain."[14] And still further:

Whence certain people attribute unknown causes to God immediately, certain to the heavens, and certain to the Devil. But an expert ought never to draw a conclusion from pure ignorance . . . and there is great doubt on the part of any expert as to whether demons exist.[15]

In the *Quodlibeta* attached to the *Contra divinatores*, Oresme repeats this theme of explaining terrestrial effects without recourse to the heavens, to God, or to demons.[16] In one place he highlights the difficulty of proving from natural effects the existence of demons in a question entitled "Whether Demons exist naturally." He admits that they do exist, but only because faith says so and, in fact, their existence cannot be proved naturally: "In regard to the question, I say that, speaking truly, demons exist, as must be believed from faith. But I assert that this cannot be proved naturally. . . . If

[13] *Ibid.*, 5ᵛ: "Ex dictis rationibus patet quod diversitas et pluralitas effectuum hic inferius plus provenit ratione materie et passivarum et causarum immediatarum magis quam superiorum."
[14] *Ibid.*, 7ᵛ: ". . . et dispositionibus naturalibus potest salvari omnis diversitas hic inferius. Igitur frustra ponuntur tales influentie particulares. . . ."
[15] *Ibid.*, 33ᵛ: "Unde quidam causas ignotas attribuunt Deo immediate, quidam celo, quidam dyabolo. Homo autem peritus nunquam ex pura ignorantia debet concludere seu incurrere maiorem nec ex parvo inconsequenti maius concedere, nunquam tibi, et cuicumque perito est magis dubium demones esse. . . ."
[16] *Ibid.*, 60ᵛ: "Ad aliud, ut superius dixi, illi qui nesciunt causas immediatas et naturales fugiunt ad demones, alii ad celum, alii ad deum, et quia talia videntur mirabilia, ideo attribuunt etc., sed hoc est falsum etc." See 80ʳ, 95ʳ, and *passim*. Incidentally, on fol. 96ᵛ he indicates that it seems to him "to believe easily is and was the cause of the destruction of natural philosophy" (". . . videtur mihi quod faciliter credere est et fuit causa destructionis philosophie naturalis").

the faith were not to pose their existence, I would say that they cannot be proved to exist from any natural effect, because all things [supposedly arising from them] can be saved naturally. . . ."[17] Although I have stressed the sharpness of the attacks of Nicole Oresme on the justice of seeking celestial influences on natural events, certainly he was not alone in his attacks, as is apparent in the equally strong attack of his younger associate Henry of Hesse (see Chapter Two, Part C, of my edition of Oresme's *De configurationibus qualitatum et motuum*); and there are no doubt others. And, in fact, Oresme's attack in the *Quodlibeta* on the concept of celestial influences was summarized in the fifteenth century (about 1478) by one Claudius Coelestinus in a work entitled *De his quae mundo mirabiliter eveniunt*, a work that was later published in 1542 by the well-known mathematician Oronce Finé. As Thorndike has pointed out (*Osiris*, Vol. 1 [1936], 631): " . . . it must have been largely through Oronce Finé's printing of the summary of Coelestinus that the views of Oresme continued to exert influence in early modern times." Indeed, this summary was translated into French at Lyon in 1557.

Before leaving our brief consideration of heavenly motions, three things ought to be remarked. The first is that formal astronomy with its Ptolemaic models and general kinematic results as inherited from antiquity continued in the fourteenth century. Hence, the *Almagest* of Ptolemy as well as the popular works of Arabic astronomy such as those of al-Farghānī and al-Battānī were studied, as were two of the best-known medieval works. These were both entitled *Theorica planetarum*: the more elementary one attributed to Gerard of Cremona (probably falsely) and the other to Campanus of Novara, the celebrated commentator on the *Elements* of Euclid. The first was subjected to commentary in the fourteenth century. And in one such commentary by an Oxford schoolman, Walter Brytte, we find the kinematic ideas of Oxford which were developed in connection with qualitative analysis, joined with the purely Greek astronomy.[18] One should further point out that there was also a sprightly development in trigonometry in the same circle, as evidenced in the works of John Maudith, Richard Wallingford, and Simon Bredon, and in France with Levi ben Gerson.[19] And, finally, there was considerable activity in the preparation of astronomical tables, much of which remains to be investigated.

---

[17] *Ibid.*, 127^v: "Ad propositum dico quod veraciter loquendo demones sunt, sicut credendum est ex fide. Sed dico quod non potest probari naturaliter, ut dixi. . . . Nisi autem fides poneret eos esse dicerem quod ex nullo effectu possent probari esse, quia naturaliter omnes possunt salvari. . . ."

[18] O. Pedersen, "The Theorica planetarum—Literature of the Middle Ages," *Classica et Mediaevalia*, Vol. 23 (1962), 228.

[19] G. Sarton, *Introduction to the History of Science*, Vol. 3 (Baltimore, 1947), 598–602, 660–61, 662–68, 673.

I am rather more interested in efforts at quantification that arose in the treatment of qualities and terrestrial motions; and we have now moved down from the heavens. I hardly need point out that quantification has a basic twofold aspect: the first is in the development of a language of quantity that can be applied to phenomena; the second is in the development of theories and techniques of measurement to produce numerical values. In its latter aspect of quantification as measurement, the fourteenth-century scientists contributed almost nothing (except perhaps in the above mentioned preparation of astronomical tables and the increasing development of practical handbooks of mensuration). In the former, of theoretical quantification, the contributions were of considerable importance, and in one way or another will occupy the remainder of this paper.

One of the initial points to mention in connection with the development of a language of quantity as applied to nature is that the medieval natural philosophers inherited the theory and substance of Euclidian proportions as the language of science, and they scarcely, if ever, transcended this language in their quantified statements of the functions they saw in nature. Hence, definitions of velocity and the like were always expressed as the ratios of like quantities, and not as metrical statements couched in algebraic form and involving proportionality constants.

Now in concentrating on the period of our discussion, we should note that within the confines of this proportionality language the fourteenth-century schoolmen made some interesting contributions. The first of these was in the development of the so-called Bradwardine rule for representing the supposed relations of force and resistance to velocity in the production of motion. The so-called Bradwardine rule or function can be expressed as follows:[20]

$$\frac{F_2}{R_2} = \left(\frac{F_1}{R_1}\right)^{\frac{V_2}{V_1}}.$$

In discussing this rule, Bradwardine made an effort to quantify more specifically than it had been done in the past the rule usually attributed by modern authors (but not by Bradwardine) to Aristotle and which he (Bradwardine) rejects. This can be written as follows:

$$\frac{V_2}{V_1} = \frac{F_2}{F_1} \text{ when } R_2 = R_1 \text{ and } \frac{V_2}{V_1} = \frac{R_1}{R_2} \text{ when } F_2 = F_1.$$

He also more neatly and mathematically expressed another rule that perhaps had a pre-history in late antiquity in a work of John Philoponus and in

---

[20] For a discussion of Bradwardine's contribution to dynamics, see my *The Science of Mechanics*, pp. 437–40.

Islam in one of Ibn Bājja. Bradwardine for the first time unambiguously gives it this form:

$$\frac{V_2}{V_1} = \frac{F_2 - R_2}{F_1 - R_1} \, .$$

This is equally rejected by Bradwardine in favor of his exponential law. Where did Bradwardine's curious "law" arise? I suggested some years ago in *The Science of Mechanics* that it had an analogue in medical or pharmacological treatments and particularly in the treatment of the degrees in medicinal compounds and simples.[21] Thus in the basic work of Alkindi *On the Degrees of Medicinal Compounds*, the author related an arithmetic ordering of degrees of intensity with a geometrical increase of the determinate qualitative powers. Hence, a medicament or a compound was "in equality" when its frigidity (i.e., its power of frigidity) was equal to that of its calidity. It was hot in the first degree when its calidity was twice its frigidity, hot in the second degree when its calidity was four times its frigidity, hot in the third degree when its calidity was eight times is frigidity, and finally, hot in the fourth degree when its calidity was sixteen times its frigidity. Hence, I proposed earlier that it was not much of a step to go from (1) an exponential relationship between qualitative degrees of intensity and qualitative powers of compounds to (2) an exponential relationship between degrees of the intensity of motion and the ratio of motive and resistive powers, particularly in view of the growing tendency from the early years of the fourteenth century to treat the intensities of qualities and the intensities of velocities in a parallel manner. My earlier suggestions have been rendered more probable by a thesis done at Princeton by Michael McVaugh,[22] except that the route of influence was more circuitous than I had imagined. It shows the influence of Alkindi, particularly on Arnald of Villanova, who appears to have conceived of Alkindi's pharmacological "law" as but one manifestation of a more general law relating intensities and powers. As he states it, however, his supposedly general law is rather a special case of Bradwardine's law (the case where the ratio of powers is continually doubled). If I spend considerable time on this clearly erroneous law, I do so only because of its widespread influence in the fourteenth and fifteenth centuries (with only occasional dissenting voices raised against it: Blasius of Parma, Giovanni Marliani, and later Alessandro Achellini). Interestingly enough, its development led to a whole new branch of mathematical proportionality dealing with fractional exponents (called by the

[21] P. 439, n. 35.
[22] "The Mediaeval Theory of Compound Medicines" (1956), pp. 228–33.

schoolmen soon after Bradwardine, "ratios of ratios"). And the language of this branch of proportionality theory as developed particularly by Oresme in his *De proportionibus proportionum* continued into the sixteenth and seventeenth centuries to the time of Kepler and Newton (Kepler's Third Law is expressed in the medieval exponential language[23]). Two aspects of Bradwardine's law make it particularly important. First, it was applied to all aspects of motion in the Aristotelian sense; to local motion, motion of alteration, and motion of augmentation. It thus had a surprisingly universal appeal. In the second place, it focused attention on instantaneous changes and led to a remarkable development in kinematics at Merton College.

In Chapters Four, Five and Six of my *Science of Mechanics in the Middle Ages* I have treated the crucial aspects of this development at Merton. It is associated with the names of four natural philosophers present at Merton between 1328 and 1350: Thomas Bradwardine, William Heytesbury, Richard Swineshead, and John Dumbleton. In brief, its essential contributions were: (1) a careful distinction of kinematics from dynamics; (2) the elaboration of precise definitions (indeed every bit as precise as those of Galileo later) of uniform speed, uniform acceleration, and the like; (3) a growing appreciation of the concept of instantaneous velocity; and (4) the elaboration and proof of a theorem describing uniform acceleration in terms

---

[23] In the "ratio of ratios" vocabulary developed in the fourteenth century, when one ratio is said to be a certain ratio of another, we are to understand that the "certain ratio" is the exponent to which the second ratio is to be raised rather than merely the arithmetic ratio in which the two compared ratios stand. Thomas Bradwardine in his *Tractatus de proportionibus,* ed. H. L. Crosby, Jr. (Madison, 1955; 2d printing, 1961), p. 126, for example, speaks of "the ratio of any two spheres as three-halves the ratio of their surfaces" ("Quarumlibet duarum spherarum proportio ad proportionem superficierum suarm eodem ordine sesquialtera comprobatur"), when he obviously intends that

$$\frac{V_1}{V_2} = \left(\frac{S_1}{S_2}\right)^{\frac{3}{2}}.$$

Similarly, Kepler, in his third law (which is, of course, entirely original with him), says that "the ratio between the periodic times of any two planets is precisely three-halves the ratio of their mean distances" (*Harmonices mundi,* Bk. V, ed. C. Frisch, Joannis Kepleri, *Opera omnia,* Vol. 5 [Frankfurt, 1964], 279; ". . . proportio, quae est inter binorum quorumcunque planetarum tempora periodica, sit praecise sesquialtera proportionis mediarum distantiarum . . ."), by which he obviously means that

$$\frac{T_1}{T_2} = \left(\frac{D_1}{D_2}\right)^{\frac{3}{2}}$$

and so he is still using the medieval exponential language. See the new edition of Oresme's *De proportionibus proportionum* (ed. Grant), where the exponential language is used throughout.

of its mean velocity. Let me say a word about the last of these contributions, since the invention of the mean speed theorem is one of the true glories of fourteenth-century science.

While we are in doubt as to the exact origin of the mean speed theorem, the earliest statement of it that we can pin down to a specific date occurs in the *Regule solvendi sophismata* of William Heytesbury, dated in 1335. In that statement of the theorem, we see the law given as an obvious truth without proof, as follows:

For whether it commences from zero degree or from some [finite] degree every latitude (i.e., increment of velocity or velocity difference), as long as it is terminated at some finite degree (i.e., of velocity), and as long as it is acquired or lost uniformly, will correspond to its mean degree. Thus the moving body, acquiring or losing this latitude (increment) uniformly during some assigned period of time, will traverse a distance exactly equal to what it would traverse in an equal period of time if it were moved uniformly at its mean degree of velocity.[24]

Other early statements of the mean speed theorem without proofs are found in the two short treatises on motion by Swineshead, which are attached to the *Liber calculationum* in a Cambridge manuscript.

However the mean speed law might have arisen at Merton, a number of interesting proofs of the theorem were given in the various works attributed to Heytesbury, Swineshead, and Dumbleton. While the relative chronological order of these proofs is by no means certain, we should not be too far wrong in supposing that the theorem was discovered some time during the early 1330's and that all of these proofs were completed before 1350.

All of these Merton proofs are of an arithmetical and logical character, while Galileo's proof of the mean speed theorem is strictly geometrical, as we shall observe. Where does the latter come from? The answer is also from the fourteenth century. The kinematic studies of the English spread by 1350 to France. And at Paris there developed a system of graphing movements by the use of a kind of co-ordinate geometry. The most complete treatment of this system—which came to be known as "latitudes of forms" or, better, "configurations of forms"—was that developed by the famous Parisian schoolman, Nicole Oresme, in two treatises, the first entitled the *Questions on the Geometry of Euclid* and the second *De configurationibus qualitatum et motuum*.

The system is interesting enough to describe briefly, even though its

[24] Clagett, *The Science of Mechanics*, pp. 262–63. The next several paragraphs depend heavily on my account of kinematics in Chapter Five of this work.

outlines are fairly well known. It was equally applied to qualities and to motions. In the case of qualities, the abscissa or base line represented the extension of the quality in a subject, while perpendiculars erected on the line of extension represented the intensities of the qualities at the various points of the subject or extension line. (See Fig. 1.) Thus, the whole two-

Fig. 1.

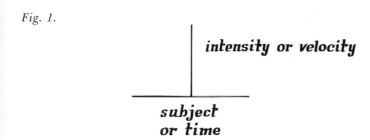

intensity or velocity

*subject
or time*

dimensional figure was used to represent the quantity of a quality. Hence, a rectangle represented a uniformly intense quality, while a right triangle represented a uniformly non-uniform quality. (See Figs. 2–4.) It should be obvious that when applied to motion the rectangle figures a uniform motion and the right triangle a uniform acceleration from rest. Now Oresme in the third part of his *De configurationibus* gave a geometrical proof of the Merton uniform acceleration theorem of the same kind as given by Galileo much later. The rectangle is employed to represent the uniform motion at the mean degree of velocity and the right triangle to represent the uniform acceleration. (See Fig. 5.) Oresme then showed by simple plane geometry that the two areas were equal. By the way, the quantity of velocity (equivalent to the area of the whole figure, or as we should say "the area under the curve") was definitely related by Oresme to the distance traversed, as it was later by Galileo in his explanation of the corollary to the theorem.

Its ultimate influence on Galileo's treatment of the law of free fall can, I believe, be shown without any doubt. I shall not try to do so here, for the details of the argument would unduly extend this paper and they are given in my forthcoming edition of the *De configurationibus*, where I have attempted to trace the step-by-step passage of the configuration doctrine from the fourteenth through the sixteenth century.[25] I should like, however, to say something here about the origins of the system with Oresme. I sug-

[25] *Nicole Oresme and the Medieval Geometry of Qualities and Motions* (Madison, in press), Chap. Two, Part B.

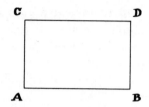

Fig. 2.—A uniform linear quality or uniform velocity

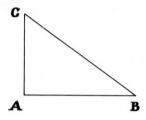

Fig. 3.—Uniformly difform quality beginning at zero degree or uniform acceleration from rest

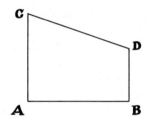

Fig. 4.—Uniformly difform quality from a certain degree or uniform acceleration beginning at a certain velocity

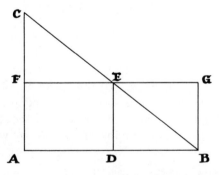

Fig. 5.—Oresme's geometric proof of the Merton rule of uniformly difform

gested some years ago that the system essentially grew out of an analogy drawn between the quantity of quality or motion and the area and dimensions of a surface.[26] At that time I pointed to the analogy as given in the brilliant *Liber calculationum* of Richard Swineshead, perhaps written a few years prior to the two main works of Oresme. Now recently I have discovered the presence of the analogy in Oresme's mind in two of his own early works: *The Questions on the Generation and Corruption of Aristotle* and *The Questions on the Physics of Aristotle*, works that are extant only in manuscript. The pertinent passage in the first work occurs in the midst of a question as to whether an indivisible could be altered:

Fourth corollary: if something were to be difform in quality, there would be more quality in the whole than in some part of it. *Whence quality is to be imagined to have two dimensions: longitude according to the extension of the subject and latitude according to intensity in degree.* Then there is a second conclusion, that if the whole quality of one body were in a point it would be infinitely intense. This is proved, for the subject is uniform. By the preceding conclusion the quality of the whole is double the quality of the half. Therefore, if by imagination the whole were placed in one half of the subject, it would be twice as intense as before. . . . It is obvious, if one uses a surface as an example, that if it is two feet long and one foot high and if it is made half as long while the total quantity of the surface remained as before, then it would be twice as high. And in the same way, if the whole quality were placed in a third part, it would be triply intense; and if in a fourth part, quadruply intense, and so on without end. Therefore, if the whole were placed in a point, it would be infinitely intense. . . . A third conclusion is that it does not follow from this that that quality would be infinite [in quantity]. This is evident in the first place because it is the same as before; it was not augmented, for, although infinitely intended, its extension has been proportionally divided to infinity, as is imaginable in regard to a surface.[27]

The basic idea, presented here so lucidly, is that we can imagine a quality as having two dimensions, longitude and latitude, the one associated

[26] Clagett, *The Science of Mechanics*, pp. 335–36.
[27] Bk. I. Quest. 20, MS Florence, Bibl. Naz. Centr., Conv. Soppr. H.ix.1628, 40v–41r: "4m corollarium: quod, si esset aliquid difformiter quale, tunc esset maior qualitas in toto quam in aliqua eius parte. Unde qualitas ymaginatur habere duas dimensiones, scilicet longitudinem secundum extensionem subiecti et latitudinem secundum intensionem in gradu. Tunc est secunda conclusio, quod si tota qualitas unius corporis esset in puncto, illa esset in infinitum intensa. Probatur quia subiectum est uniforme. Per precendentem conclusionem qualitas totius est dupla ad qualitatem medietatis. Igitur si per ymaginationem tota poneretur in una medietate subiecti, esset in duplo intensior quam ante. Patet statim quia aliter non esset tanta qualitas sicud ante, quia si ipsa esset equaliter (41r)

with the extension of the subject of the quality and the other with its intensity expressed in degrees. These, of course, are the very names that Oresme adopts in his fully developed graphing system. Because of this imagined two-dimensionality of the quality, we can accordingly imagine its quantity by means of a surface, with the quantity of the surface varying in the same way as the quantity of the quality. That Oresme also thought of the surface analogy as applying to motion is clear from a passage in his *Questions on the Physics*:

This then is the first conclusion concerning a motion qualifiedly infinite: A motion infinite in velocity can be produced in a finite time. Proof: It is possible for something to be moved during the proportional parts of an hour, first with some velocity, then twice as fast, then three times, four times, and so on, this being accomplished by the diminution of resistance to infinity. Then I infer by way of corollary that such a motion would be qualifiedly infinite, for it would only traverse a finite space, namely four times the space traversed in the first [proportional] part [of the hour]. One could concede this to be demonstrable as follows. It having been posited that in the first part it traverses the space of a foot, then if it were moved with such a degree through the remainder [of the hour] it would traverse two feet in the whole time. Again, if with the latitude acquired in the second part it would be moved through that second part and the rest [of the hour], it would traverse one foot. And if it were moved with the latitude of the third part through that part and the rest [of the hour], it would traverse one half foot, and then one half of one half, and so on in this way continually; it traverses [all together] precisely as much as that posited in the case; hence it traverses four feet. . . . And if upon a line of two feet in length, which is divided into proportional parts, one proceeded in that way so that upon the first part there would be a surface one foot in altitude, and [on the second one of two feet, and] on the third one of three feet, and so on continually, then this [total] surface would be equal to a surface four feet in length and one foot wide.[28]

---

intensa, cum ipsa sit in duplo minus extensa, sequitur quod esset in duplo minor, quod est contra ypothesim. Patet in exemplo de superficie, quod si sit longa duobus pedibus, alta uno pede, et si fiat minus longa in duplo et maneat tanta sciud ante, esset in duplo magis alta. Et eodem modo si qualitas totius poneretur in parte ⟨tertia⟩, esset in triplo plus intensa; et si in 4ª, in quadruplo, et sic sine fine. Igitur si tota poneretur in puncto, ipsa esset in infinitum intensa, quod est propositum. Et eodem modo argueretur, si subiectum esset difforme et per ymaginationem posset reduci ad uniformitatem. . . . Tertia conclusio est quod ex hoc non sequitur quod illa qualitas esset infinita. Patet primo quia est eadem quam prius; erat non augmenta quia licet sit in infinitum intensa tamen eius extensio est in infinitum divisionata proportionaliter, sicud patet ymaginari de superficie."
[28] *Questiones super libros physicorum*, Bk. VI, Quest. 8, Seville, Bibl. Col. 7.6.30, 71r, c. 2: "Tunc de motu infinito secundum quid est prima conclusio, quod infinitus motus in velocitate potest fieri tempore finito. Probatur, quia possible est quod secundum partes

This application of surfaces to illustrate a line of reasoning about velocities varying in time reappears in more formal dress in Chapter III. viii of the *De configurationibus*, which I shall discuss later. But in the *Questions on the Physics*, it is being given as a geometrical illustration or analogy to a preceding verbal argument. I should also point out there seems to be no further use or treatment of the nascent configuration doctrine in the *Questions on the Physics*. I am thus reasonably confident that these questions as well as those on the *De generatione* date from a period prior to the composition of the *De configurationibus* and even prior to his *Questions on the Geometry of Euclid*, where he first outlines the configuration doctrine in detail. This expression of the analogy as present in the earlier works of Oresme seems then to be the last step prior to a full exposition of the system, and, as I have said, Oresme's first full exposition came in the *Questions on the Geometry of Euclid*, written some time close to 1350, I would suppose as a part of his arts teaching assignments. While I shall not give here a lengthy exposition of the doctrine as outlined in the *Questions* or distinguish the treatment adequately from his later exposition in the *De configurationibus qualitatum et motuum*, I should like to share with you my discovery of a rather remarkable coincidence in Oresme's treatment of the rule of uniformly difform with that of Galileo in the *Discorsi*. In order to do this, let me say a word about the use of the configuration doctrine by Galileo.

Galileo's use of the configuration doctrine is most fruitfully present in his proof of the fundamental uniform acceleration theorem (Theorem I, Proposition I) in the "Third Day" of his *Discorsi e dimonstrazioni matematiche intorno a due nuove scienze*,[29] where Galileo states the Merton College acceleration theorem in slightly different form and gives a geometric proof employing a right triangle to represent uniform acceleration from rest and a rectangle to represent uniform motion at the speed of the middle

proportionales hore aliquid moveatur aliqua velocitate, deinde duplo velocius, deinde triplo, et quadruplo, et cetera, et hoc propter diminutionem in infinitum resistentie. Tunc infero corollarie quod talis motus esset infinitus secundum quid quia non pertransiret nisi spacium finitum, videlicet quadruplum ad pertransitum in prima parte; sicut posset concedere demonstrari, quia posito quod in prima parte pertranseat spacium pedem, tunc sit tali gradu moveretur per residuum pertransiret duplos in toto tempore; et iterum si latitudine asquisita in secunda parte moveretur per ipsam et per residuum pertransiret unum pedem; et si latitudine tertie moveretur per ipsam et per residuum pertransiret dimidium pedem, deinde medietatem medietatis, et sic semper et continue, precise tantum pertransit in casu posito; quare pertransit 4^{or} pedes. . . . Et si super lineam bipedalem divisam per partes proportionales fieret illo modo: super primam partem esset superficies pedalis altitudinis [et super secundam bipedalis] et super tertiam tripedalis et sic semper, tunc ista superficies equivaleret uni superficiei 4^{or} pedum in longum et unius in latum."
[29] *Le Opere, ed. naz.*, Vol. 8 (Florence, 1898), 208–12, where the first two theorems and the first corollary to the second theorem are given. For an English translation, see

instant of the period of acceleration. Theorem I states: "The time in which a certain space is traversed by a moving body uniformly accelerated from rest is equal to the time in which the same space would be traversed by the same body traveling with a uniform speed whose degree of velocity is one-half of the maximum, final degree of velocity of the original uniformly accelerated motion." This theorem is the basis of the proof of his celebrated second theorem: "If some body descends from rest with a uniformly accelerated motion, the spaces traversed in any times at all by that body are related to each other in the duplicate ratio of these same times, that is to say, as the squares of these times." This theorem was used for the proof of its first corollary, which held that in uniform acceleration from rest the spaces traversed in any number of equal and consecutive time periods starting from the first instant of motion "will be related to each other as the odd numbers beginning with unity, i.e., 1, 3, 5, 7. . . ." And so we see that Galileo has given three forms of the acceleration theorem. Now let us step back to Oresme's *Questions on the Geometry of Euclid*. The *Questions* first includes the conventional form of the Merton rule which measured a uniformly difform quality or motion by its middle degree:

The penultimate [conclusion] is that from this latter together with the aforesaid it can be proved that a quality uniformly difform is equal to the middle degree, i.e., that it would be just as great in quantity as if it were uniform at the middle degree. And this can be proved as for a surface.[30]

Although Oresme probably gave no proof of the theorem here, there is one manuscript which includes a garbled proof.[31] The proposition is repeated in a somewhat different form in Question 15:

Then the first proposition is that it is impossible for *b*, which is uniformly difform to no degree, to have as much quality as *a* [which is equal in subject and is

Clagett, *The Science of Mechanics*, pp. 409–16. A discussion of the Merton theorem with the appropriate literature has been given in the same work, Chaps. 5 and 6, and pp. 630–31n, 646–47, 649, 654–55. Note that the medieval form of the theorem usually emphasized that the spaces traversed in equal times were equal, while Galileo's theorem states that the times for the traversal of the equal spaces are equal, a change in wording without great mathematical significance.

[30] Nicole Oresme, *Quaestiones super Geometriam Euclidis*, Quest. 10, ed. H. L. L. Busard (Leiden, 1961), p. 28, ll. 8–11. (Note: I have given a new edition and English translation of Questions 10–14, and the beginning of 15, in Appendix I of my *Nicole Oresme and the Medieval Geometry of Qualities*. There are no significant changes in any of the passages quoted below, but in the full proofs there are numerous changes and the new text should be consulted, if possible.)

[31] See n. 6 of my English translation of Question 10 in the above cited Appendix, where I discuss the corrupt proof appearing in MS Seville, Bibl. Colomb. 7.7.13.

uniform in quality], unless it begins from a degree double [the intensity of a and ends with no degree].[32]

For this proposition, one manuscript gives the well-known diagram of a right triangle and its equivalent rectangle constructed on the same base (Seville, Bibl. Colomb, 7.7.13, 109^r), i.e., the figure that appears in *De configurationibus*, Chapter III.vii. (See Fig. 5.) Oresme in the *Questions* also recognized that the proposition held for uniform deceleration to rest:

If a is moved uniformly for an hour and b is uniformly decelerated in the same hour by beginning from a degree [of velocity] twice [that of a] and terminating at no degree, then they will traverse equal distances, as can be easily proved.[33]

Since Oresme stresses here that it is equality of distance (in the same time) that is involved, it is evident that when talking about velocities, he conceives of the areas of the surfaces as representing total distances traversed. It is then clear that Oresme, even in the earlier work, understood what was to be the substance of Galileo's first theorem and, as we pointed out, Oresme later added a formal geometric proof in his *De configurationibus*, Chapter III. vii.

But in addition to the Merton rule for uniformly difform qualities and motions, Oresme has given in his early work a proposition that is formally equivalent to Galileo's second theorem relating the distances to the squares of the times, a proposition usually considered as entirely original with Galileo. It is true, however, that Oresme's proposition is applied to qualities rather than directly to velocities. Oresme asserts:

The second is that, in the case of a subject uniformly difform [in quality] to no degree, the ratio of the whole quality to the quality of a part terminated at no degree is as the square of the ratio of the whole subject to that part [of the subject]. This is evident in the first place because, by the preceding [proposition] such triangles and such qualities [which the triangles represent] are similar. But the ratio of similar triangles is as the square of the ratio of a side to a corresponding side, by VI.17 [of the *Elements*], but the extension of the subject is as the side of the triangle designating the quality. From this, the proposition is evident. . . . A third argument for it [i.e., the proposition] is, that by the first question [i.e., Question 10], it is obvious that any such quality would be equal to its middle degree with respect to intensity. Therefore, the [equivalent] intensity of the whole [uniformly difform quality] is twice the [equivalent] intensity of its half, and also the extension [of the whole is twice] the extension [of its half].

[32] Ed. Busard, p. 42, ll. 6–8.
[33] *Quaestiones,* ed. Busard, Quest. 13, p. 37, ll. 17–20.

Hence, by the second question [i.e., Question 11, giving a rule for comparing the quantities of two uniform qualities one of which is both twice as intense and twice as extended as the other], the ratio of the qualities is as the square of the ratio of subject to subject.[34]

Oresme later in the same question notes that one can speak in the same way "of velocities with respect to time." Hence, if such a transfer to velocities is made so that the quantities of the velocities, i.e., their total distances traversed (e.g. represented by triangles ABC and ADE in Fig. 6), are considered in the place of quantities of qualities, and their times (e.g. represented by lines AB and AD) are considered in place of subjects of qualities, we will have Galileo's second theorem. Furthermore, one of Oresme's arguments in proof of the proposition is based on the Merton rule for uniformly difform, just as Galileo's proof of his second theorem is based on his expression of the Merton rule found in his first theorem.

Fig. 6.

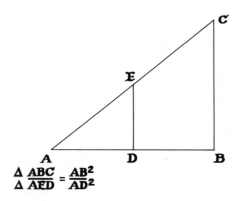

$$\frac{\Delta\ ABC}{\Delta\ AED} = \frac{AB^2}{AD^2}$$

Finally, it should be remarked that Oresme also has given in the *Questions* a proposition equivalent to Galileo's corollary to Theorem II holding that in the case of uniform acceleration from rest the distances traversed in equal, consecutive time periods are as the odd numbers: 1, 3, 5, 7, . . . Again, it should be noted that Oresme has framed the proposition in terms of qualities:

The second conclusion is that, with a subject so divided [into equal parts] and with the most remiss part designated as the first part, the ratio of the partial

[34] *Ibid.*, Quest. 13, p. 36, ll. 20–27; l. 32 to p. 37, l. 4.

qualities, i.e., their mutual relationship, is as the series of odd numbers where the first is 1, the second 3, the third 5, etc., as is evident in the figure.[35] (See Fig. 7.)

Fig. 7.

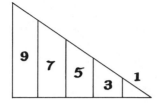

Oresme's proof, it will be clear to the reader consulting the full text, goes back ultimately to the proposition relating the quantities of the qualities to the squares of the subject lines. Similarly, Galileo proved his corollary by means of his Theorem II relating distances to the squares of the times. One must presume that Oresme would also have recognized that his proposition held for the distances traversed, since throughout the work he stresses the applicability of his propositions to velocities as well as qualities. And, in fact, the first step toward Oresme's odd-number theorem was taken by the Merton College authors in the context of velocities when they proved that a body uniformly accelerated from rest traverses three times as much space in the second half of the time as in the first.[36]

In view of the fact that Oresme drew much the same consequences from the Merton rule as did Galileo, and in view of the further fact that the order and substance of proofs in the two authors is essentially the same, we might well ask whether Galileo read Oresme's *Questions*. For the present, the answer certainly must be in the negative. There is no evidence of Galileo's having used any Oresme manuscripts, and it must be remembered that Oresme's *Questions on the Geometry of Euclid* was not published until the twentieth century. I hasten to add, however, that while I cannot connect this particular work with Galileo, there is no doubt in my mind that Galileo was the heir, in at least an indirect way, of the medieval configuration doctrine. And, in fact, I believe that one can make a fair case for identifying the particular published works from which he drew his general knowledge of the system and his particular knowledge of the mean speed theorem with its geometric proof.

[35] *Ibid.*, Quest.. 14, p. 38, ll. 29–32.
[36] Clagett, *The Science of Mechanics,* p. 266.

In centering my attention on the development of kinematics with its geometric representation by Oresme, I should also point out that as a part of this evolution of kinematics there appears a remarkable treatment of the summation of infinite series. This appears to be almost wholly a medieval development and largely a product of the fourteenth century. Again, the key names are those at Merton College, such as, Swineshead, and Oresme at Paris. It may surprise some of you that the treatment of the summation of infinite series had its beginning in the consideration of Aristotle's views of the infinite rather than in Greek mathematical works. That is to say, it arose in a strictly philosophical context. The point of departure seems to have been Book III, Chapter 6 (206^b, 3–12), of the *Physics* of Aristotle, where Aristotle says that in a certain way an infinite according to addition is the same as an infinite according to division.[37] The example given is of a finite magnitude, which is divided successively according to a given ratio (in medieval terminology "divided according to proportional parts"). And so the parts of the line produced by this division when added together produce the finite magnitude. It is but a small jump from what the passage says to the conclusion that, if the division is completed to infinity, the whole original finite quantity is produced by the addition of an infinity of parts. In his exposition of this passage, Aquinas gives, as a specific example of this division, a cubit line divided first into halves, then one of the halves into halves, one of the quarters into halves, and so on to infinity.[38] Hence, if one thought of the infinite division as completed, we would then have by addition the following infinite series:

$$(1) \qquad \tfrac{1}{2} + \tfrac{1}{4} + \tfrac{1}{8} + \ldots + \tfrac{1}{2}^n + \ldots = 1$$

I should add, however, that neither Aristotle nor Aquinas seems to have actually thought of the infinite series as completable, since the division to

[37] See the Medieval Latin translation of Moerbeke accompanying Thomas Aquinas, *In octo libros physicorum Aristotelis expositio*, ed. P. M. Maggiòlo (Turin, 1954), p. 183 (Text. No. 59): "Quod autem secundum appositionem idem quodammodo est et quod est secundum divisionem. Infinitum enim secundum appositionem fit e contrario: secundum quod enim divisum videtur in infinitum sic appositum videbitur ad determinatum. In finita enim magnitudine si accipiens aliquis determinatum, accipiet eadem ratione, non eandem aliquam magnitudinem ratione accipiens, non transibit finitum. Sin vero sic augmentet rationem, ut semper eandem aliquam sit accipere magnitudinem, transibit finitum, propter id quod omne finitum absumitur quolibet finito. Aliter quidem igitur non est, sic autem est infinitum, potentia et divisione. Et actu autem est, sicut diem esse dicimus et agonem; et potentia sic sicut materiam, et non per se sicut finitum. (Text No. 60) Et secundum appositionem igitur sic infinitum potentia est, quod idem dicimus quodammodo esse ei quod est secundum divisionem: semper quidem enim aliquid ipsius extra est accipere."
[38] *Ibid.*, p. 185, ". . . puta si a linea cubitali accipiat medietatem, et iterum a residuo medietatem; et sic in infinitum procedere potest. . . ."

II

infinity was only potential rather than actual. However, by the time of the fourteenth century, the schoolmen were blithely assuming the summation of this and other more complicated series. Furthermore, it is important to realize that most of the treatment of the more complicated series rests on the assumption of the summation of this first simple series. To show this to be so, let us take the most popular of the series summed by the schoolmen of the fourteenth, fifteenth, and sixteenth centuries:

$$(2) \qquad 1 + \tfrac{1}{2} \cdot 2 + \tfrac{1}{4} \cdot 3 + \ldots + \tfrac{1}{2}^{n-1} \cdot n + \ldots = 4$$

It should be noted that this was the series which we have already mentioned as appearing in geometric form in Nicole Oresme's *Questions on the Physics of Aristotle*. Let us now look at the more complete geometric treatment of it given by Oresme in his *On the configurations of qualities and motions* (see Fig. 8):

Fig. 8.

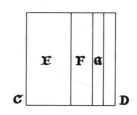

III.viii. On the measure and intension to infinity of certain difformities. A finite surface can be made as long as we wish, or as high, by varying the extension without increasing the size. For such a surface has both length and breadth and it is possible for it to be increased in one dimension as much as we like without the whole surface being absolutely increased so long as the other dimension is diminished proportionally, and this is also true of a body. For example, in the case of a surface, let there be a surface of one square foot in area whose base line is AB; and let there be another surface, similar and equal to it, whose base line is CD. Let the latter surface be imagined to be divided on line CD to infinity into parts continually proportional according to the ratio of 2 to 1, with its base divided in the same way. Let E be the first part, F the second, G the third, and so on for the other parts. Therefore, let the first of these parts, namely E, which is half the whole surface, be taken and placed on top of the first surface towards the extremity B. Then upon this whole let the second part, namely F, be placed, and again upon the whole let the third part, namely G, be placed, and so on for the others to infinity. When this has been done, let base line AB be imagined as being divided into parts continually proportional according to the ratio 2 to 1, proceeding toward B. And it will be immediately evident that on the first proportional part of line AB there stands a surface one foot high, on the second a surface two feet high, on the third one three feet high, on the fourth four feet high, and so on to infinity, and yet the whole surface is only the two [square] feet [in area] previously given, without augmentation. And consequently the whole surface standing on line AB is precisely four times its part standing on the first proportional part of the same line AB. Therefore that quality or velocity which would be proportional in intensity to this figure in altitude would be precisely four times the part of it which would be in the first part of the time or the subject so divided. For example, let the first part (towards extreme A) of the proportional parts divided along AB according to the ratio 2 to 1 be a certain amount white or hot, the second twice as white [intensively], the third three times as white, the fourth four times, and so on to infinity on both sides according to the [natural] series of [whole] numbers. Then from the prior statements it is apparent that the total whiteness of line AB is precisely four times the whiteness of the first part; and it would be the same for a surface [whiteness], or for a corporeal whiteness, if it were increased in intensity in a similar fashion. In the same way, if some mobile were moved with a certain velocity in the first proportional part of some period of time, divided in such a way, and in the second part it were moved twice as rapidly, and in the third three times as fast, in the fourth four times, and increasing in this way successively to infinity, the whole velocity would be precisely four times the velocity of the first part, so that the mobile in the whole hour would traverse precisely four times what it traversed in the first half of the hour; e.g., if in the first half or proportional part it would traverse one foot, in the whole remaining period it would traverse three feet and in the total time it would traverse four feet. . . .[39]

[39] *The Science of Mechanics*, pp. 380–81. A slightly different new edition of this appears in my *Nicole Oresme and the Medieval Geometry of Qualities and Motions*, and it is from

It is evident that what Oresme has done by piling up the surfaces is to indicate that the original series (2) is equivalent to the following series:

(3) $2 + [1 + \frac{1}{2} + \frac{1}{4} + \ldots + \frac{1}{2}^{n-1} + \ldots] = 4$

And indeed the subseries in brackets is not "proved" in this work, it being merely stated that one of his square feet is divided into proportional parts according to the ratio of 2 to 1. But he has given a proof of this in his earlier *Questions on the Geometry of Euclid*,[40] where a proof is expounded for the general series:

(4) $\frac{a}{b} + \frac{a}{b}\left(1 - \frac{1}{b}\right) + \frac{a}{b}\left(1 - \frac{1}{b}\right)^2 + \ldots + \frac{a}{b}\left(1 - \frac{1}{b}\right)^{n-1} + \ldots = a$

$\frac{a}{b}$ being the first aliquot part of a to be removed; and series (1), i.e.,

$\frac{1}{2} + \frac{1}{4} + \frac{1}{8} + \ldots + \frac{1}{2}^n + \ldots = 1$,

is shown to be a corollary of the general series.

Now series (2) summing at 4, with which I opened this discussion, was also given by Richard Swineshead in his *Liber calculationum*, with the objective of determining the average intensity of a quality whose subject has been divided into proportional parts of increasing intensity.[41] And

_____

this rather more complete text that my translation has been made. See H. Wieleitner, "Über den Funktionsbegriff und die graphische Darstellung bei Oresme," *Bibliotheca Mathematica*, 3. Folge, Vol. 14 (1913–14), 231–33.
[40] *Quaestiones*, ed. Busard, Quest. 1, p. 2, ll. 23–34; p. 3, ll. 1–3; Quest. 2, p. 5, ll. 9–15. See the review of Busard's edition of the *Quaestiones* by John Murdoch in *Scripta mathematica*, Vol. 27 (1964), 68.
[41] "De difformibus," in the *Calculationes* (Pavia, 1498), pp. 14–15 (cf. MSS Cambridge, Gonville and Caius 499/268, 168ʳ⁻ᵛ; Paris, BN lat. 6558, 6ʳ⁻ᵛ; Pavia, Bibl. Univ. Aldini 314, 5ᵛ–6ʳ): "Prima tamen opinio de qualitate difformi, cuius utraque medietas est uniformis, potest sustineri, scilicet quod corresponderet (*C*, respondeat *Ed*) gradiu medio inter illas qualitates. Et fundatur argumentum super illo: in duplo plus facit qualitas extensa per totum subiectum ad totius denominationem quam si sola (*C*, tota *Ed*) per medietatem extenderetur, quod arguitur sic. Signetur *a* quod habeat caliditatem ut 4 per totum. Tunc totum erit (*C*, est *Ed*) calidum per totam caliditatem ut 4. Sed una medietas illius tantum facit ad denominationem totius (*C*, om. *PEd* subiecti) sicut alia medietas. Igitur tota illa qualitas in duplo plus denominat totum quam una eius medietas (*C*, pars sive medietas *Ed*) totum denominat, quod fuit probandum. Ex quo sequitur quod denominatio totius subiecti per qualitatem extensam per medietatem subiecti (*C*, om. *Ed*) solum est subdupla ad illam qualitatem, quia in duplo minus dominat totum quam medietatem illam per quam extenditur, et illam medietatem denominat gradu suo (*C*, summo *Ed*). Igitur totum per illam qualitatem denominatur gradu subduplo ad illam qualitatem. Et si extenderetur (*C*, extendatur *Ed*) per quartam totius, tantum tunc denominaret totum gradu subquadruplo ad illam qualitatem. Et sic correspondenter sicut proportionaliter extenditur per minorem partem quam est totum, ita totum denominat (*CP*, nominatur *Ed*) gradu remissiori quam partem (*C*, pars *Ed*) per quam illa (*C*, om. *Ed*) extenditur. . . . Contra quam positionem et eius fundamentum arguitur sic, quia sequitur quod si prima pars proportionalis alicuius esset aliqualiter intensa, et secunda in duplo intensior, et

299

Swineshead's treatment was made the object of a Renaissance commentary by the Portuguese master at Paris, Alvarus Thomas, in his *Liber de triplici motu* (Paris, 1509), where the series is proved in a more general form,[42] with the division into proportional parts made according to any ratio and not just the ratio of 2 to 1. Incidentally, this brilliant tract illustrates exceedingly well the continuity of the discussion of infinite series between the Middle Ages and the Renaissance.

But let us turn away from the Swineshead type of expression and proof to another fourteenth century proof appearing in one copy of a short philosophical tract entitled from its incipit *A est unum calidum*,[43] and perhaps composed by Johannes Bode. The author states the conclusion of the series as applied to velocities in almost exactly the same way as Oresme did in the last part of Chapter III.viii, which I have just quoted. In his proof the author ingeniously transforms series (2) into the following series:

$$(5) \qquad 1 + 1 + 1 + [\tfrac{1}{2} + \tfrac{1}{4} + \tfrac{1}{8} + \ldots + \tfrac{1}{2^n} + \ldots] = 4$$

---

tertia in triplo, et sic in infinitum, totum esset eque intensum precise sicut est secunda pars proportionalis, quod tamen non videtur verum. Nam apparet (*add. C* ex illa conclusione) quod illa qualitas est infinita, ergo si sit sine contrario infinite denominabit suum subiectum. . . . (15, c. 2) . . . Ad argumentum in oppositum negatur consequentia: hec est qualitas infinite intensa, ergo infinite totum subiectum denominat, quia illo modo extenditur quod infinite modicum faciet qualitas illa infinita respectu illius subiecti. Nam qualitas quarte partis proportionalis est in duplo intensior quam qualitas secunde partis proportionalis et est per (*CP, om. Ed*) subiectum subquadruplum (*PEd*, subduplum *C*) et ideo in duplo facit minus (*add. Ed* scilicet quam secunda *quod om. CP*). Si enim quarta pars proportionalis esset in octuplo (*EdP*, duplo *C*) intensior quam prima (*EdP*, secunda pars *C*), sicut est in octuplo (*EdP*, quadruplo *A*) minor illa [in extensione], tunc tantum faceret ad totius denominationem sicut prima (*EdP*, secunda *C*). Sed quarta pars, ut constat, est nunc in duplo minus intensa quam tunc esset, quia nunc est in duplo remissior quam tunc esset. Ergo nunc quarta pars proportionalis in duplo minus facit in comparatione ad totum quam facit prima pars proportionalis et prima tantum facit comparatione ad totum ut secunda, ut patet. Igitur quarta pars proportionalis in duplo minus facit quam secunda ad totius denominationem (*C*, intensionem *APEd*), et tamen sua qualitas est in duplo intensior. Et sic discurrendo, quelibet qualitas extensa per partem posteriorem minus facit quam qualitas extensa per partem priorem, vocando partes priores que sunt propinquiores extremo ubi partes maiores terminantur. Et hoc est verum de omnibus partibus proportionalibus *a* nisi de prima et secunda que equaliter faciunt ad totius denominationem." I have made a few corrections from the three manuscripts (*C*-Cambridge; *P*-Paris; *A*-Pavia, Aldini) but only those which help to emend the edition so that it would make better sense. I have altered the punctuation freely.

[42] This edition is without pagination, but see Sig. p. 4v, c. 2–q 1r, c. 2. The full Latin text is given in my *Nicole Oresme and the Medieval Geometry*, Commentary to Chap. III.viii.

[43] The Latin text is given by H. L. L. Busard, "Unendliche Reihen in *A est unum calidum*," *Archive for History of Exact Sciences*, Vol. 2, No. 5 (1965), 394–95. I have also given the text independently with a somewhat different reading in several places in the *Nicole Oresme and the Medieval Geometry*, Commentary to Chap. III.viii.

Hence, since the subseries sums at 1, the total sums at 4, and he has his proof. The ingenious method of transformation is first applied to the velocity of the third proportional part of the time and then successively to each succeeding part. It is first noted that in the third part, and indeed in every succeeding part, the time span of that part is equal to the time remaining. The implication of this remark is that there are just as many instantaneous velocities in the third part as in the time remaining. And so, if we take from all the remaining velocities one degree and add it to velocities in the third proportional part, then the velocity during the third proportional part will be uniform at 4 (instead of at 3), and the uniform velocity of each of the remaining velocities will be reduced by one. Since the uniform velocity in the third proportional part is now 4 and so twice the velocity of the second part, and since it is half as extended in time, it is evident that in the third part of the time just as much space will be traversed as in the second; namely, a space of 1. Similarly the velocity of the fourth proportional part, having been reduced from its original 4 to a velocity of 3 by the increase of the velocity of the third proportional part, is now restored to 4 by borrowing a degree from all of the succeeding velocities. Hence, with its velocity now restored to 4 (and, hence, equal in velocity to that of the third part) and with a time period half that of the third part, the mobile in the fourth part traverses a distance half of that traversed in the third part. Similarly, the velocity of the fifth part, having been reduced by two borrowings to 3, is now raised to 4 by borrowing one degree from all the succeeding velocities. Hence, with a velocity now equal to that of the fourth part and with a time period half that of the fourth, the mobile will traverse in the fifth part half as much space as that traversed in the fourth. In the same manner, all of the succeeding velocities are, one by one, brought up to a velocity of 4 by successive borrowings, and since each part is one half that of the preceding one in time span, the space traversed in each proportional part of the time will be half that traversed in the preceding part. Hence, the transformation of the series is complete and the proof is obvious.

This proof was not without its influence in the Renaissance, since it was repeated (although with some minor differences) by the fifteenth-century Florentine Bernardus Torni, in his commentary to Heytesbury's *Regule solvendi sophismata*, published with that latter text in Venice, 1494.[44] In fact, Torni praises Oresme as the one who stimulated his consideration of the conclusion, but it was the proof from the *A est unum calidum* that he takes over. He has made one addition, namely, a proof of the summation

[44] Folios 76v–77r. The full text is given in *Nicole Oresme and the Medieval Geometry*, Commentary to Chap. III.viii.

of series (1); that is, the subseries in the transformed series of the *A est unum calidum*: $\frac{1}{2} + \frac{1}{4} + \frac{1}{8} + \ldots + \frac{1}{2^n} + \ldots = 1$. His proof is based on Euclid V.13 (Greek text, V.12). Incidentally, Torni's treatment was also influential on the group of schoolmen at Paris in the early fifteenth century, perhaps on Alvarus Thomas, but certainly on Juan de Celaya, Luiz Coronel, and Jean Dullaert.[45]

Before leaving this rather exciting development of infinite series, let me underline that the discussion of all of these series occurs as part of the discussions of qualities and velocities and so as part of the treatment of hypothetical physical situations. Furthermore, it is worth pointing out that Swineshead and others actually employed the summation of infinite series in a very subtle way to prove such crucial theorems as the mean speed theorem and, hence, the use of such series was already improving the mathematical facility of the natural philosophers.[46]

In my brief search for trends in the fourteenth century—those influential and those not so influential—I have left some glaring omissions. Hence, I have not even mentioned the interesting progress made in statics.[47] This omission is partly justified in that the main medieval contributions to a quantified statics were made in the thirteenth rather than the fourteenth century and in essentially Hellenistic terms, although the author of one commentary on Jordanus' *Elementa de ponderibus* known as the *Aliud commentum* showed, in a way more clear than any before him, that Jordanus' proof of the law of the lever rested on the principle of virtual displacements. I have also barely touched upon the dynamical discussion of *impetus* and its rudimentary quantification in the hands of Jean Buridan, with the impetus described as being proportional to prime matter and velocity.[48] Its ultimate influence on sixteenth-century authors[49] is much more debatable than the kinematic doctrines here emphasized. Furthermore, it has been so often treated that I forbear further discussion. Nor have I commented on certain developments in mathematics in the fourteenth century, such as the increasing use of the Moerbeke translations of Archimedes at Paris and, in particular, their influence on the French mathematician Johannes de Muris.[50] (Incidentally, this same author also produced the most mature handbooks of mensuration and calculation in the fourteenth cen-

[45] Duhem, *Léonard de Vinci*, Vol. 3, 546–48.
[46] *The Science of Mechanics*, pp. 295–96.
[47] *Ibid.*, Chap. 2.
[48] *Ibid.*, pp. 522–25.
[49] *Ibid.*, pp. 653–57.
[50] For a preliminary summary of this growing use of Archimedes, see my *Archimedes in the Middle Ages*, Vol. 1 (Madison, 1964), Chap. I. Volume 2 of this work will treat the matter in more detail.

tury.)[51] My only hope is that I have left you with the impression of a fertile upheaval in natural philosophy in the fourteenth century that augured well for the scientific quickening that followed. And if, like Swineshead, you are inclined to see the clarification of change in the finding of means, perhaps you will be successful in assaying the scientific thought of the Renaissance in terms of a mean between the trends I have stressed for the fourteenth century and those which Mr. Drake will discuss as representing Galileo.

[51] These are his *Quadripartitum numerorum*, which gave numerous formulas that apply to physical science, all with numerical examples, and his *De arte mensurandi*. Both deserve critical editions.

III

Richard Swineshead
and Late Medieval Physics

I THE INTENSION AND REMISSION OF QUALITIES (1)

A. The Rise of the Quantitative Treatment of Qualities.

Certainly one of the most significant developments in the rise of early modern physics was the quantification of physics. It has usually been assumed that this quantification or mathematicization took place in the sixteenth and seventeenth centuries under the influence of the introduction of the mathematics of ARCHIMEDES and its use and application by GALILEO GALILEI. Now there is no question that it was the spread of the Archimedean works and spirit which did much to shape the form that the mathematical investigation of natural phenomena took. But it has become more apparent in the last generation that the 16th century interest in the mathematical if not experimental treatment of natural questions is at least in part a product of the initial investigations along that line undertaken in the Universities of Oxford and Paris in the fourteenth century. This interest in the quantification of the physical at these universities was manifested particularly by the treatment of qualities quantitatively, a treatment that also embraced a new interest in the quantitative study of local movement. For the first time in history natural philosophers were seriously

---

(1) This essay is the first of a series on the role of Swineshead in medieval physics. It has grown out of an edition of Swineshead's *Calculationes* which I am preparing with the gratefully acknowledged assistance of fellowships (1946, 1950-1) from the John Simon Guggenheim Memorial Foundation. For section A. cf. A. Maier's brilliant *Die Problem der intensiven Grosse*, 1939, which I read too late for use here.

thinking about kinematics, the study of movement in terms of the dimensions of space and time.

This essay is primarily concerned with certain opinions of RICHARD SWINESHEAD, one of the Oxford logicians who unhesitatingly accepted a quantitative treatment of qualities. But by way of introduction we should briefly note how the quantitative approach to qualities arose.

From the latter part of the thirteenth century there began to be discussed and circulated at least four opinions which hoped to explain differences in qualitative intensities. Three of these positions, we shall see, rejected any basic similarity between quantity and quality, while the fourth insisted on such a similarity. The discussion of intensity variations in qualities, which the schoolmen called intension and remission of qualities, had its medieval origin in (1) scholastic efforts to discover how the virtue charity increased in a person, whether or not it was by the addition of one charity to another, and kindred questions. On the other hand, it had its ancient sources in (2) statements found in the *Categories* of ARISTOTLE and the *Commentary on the Categories* of the sixth-century Neoplatonist, SIMPLICIUS.

We can conveniently start our background investigation with AQUINAS who appears to have been one of the first of the Western schoolmen to have made use of SIMPLICIUS' *Commentary* (translated by his friend, WILLIAM MOERBEKE) and who therefore is one of the first to treat variations in qualitative intensities in the light of a variety of opinions from Greek antiquity.

As the result of his mature consideration found in the *Summa theologiae* ST. THOMAS advanced the opinion (1) that intensive increases in quality, e.g., where we say something is " more " or " less " white, result from the *varying participation of a subject in a given, unchanged quality.* Thus intension and remission do not originate in the quality or form, but in the subject and its varying disposition for a given quality. Since this is so, THOMAS goes on to assert (2) that intensive increases in habits and forms (and thus charity) *do not take place by the addition of one part of the form or habit to another,* even though extensive increases, e.g., where we say " little " or " great " whiteness, do take place by addition. This theory is very lucidly outlined in two questions of the *Summa theologiae* from which I shall select illustrative passages, the first of which appear in his discussion of the increase of habits (2) :

Now the perfection of a form may be considered in two ways : first, according to the form itself : secondly, according to the participation in the form by the subject. In so far as we consider the perfections of a form according to the form itself, thus the form is said to be *little* or *great* : for instance, great or little health or science. But in so far as we consider the perfection of the form according to the participation in it by the subject, it is said to be *more* or *less :* for instance, more or less white or healthy...

For we said that increase or decrease in forms which are capable of intensity and remission happen, in one way, not on the part of the form itself considered in itself, but through a diverse participation in it by the subject. Therefore such increases of habits and others forms is not caused by the addition of form to form, but by the subject participating more or less perfectly in one and the same form. (3)

Much the same sort of thing is said in a later question where the problem of the increase in charity is considered (4) :

For since charity is an accident, its being *(esse)* is to be in something. So that an essential increase of charity means nothing else but that it is yet more in its subject, which implies a greater radication in its subject... Hence charity increases essentially, not by beginning anew, or ceasing to be in its subject, as the objection implies, but by beginning to be more and more in its subject...

It follows (5) therefore that if charity be added to charity, we must presuppose a numerical distinction between them which follows a distinction of subjects : thus whiteness receives an increase when one white thing is added to another, although such an increase does not make a thing whiter (this is the same distinction above between " little, great " and " less, more.") This, however, does not apply to the case in point, since the subject of charity is none other than the rational mind, so that such like an increase of charity could only take place by one rational mind being added to another; which is impossible. Moreover, even if it were possible, the result would be a great lover, but not a more loving one. It follows, therefore, that charity can by no means increase by the addition of charity to charity, as some have held to be the case.

---

(2) THOMAS AQUINAS, *Summa Theologiae*, I-II, quaest. 52, art. 1 (Edition of the Ottawa Institute of Medieval Studies, 1941, 982b; translation of the English Dominicans, 1st part of second part, p. 38).

(3) *Ibid.*, I-II quaest. 52, art. 2 (*edit. cit.*, 984b; *transl. cit.*, p. 43).

(4) *Ibid.*, II-II, quaest. 24, art.4 (*edit. cit.*, 1532a; *transl. cit.*, 2nd part of second part, p. 286).

(5) *Ibid.*, II-II quaest. 24, art. 4, art. 5 (*edit. cit.*, 1533a; *transl. cit.*, pp. 287-288). It should be noted that DUHEM completely misrepresents AQUINAS' position on this question, at least so far as what AQUINAS has said in the *Summa Theol.* P. DUHEM, *Études sur Léonard de Vinci*, Paris, 1913, vol. 3, pp. 317-318.

Accordingly charity increases only by its subject partaking of charity more and more, i.e., by being more reduced to its act and subject thereto. For this is the proper mode of increase in a form that is intensified, since the being of such a form consists wholly in its adhering to its subject. Consequently, since the magnitude follows on its being, to say that a form is greater is the same as to say that it is more in its subject, and not that another form is added to it ; for this would be the case if the form, of itself, had any quantity, and not in comparison with its subject.

The source of THOMAS' theory, it would appear, lies in the statements made in the *Categories* of ARISTOTLE and the *Commentary* of SIMPLICIUS (6). We can be fairly confident of this because THOMAS includes in the discussion of the increase of habits a very close paraphrase of a summary passage from SIMPLICIUS (7) :

In this way, then, there were four opinions among philosophers concerning the intensity and remission of habits and forms as SIMPLICIUS related in his *Commentary on the Categories.* For PLOTINUS and other Platonists held that qualities and habits themselves were susceptible of more and less, for the reason that they were material, and so had a certain indetermination because of the infinity of matter. Others, on the contrary, held that qualities and habits of themselves were not susceptible of more and less; but that the things affected by them are said to be more and less, according to a diversity of participation : that for instance, justice is not more or less, but the just thing. ARISTOTLE alludes to this opinion in the *Categories.* The third opinion was that of the Stoics and lies between the two preceding opinions. For they held that some habits are of themselves susceptible of more and less, for instance, the arts; and that some are not, as the virtues. The fourth opinion was held by some who said that qualities and immaterial forms are not susceptible of more and less, but that material forms are.

A second opinion relative to intensive increases was that extensively developed by HENRY OF GHENT (8). This view held that intensive augmentation of a qualitative form *(augmentatio in forma infusa)* does *not* take place from varying participation of the subject in the quality or from variations of matter, but only from the form itself, which has in its essence parts. That is to say, it is in the essential character of the quality to have different intensive

---

(6) *Categories*, 10b-11a; SIMPLICIUS, *In Categ.* Chap. 8 (Prussian Academy edition, vol. *8*, pp. 283-5).

(7) AQUINAS, *op. cit., loc. cit.,* in note 2.

(8) HENRY OF GHENT, *Quodlibeta,* quodlib. 5, quaest. 18 (Edition of 1518). Cf. DUHEM, *op. cit.,* p. 319. I have also used DUNS SCOTUS discussion of HENRY's opinion, *In lib. Sententiarum,* Bk. 1, distinctio XVII, quaest. 6 (*Opera omnia*, vol. *5-2*, Lyon, 1639, pp. 988-995).

parts. The actual intension results when each new part passes over from potentiality to act, or as is sometimes said, when the new part is " extracted " from potentiality to act. But this theory, like the first one, rejected any increase by an apposition of part to part of the same species, i.e., by any strictly part to part addition.

Still a third opinion which rejected any part by part addition of form in intension and remission is the theory intimately associated with the name of GODEFROID DE FONTAINES (9), and was later given the authoritative support of WALTER BURLEY in his treatise *De intensione et remissione formarum* (10). In the preliminary form of this theory as outlined by GODEFROID it was held that in the case of an intensively increased form or quality, e.g., when there is " more " charity, no identical numerical part of the preexisting less intensive quality remains. The preexisting " individual " *(individuum)* is destroyed and replaced by a more perfect individual which does not contain that preexisting individual as a numerical part of it and in fact is absolutely distinguished from it. BURLEY embraces this opinion and develops it further. He concludes that in every formal motion, i.e., intensive increase in qualitative forms, something completely new is acquired, and this is a form. And so in such a formal movement, the whole preceding form from which the movement begins is destroyed and a totally new form *(una forma totaliter nova)*, non-existent in the subject before, is acquired. Since, then, there is a whole series of distinct forms involved in intension, BURLEY maintained that " no form is intended or remitted, but rather the subject is intended and remitted according to form." This whole position has obviously been influenced by the Franciscan doctrine of the plurality of substantial forms. It suggests rather a plurality of qualitative forms.

All of the preceding positions, we have seen, rejected the increase in the intensity of a quality by means of a part to part addition. But as we can note from the statements of AQUINAS above, there were some people in his day who thought charity could increase

(9) GODEFROID DE FONTAINES, *Quodlibeta*, quodlib. 7, quaest. 7, unpublished, but summarized by SCOTUS, *op. cit., edit. cit.*, Bk. I, dist. XVII, quaest. 4 (p. 976). Cf. DUHEM, *op. cit.*, pp. 327-328.

(10) WALTER OF BURLEY, *De intensione et remissione formarum*, Venice, 1496, ff. 1-15v, for passage quoted, f. 10v, c. 1.

136

by the addition of part to part, form to form. AQUINAS himself reports the common distinction between two kinds of quantity, dimensive or corporeal *(quantitas dimensiva, corporalis)* and virtual *(quantitas virtualis)* (11). But as we remember, he believed that while corporal quantity, with respect to a form, increased by the addition of parts, intensive or virtual quantity did not so increase by addition, but by the varying participation of the subject in the form. And so AQUINAS rejected any attempt to treat these two quantities as similar.

RICHARD MIDDLETON, who wrote a commentary on the *Sentences* probably a little after 1281, retains the distinction between corporal quantity, which he calls quantity of mass *(quantitas molis)*, and virtual quantity (12). The former is measured by the number of objects submitted to the action of the power and thus ressembled discontinuous quantity, the latter by the intensity of the act produced in a given object, and so resembles continuous quantity. In spite of this distinction, however, he believes that intensity or quantity of force can be increased by addition in a manner similar to increase in quantity of mass. He posits that just as adding one quantity of mass to another produces a greater mass, so the addition of one degree of a quantity of force to a preexisting one produces something greater in force.

This doctrine seems to have been supported by DUNS SCOTUS (13), who indicates very briefly, after having rejected the doctrines of GODEFROID and THOMAS and in the course of refuting the opinion of HENRY OF GHENT, that " there is no extraction (of a new part from potentiality to act)." " But I say that there is a new reality added to the preexisting one. This reality is like parts or non-quidditative degrees, which are individual and existing." One of his later commentators JOANNES PONCIUS interprets this passage as concluding that intensive augmentation " takes place by the *addition* of non-quidditative parts " (14). He goes on to say that " this doctrine of Scotus is commonly held

(11) AQUINAS, *op. cit., edit. cit.*, 1531b, 1532a.

(12) RICHARD OF MIDDLETON, *Super quatuor libros Senetentiarum*, Bk. 1, dist. XVII, art. 2, quaest. 1, Brixiae, 1591, vol. *1*, p. 162. Cf. DUHEM, *op. cit.*, pp. 330-331.

(13) DUNS SCOTUS, *op. cit., edit. cit., loc. cit.* in note 6, p. 990.

(14) *Ibid.*, p. 991.

by all those who say that the intension of a quality is accomplished by the addition of degree to degree whether these degrees are homogeneous or heterogeneous." Certainly there is no doubt that SCOTUS's " faithful student " JOANNES DE BASSOLIS adopted the concept of intension by degree to degree addition (15). He repeats and answers an old argument brought up against part to part addition, namely that if we add tepid water to tepid water both at the same degree of heat, the resulting mixture is not hotter. This, JOANNES answers, is because we are adding subject to subject and getting an increase in the quantity of mass, but if we could put the two quantities of heat in the same subject so that there would be no extensive increase of the quality in the subject, then we would get an increase in intensity. Many schoolmen in the 14th and 15th centuries picked up this distinction between " quantity of heat " and " intensity of heat " and used it in determining the ultimate effectiveness of a heat agent in a heat action (16).

The doctrine of intension by gradual additions was taken up

---

(15) JOHN OF BASSOLS, *In quatuor Sententiarum libros*, Bk. I, dist. XVII, queast. 2; Paris, 1579, ff. 114-117.   Cf. DUHEM, *op. cit.*, pp. 335-339.

(16) M. CLAGETT, *Giovanni Marliani and Late Medieval Physics*, New York, 1941, pp. 34-39. It is rather interesting to note that the distinction between quantity of mass and virtual or intensive quantity which was applied to heat actions may, in my opinion, have been the source of the quantitative description of the *impetus* which the schoolmen believed accounted for the continuance of projectile movement. Although the theory of an impressed force continuing the movement of the projectile is very old, going back at least to John Philoponus in the sixth century, A.D., the first quantitative description of impetus as varying with the quantity of prime matter in the projectile and the velocity imparted to the projectile is introduced by JEAN BURIDAN (See M. CLAGETT, " Some General Aspects of Medieval Physics." in *Isis, 39* (1948), pp. 40-41). But what was the source of this new definition? And why was it expressed in terms of the quantity of prime matter and velocity? My opinion is that it goes back to the fundamental distinction noted with respect to qualities, namely quantity of mass and intensive quantity. This is borne out by realizing that when local motion was treated at the same time as qualitative changes, the velocity of local motion was equivalent to intensity in qualitative changes, and both were expressed in terms of degrees *(gradus)*. Now the similarity between quantity of mass *(quantitas mole)* and quantity of prime matter *(quantitas prime materie)* is obvious. So just as consideration of both quantity of heat and intensity were thought to be necessary in measuring the effectiveness of an agent in a heat action, so the similar factors, quantity of prime matter and velocity were thought to measure the effect of a moving projectile.

138

with great vigor at Oxford. Nor is this surprising, for Oxford, even from the 13th Century was known for its nourishment of mathematics, and this attempt to treat qualities quantitatively was picked up with remarkable alacrity if not always with equal clarity. THOMAS BRADWARDINE's efforts in dynamics and kinematics in his *Treatise of the Proportions of Movements* in 1328 no doubt constituted the most important initial step in that movement, and the dependence of later authors on him is evident and acknowledged (17).

One of the principal leaders at Oxford in the school who recognized the value of treating the qualitative intensions quantitatively was RICHARD SWINESHEAD (Suiseth). And while I shall reserve questions of biography to a later article, it seems fairly certain that RICHARD SWINESHEAD was a fellow of Merton College about 1340 and that his principal work, later known as the *Calculationes*, must date from about this time (18). And of all of the English philosophers of Oxford at this time who were interested in questions of qualities and movements, men such as THOMAS BRADWARDINE, JOHN DUMBLETON, WILLIAM HEYTESBURY, the anonymous author of the *Six Inconveniences*, et. al. (19), SWINESHEAD seems to have gained greatest renown in later centuries for this kind of activity   Fifteenth century schoolmen in Italy like ANGELUS DE FOSSAMBRUNO, GIACOMO DA FORLI, PAUL OF VENICE, and GIOVANNI MARLIANI, call him ,,the Calculator" with obvious respect, and point to him as the great authority in this quantitative physics that was growing up. It is also true that Italian humanists of the same century failed to appreciate the subtleties of his thought and in fact appropriated his name in coining the derisive term for scholastic inanities,

---

(17) Thus, for example, see below where SWINESHEAD mentions BRADWARDINE. Most of the treatises dealing with local motion in the fourteenth century refer to BRADWARDINE. See CLAGETT, *Giovanni Marliani, etc.*, chap. VI.

(18) For the essential points in the biography of SWINESHEAD consult the DNB; CLAGETT, *Giovanni Marliani*, Appendix; and L. THORNDIKE, *The History of Magic and Experimental Science*, vol. 3, New York, 1934, pp. 370-385.

(19) Some of the basic ideas of the Oxford schoolmen are treated by P. DUHEM, *op. cit.*, vol. 3, pp. 405-481. For the Italian authors mentioned in the next sentence of the text, consult the same work, pp. 481-510, and also CLAGETT, *Giovanni Marliani*, by referring to the index.

suisetica (20). But in the sixteenth century he was admired by men of the caliber of SCALIGER and CARDAN (21), and no less a figure than LEIBNIZ pleaded for the editing of the Calculator's work, at the same time describing him as the man " who introduced mathematics into scholastic philosophy " (22).

Granting that qualities may be treated quantitatively as many of the English school did, it became a fundamental question as to how to measure the intension and remission of these qualities to use scholastic vocabulary, or to put it in another way, how to measure the progressive or regressive alteration in the intensity of qualitative forms (or the alteration of velocity in local motion.). This is the question which SWINESHEAD raises and seeks to settle in the first tractate of his *Calculations*. It is this tractate which is ordinarily known as the *Intension and Remission of Qualities*, or along with some of the succeeding tractates, as *Intension and Remission of Forms* (23). It is this first short tractate which I should like to analyse briefly in this article.

Before starting this analysis, it ought to be noted that the expression, latitude, which has an interesting history toward the end of the thirteenth and in the early fourteenth century, now clearly is used by SWINESHEAD in the sense of a range of alteration in intensity from one degree of intensity to another, or in the case of local motion a positive or negative increment in velocity, i.e., an increment from one degree of velocity (equivalent to in-

(20) ERMALAO BARBARO, *Epistolae, orationes, et carmina*, edited by Vittore Branca, Florence, 1943, pp. 23, 78. Cf. the statement put in the mouth of NICCOLÒ NICCOLI by LEONARDO BRUNI, L. THORNDIKE, *University Records and Life in the Middle Ages*, New York, 1944, p. 269.

(21) THORNDIKE, *op. cit.*, in note 18, vol. 3, p. 373. Cf. W. G. TENNEMAN, *Geschichte der Philosophie*, vol. 8, Leipzig, 1808, pp. 904-905, with appropriate references in note 70 to CARDAN and SCALIGER.

(22) THORNDIKE, *op. cit.*, vol. 3, p. 370. Cf. LOUIS COUTURAT, *Opuscules et fragments de Leibniz*, Paris, 1903, p. 340 : Fuit enim aliquis Johannes Suisset, dictus Calculator, qui circa motus et qualitatum intensiones in media metaphysicorum regione mathematicum sine exemplo agere coepit. See also pp. 177, 191, and 330.

(23) In all of the various manuscripts the incipits of the first treatise in the Calculationes reveal that it is concerned with intension and remission of a quality. However the Paris manuscript (BN, Fonds latin, no. 6558) and later authors give the title of *Intensio et remissio formarum* to the first few treatises which concern various aspects of intension and remission.

III

140

stantaneous velocity) to another. Or, we might think of a body where there is a qualitative latitude when the adjacent parts of that body vary in heat intensity. We would say, then that there is a latitude of hotness or calidity in that body from such or such a minimum degree of calidity (or zero degree of calidity) to such or such a maximum degree *(gradus summus)*. Other scholastic terminology will be explained in the context of the discussion (24).

B. The *Calculations* (25); Tractatus I : The Measurement of the Intension and Remission of a Quality.

The Calculator informs us at the outset that there are several opinions as to how intension and remission are to be measured. But before taking these up he wishes to point out (26)

intension is accepted in two ways. In one way it is spoken of as the alteration by means of which a quality is acquired—and speaking thus intension is motion. In the other way it is referred to as the quality by means of which something is intended, e.g., a hot body is said to be intended by calidity.

Though he seems to favour the second method, as matter of fact the first method of accepting intension as motion seems to be implied on occasions in later sections (27).

---

(24) Since the question of the terminology used in the fourteenth century physical treatises is often difficult I can recommend two fourteenth century works which offer definitions : The anonymous introduction found attached to THOMAS BRADWARDINE, *Proportiones* included among the several treatises in B. POLITI, *Quaestio de numero modalium* etc., Venice, 1505, ff. 9r-10r. Compare also in the same edition, GIOVANNI CASALI, *De velocitate motus alterationis*, f. 59v *et seq.*

(25) The Latin text of the *Calculationes* used in these footnotes has been prepared primarily from the first one of the following manuscripts and editions : University of Pavia, Aldini codex no. 314, ff. 1-83 (abbreviated below as A); Cambridge Library, Gonville & Caius Ms. 499/268, ff. 165r-215r (abbreviated as B); Paris, Bibliothèque Nationale. Fonds latin no. 6558 (abbrev. as C); Rome, Biblioteca Angelica, Inventario no. 1963, (abbrev. as D); Editio princeps, Padua ca. 1477 (Abbrev. as EP); Editio secunda, Pavia, 1498 (abbrev. as ES). Since the completed text will be published later the variant readings have been ommited in all but a few cases.

(26) Ms. A, f. 1r : Penes quid habent intensio et remissio qualitatis attendi plures sunt opiniones. Pro quo tamen primo est notandum quod intensio dupliciter potest accipi. Uno modo pro alteratione mediante qua qualitas acquiritur. Et sic loquendo intensio est motus. Alio modo dicitur intensio qualitas mediante qua aliquid est intensum sicut calidum est intensum mediante caliditate et sic proportionaliter de remissione est dicendum.

(27) A, f. 1r : De intensione et remissione secundo modo dictis (1) ad presens est locutio. Variant : (1) A seems to have *dicta*. B, C, D, EP, ES have *dictis*.

With this initial distinction made the Oxford logician then states three opinions regarding the measurement of intension and remission. As the discussion develops we shall see that he favours the third opinion (28) :

> The first position posits that the intension of any quality is measured by the proximity to the most intense degree of its latitude. Remission in this position is measured by the distance from the most intense degree.

It will be clear when we examine the arguments used by SWINESHEAD against this position that he thinks it represents intensity growing as the line which joins the intensity with the maximum degree decreases. Remission on the other hand grows as that line increases in length.

After stating the first position the Calculator immediately states the next two (29) :

> The second position posits that intension is measured by the distance from zero degree of the latitude and remission with the distance from the most intense degree of its latitude. The third position holds that intension is measured by the distance from the zero degree and remission with the proximity to the zero degree.

That these were not all of the possibilities in the course of this discussion is demonstrated by one of the later Italian authors who concerns himself with this question, PIETRO POMPONAZZI, in his *De intensione et remissione formarum*. After mentioning the three opinions we have already quoted, he notes that the third opinion is that of the Calculator and then proceeds to describe a fourth opinion which held that intension is measured by proximity to the maximum, as in the first position, but that remission is measured by proximity to the zero degree (30).

SWINESHEAD's criticism of the first position makes interesting

---

(28) A, f. 1r : Prima positio ponit quod intensio cuiuslibet qualitatis attenditur penes appropinquationem gradui intensissimo (1) illius latitudinis; remissio tunc penes distantiam a gradu intensissimo (2).   Variants: (1) EP and ES read *summo;* (2) EP has *perfectissimo*, ES, *summo.*

(29) A, f. 1r : Secunda positio ponit quod intensio attenditur penes distantiam a non gradu et remissio penes distantiam a gradu intensissimo illius latitudinis. Tertia positio ponit quod intensio attenditur distantiam a non gradu et remissio penes accesum versus non gradum.

(30) PIETRO POMPONAZZI, *De intensione et remissione formarum*, etc., Venice, 1525, f. 2r.

III

142

use of arguments involving the infinite. He claims that if we accept this position then it would follow that the maximum degree of intensity in any hot body would be infinite (31).

This is because if some calidity were increased in intensity to the maximum degree, then that calidity will be a certain amount intense, then doubly intense, then quadruply intense, and thus to infinity because it will be a certain amount near to the maximum degree, then doubly nearer, quadruply nearer, and thus to infinity, and proportionally as it will be nearer to the maximum degree, so according to this position it will be more intense than any time before. Therefore, this calidity will be intended toward infinity before the end, and in the end it will be more intense than any time before. Hence the maximum calidity will be infinitely intense, which was to be proved.

The substance of this argument will perhaps emerge more clearly if we repeat it using modern symbolization. To prove : The first position is false. Proof :

(1) Assume the first position to be true, representing it as follows :
(2) Let $x = f(1/y)$, where x is the intensity and y is the magnitude of a line drawn from the degree of intensity to the maximum degree.
(3) Then as we increase x towards the maximum degree, y becomes smaller, and will, for example, become one-half as long, one quarter as long, getting as small as you like.
(4) Now y must become zero at the maximum degree.
(5) But from (2) when y is zero, then x must be infinite.
(6) Thus the maximum degree must always be infinitely intense when we assume (2) and (4).
(7) But since all maximal degrees are not infinitely intense, the position is false.

SWINESHEAD is pointing out, then, the difficulty involved in trying to measure an increase by a decreasing function which must be zero when that increase is to some finite maximum.

A similar type of argument is presented in his next criticism of the first position. For he claims that if we accept the first

---

(31) A, f. 1r : Ex illa [positione] sequitur quod gradus summus est infinite intensus, quia intendatur (1) aliqua caliditas ad summum, tunc illa caliditas erit aliqualiter intensa, et in duplo intensior, et in quadruplo (2), et sic in infinitum, quia aliquantulum propinqua erit ista caliditas gradui summo, et in duplo propinquior, et in quadruplo, et sic in infinitum. Et proportionaliter sicut erit propinquior gradui summo, ita iuxta illam positionem erit intensior. Ergo in infinitum intensa (3) erit hec caliditas ante finem et in fine erit intensior quam unquam ante. Ergo caliditas summa erit infinite intensa, quod fuit probandum (4). Variants : (1) ES adds *in hora*; (2) B has *triplo*, EP and ES omit; (3) B has *intensior*; (4) A omits *quod fuit probandum.*

position, then there will not be any degree two times less intense than the mean degree of the latitude. The argument is proved as follows (32) :

No degree is distant from the maximum more than double the mean degree between the maximum degree and the zero degree, the mean being equally distant from the extremes. Since no degree is as distant from the maximum degree as the zero degree it follows that the zero degree is doubly more distant from the maximum degree than the mean of the whole latitude. Therefore, no degree is doubly less intense than the mean degree of the whole latitude. The *consequens* is false because there is some degree a certain amount intense, and some degree half as intense, and some a quarter as intense, and thus to infinity, just as some quantity is a certain amount, and some quantity is half as much and another quantity is one fourth as much, and thus to infinity. Therefore the position is false.

Much of this argument is directed as the previous one at the difficulties that arise when we try to represent the intensity by the linear distance to the maximum, i.e., when we attempt to correlate a quantity that rises to a definite finite limit with a magnitude that can, as it approaches zero, be any quantity as small as you wish.

I suggest that this second argument might be restated as follows :
(1) Assume the position, $x = f(1/y)$ where x is the intensity and y is a line distance from any intensity to the maximum degree of intensity.
(2) Accept the common belief that in a latitude of intensity there is maximum degree and a minimum or zero degree with a mean degree equally distant from the extremes.
(3) Assume the lower extreme of the latitude is zero and further assume that the mean degree is a distance a from the maximum degree.
(4) Restating (2), when the intensity is zero, the distance from the maximum will be $2a$.

---

(32) A, f. 1r : Item ex illa positione sequitur quod nullus est gradus in duplo minus intensus quam est medius gradus, scilicet totius latitudinis. Hoc probatur sic : Nullus gradus per in duplo plus distat a gradu summo (1) quam gradus medius inter gradum summum et non gradum, eo quod medium est quid equaliter distat ab extremis. Cum ergo nullus gradus per tantum distet a gradu summo sicut non gradus distat a gradu summo, sequitur quod (2) non gradus per in duplo plus precise distat a gradu summo quam distet gradus medius totius latitudinis caliditatis a gradu summo. Ergo nullus gradus est in duplo minus intensus quam est gradus medius totius latitudinis caliditatis. Consequens est falsum, quia aliquis gradus est aliqualiter intensus, et in duplo minus intensus est aliquis gradus, et in quadruplo, et sic in infinitum, sicut aliqua quantitas est aliqualiter magna, et aliqua in duplo brevior, et aliqua in quadruplo brevior, et sic in infinitum. Ergo positio falsa. Variants : (1) B, C have redundant phrases after *summo*, here omitted; (2) A has *quia* for *sequitur quod*.

III

144

(5) But in assuming (1), by the time the intensity x is zero, y will have gone through a whole series of value beyond $2a$ to infinity.

(6) Now we know that intensity can have all the decreasing values to zero. But according to (4) $2a$ represents the maximum value we can accord to y. Thus if we follow (1) at the same time that we accept (4), then we cannot go through the decreasing values of intensity to zero because we would very soon be beyond our limit $2a$. And if we could not go through values to $x=0$, then there would not be any degree doubly less intense than the mean degree.

(7) It is clear that (4) and (1) are contradictory. But if (4) is accepted as based on the common experience expressed in (2), and thus is true, then (1) is false, and the position is false. Q.E.D.

The third argument directed by SWINESHEAD against the first position concludes that out of this first position it follows : (33)

that any degree of motion you wish is of infinite remission, because every degree of motion is infinitely distant from an infinite degree of motion. Since there is no maximum degree beneath the infinite degree, it follows that any degree at all is infinitely distant from the maximum degree of its latitude. The *consequens* is false, and therefore, so is the antecedent.

This argument goes back to the first argument against this position, namely, that if intension is measured by nearness to the maximum degree, then any maximum degree is infinitely intense. Now this third argument says that if remission is measured by distance *from* the maximum degree, then every degree of remission is infinitely remiss because it must be infinitely distant from an infinite maximum degree.

In the course of the next or fourth argument SWINESHEAD introduces the concept of the infinitely small *(infinitum modicum)*. Before taking up his argument, let us see in what sense he is using the term *infinitum* in this expression. From at least the time of PETER OF SPAIN's *Summulae logicales* in the thirteenth century, it was customary for the schoolmen to distinguish between what they called *categorematic* and *syncategorematic* infinites (34). The

(33) A, f. 1r : Item sequitur quod quilibet gradus motus est infinite remissionis, quia omnis gradus motus per infinitum distat a gradu infinito motus. Cum ergo nullus sit gradus intensissimus citra gradum infinitum motus (1), sequitur quod quilibet gradus motus per infinitum distat a gradu intensissimo sue latitudinis sed consequens est falsum; ergo et antecedens. Variant : (1) A omits.

(34) PETER OF SPAIN, *Summulae logicales*, edited by J. P. MULLALLY, Notre Dame, Indiana, 1945, p. 118 : Tertia distinctio est quod " infinitum " capitur dupliciter : uno modo capitur categorematice, significative ut est terminus communis et sic significat quantitatem rei subiectae vel predicatae, et signum ut cum

categorematic " infinite " was a general term signifying an actual quantity without limit or end. For example : " The world is infinite (i.e., unbounded in extent)." The syncategorematic " infinite " was a distributive term signifying a quantity larger (or smaller) than any quantity you please. Thus the expression " the infinitely small is a part of a continuum " signifies that a quantity smaller than any quantity you please is a part of a continuum. There is little doubt that when SWINESHEAD uses the expression *infinitum modicum* he is employing *infinitum* in the syncategorematic sense and thus means by that phrase in the following and succeeding arguments " a magnitude smaller than any magnitude you please." But let us turn to SWINESHEAD's arguments (35) :

---

dicitur : " Mundus est infinitus; " alio modo capitur syncategorematice, non prout dicit quantitatem rei subiectae vel praedicativae, sed inquantum se habet subiectum in ordine ad praedicatum, et sic est distributio subjecti et signum distributivum. Cf. EDWARD STAMM, " Tractatus de Continuo von Thomas Bradwardine," in *Isis*, vol. 26 (1936), pp. 19-20 where BRADWARDINE is quoted as saying: " Infinitum cathetice (cathegorematice)... est quantum sine fine..., magnum vel multum sine fine seu non finitum... Infinitum syncathetice (syncathegorematice)... est quantum finitum et maius isto, et finitum maius isto maiori, et sic sine fine ultimo terminante, et hoc est quantum et non tantum quin maius..." The words in brackets are my own. Cf. also P. DUHEM, *op. cit.*, vol. 2, pp. 22-24, appendix E, pp. 368-408; vol. 3, pp. 274-276. In this last passage he quotes from the sixteenth century schoolman, SOTO, the example of the syncategorematic infinite I have given : Infinita pars est pars continui.

(35) A, f. 1r : Item ex illa positione sequitur quod quaelibet caliditas citra summam est infinite remissionis, quia vel est gradus summus infinite intensus vel finite intensus. Si infinite intensus, ergo omnis gradus citra summam per infinitum distat a gradu summo. Consequentia patet per hoc ´quod omnis gradus finitus per infinitum distat a gradu infinito. Ergo cum penes appropinquationem gradui summo iuxta positionem habeat intensio caliditatis attendi (1) sequitur quod omnis gradus caliditatis sit infinite remissus. Si gradus summus sit finite intensus, tunc sit *a* unus gradus remissus. Tunc sic infinite propinquior est aliquis gradus gradui summo quam est *a* gradus, quia per infinitum modicum distat aliquis gradus a gradu summo, ut satis patet. Ergo si penes appropinquationem gradui summo intensio qualitatis vel gradus habeat attendi, sequitur quod in infinitum intensior est aliquis gradus quam est *a* gradus. Ergo cum quiiibet gradus citra summum et etiam gradus summus sit solum finite intensus, sequitur quod *a* gradus est infinite remissus. Sed illud est falsum. Ergo positio falsa. Et quod conclusio sit falsa patet, quia si quilibet gradus foret infinite remissus, nullus gradus foret aliqualiter intensus et per consequens nullus gradus foret alio intensior, quid est impossibile. Variants : (1) ES adds phrase *et penes distantiam a gradu summo remissio*.

146

It also follows from that position that any degree of calidity (hotness) less than the maximum is of infinite remission, since the maximum degree is either infinitely intense or finitely intense. If it is infinitely intense then every degree beneath the maximum is an infinite distance from the maximum degree. The *consequentia* is obvious from the fact that every finite degree is infinitely distant from an infinite degree. Hence, since according to this position intension of calidity is measured by its proximity to the maximum degree, it follows that every degree of calidity is infinitely remiss.

Now if the maximum degree is finitely intense and *a* is a remiss degree, there is some degree infinitely nearer to the maximum degree than is *a* because some degree is distant by an infinite modicum from the maximum degree, as is obvious enough. Therefore, if the intension of a quality or a degree is to be measured by its propinquity to the maximum degree, it follows that some degree is infinitely more intense than is *a* degree. Hence, since any degree at all beneath the maximum as well as the maximum degree is only finitely intense, it follows that *a* degree is infinitely remiss. But that is false, hence the position is false.

As in the preceding cases it would, I believe, be of some use to translate this into more modern terminology, trying to get at what SWINESHEAD means. The argument appears to be as follows :

(1) Assume as an expression for intension in the first position, $x = f(1/y)$, with x as intensity and y a line distance from the degree intensity *to* the maximum degree. But also assume that $z = f(u)$ where z is remission and u is a line distance *from* the maximum degree of intensity.

(2) Now by the argument the maximum degree of intensity is either infinite or finite.

(3) If the maximum degree is infinite, then every value of u will be infinite, since every value of u would be measured from a terminus an infinite distance away.

(4) Hence every value of z will also be infinite.

(5) But suppose that the maximum degree of intensity were finite. Then take any value of intensity a less than the maximum.

(6) Now there can always be some value of intensity infinitely nearer the maximum than a because there can be some value an infinitely small distance from maximum, infinitely nearer and infinitely smaller being used in a syncategorematic sense.

(7) If by (1) intension is really measured by proximity to the maximum and by (6) there is some value infinitely nearer to the maximum than is a, then there is some degree infinitely more intense than a.

(8) But since by (5) we assumed that the maximum is finite and thus any degree below the maximum is finite, and yet at the same time by (7) show that some degree below the maximum is infinitely more intense than a, hence any degree a must be infinitely remiss, i.e., a is infinitely removed from some degree below the maximum and thus is infinitely removed from the maximum.

(9) But a is any degree of calidity and is in actuality not infinitely remiss. Hence the position is false. Q.E.D.

It is clear that steps (6) and (7) are the crucial parts of the argument. If we were representing intension by a line drawn from an initial zero terminus, then we could not conclude that because there was some degree infinitely nearer in a syncategore-matic sense to the maximum, there must be a degree infinitely more intense than *a*. This latter conclusion could only follow if we are representing increasing intensity by a line drawn to the maximum, which line is decreasing in length. In such a situation the line could be reduced to a quantity as small as we like. Hence, in substance this argument is similar to the preceding ones. The last argument against the first position is of the same nature and I shall omit any discussion of it here.

Having argued against the first position, the Calculator now proceeds to argue against the second, or at least the second part of the second position, namely that remission is to be measured from the maximum degree of intensity.

The first argument against the second position is similar to the second argument against the first position (36) :

From the second position it follows that no degree is twice as remiss as the mean degree since no degree is two times farther from the maximum degree than is the mean degree, as is clear enough. The *consequens* is false because there is some degree twice as remiss, some degree four times as remiss, and so on, just as there is some quantity two times less than any given quantity, four times less, and so on to infinity.

Although it is difficult to follow with complete assurance the abbreviated argument presented by the Calculator, I believe his intention is the following :

(1) Assume $z = f(u)$ where z is remission and u is a line distance from the maximum degree of intensity.
(2) It is commonly accepted that in a given latitude or range of intensity there is a maximum degree, a minimum degree (either zero or relatable to zero), and a mean degree equidistant from the maximum and the zero degrees.
(3) Now suppose that the mean degree is a distance from a finite maximum degree. The zero degree or lower limit of the latitude then would be a distance of

---

(36) A, f. 1r : Ideo ponatur secunda positio contra quam sic arguitur : Ex illa sequitur quod nullus gradus est in duplo remissior (1) medio, quia nullus gradus per in duplo plus distat a summo quam gradus medius, ut satis constat. Et consequens est falsum, quia aliquis gradus est in duplo remissior, alius in quadruplo, et sic de aliis, sicut quacumque quantitate data est aliqua quantitas in duplo brevior, et in quadruplo brevior, et sic in infinitum. Variant:(1) C adds *alio vel*.

2a from the maximum. And in measuring according to (1), the zero degree of intensity would be identifiable with a remission twice that of the mean. Since the zero degree is the lower limit, then according to this position *2a* represents the largest possible distance from the maximum.

(4) But in the terms in which remission is universally conceived, there can be no limit to the relational comparison of remission, i.e., something can be twice as remiss as a given degree, four times as remiss, and so on to infinity, where presumably intensity would be zero.

(5) But if we accept the position assumed in (1), we cannot increase remission at will to the point where intensity is zero, but would rather be limited by some arbitrary value *2a*.

(6) Thus since we could not arrive at an intensity of zero by using (1), we could not have a degree twice as remiss as defined by the identification of zero degree of intensity with a degree twice as remiss in step (3).

(7) Thus (1) and (4) are contradictory, and since (4) is generally accepted, (1) must be false. Hence the position is false. Q.E.D.

The only difficulty in my interpretation of the argument would appear to be my assumption that Calculator means in (6) that it is the degree twice as remiss as the mean that must be identifiable with zero degree of intension in this position and yet which cannot be identifiable with it because, accepting the common opinion of remission, we can never reach zero degree using this second position. But I believe that this is just what he is trying to say. Let us put it another way. According to this second position, the zero degree of intensity must be identifiable with the distance twice the mean (step (3)). But at the same time common opinion tells us that we can use the expressions twice as remiss, four times as remiss, and so on as we are approaching the highest possible remission which is identifiable with zero intensity. Hence, we cannot reconcile this common opinion which identifies zero degree of intensity with an unlimited remission and the opinion of this second position which tries to identify zero degree of intensity with twice the remission of the mean degree, but which never even yields a degree twice as remiss as the mean degree, because so long as it accepts any part of the common opinion it cannot reach the zero degree. Then since the common opinion is held to be true and since it contradicts the second position, the second position must be false.

Omitting two of the arguments against the second position, we can make passing reference to the last of these arguments since again we have use of the concept of the infinitely small. Its translation into modern terminology is quite similar to the fourth argument against the first position (37) :

Let *a* be a degree of calidity which is distant from the maximum by a certain latitude. This latitude is divided into proportional parts toward the maximum. Then by an infinitely less amount would some part be distant from the maximum degree. Since, therefore, remission is to be measured by the distance from the maximum degree, it follows that some part is infinitely less remiss than is *a* degree and any part at all is more remiss than the maximum degree. Therefore, *a* is infinitely remiss. The *consequens* is false. Hence the position is false.

Again rearranging this argument we have the following steps :

(1) Assume that remission is measured by distance from the maximum, *m*.
(2) Take any degree *a*. Then from (1) its remission is represented by line *ma*.
(3) Now *ma* can be divided up into proportional parts 1/2, 1/4, 1/8,... to infinity.
(4) Some part will be infinitely nearer to *m*, than is *a*, infinitely nearer in the syncategorematic sense.
(5) Hence that part will be distant from the maximum by an infinitely less amount.
(6) Then that part is infinitely less remiss than *a* since remission is measured by distance from the maximum.
(7) Any part at all is more remiss than the maximum.
(8) Hence, any degree *a* is infinitely remiss (from (6)).

It seems to me that when the Calculator proceeds from statement (6) to that noted in (8), he is shifting from the infinitely small to the infinitely large in an illegitimate manner, a mistake which he had not made in the fourth argument against the first position, because that position had assumed tacitly that the infinitely large was function of the infinitely small.

Before passing to the third position which the Calculator accepts, I should like to observe that he does not bring out sharply one fundamental criticism against using the maximum as a reference point, that namely, the latter must vary from measurement to measurement, from problem to problem. But of course this

---

(37) A, f. 1v : Sit ergo *a* unus gradus (1) caliditatis qui per certam latitudinem distet a summo, que latitudo dividitur in partes proportionales versus summum. Tunc per infinitum minus distat aliqua pars a gradu summo quam distet *a* gradus, quia in infinitum propinquior est aliqua illarum partium gradui summo. Cum ergo remissio habeat attendi penes distantiam a gradu summo, sequitur quod in infinitum minus remissa est aliqua illarum partium quam est *a* gradus et quelibet illarum partium est remissior quam est gradus summus. Ergo infinite remissus est *a* gradus. Consequens est falsum; ergo positio falsa. Variant : (1) A has *una latitudo ;* others *gradus.*

150

would be an argument that would appeal to an experimental investigator who sees the benefits of standards, while the Calculator is primarily concerned with the logical difficulties inherent in certain quantitative ways of representing qualitative changes.

Having disposed of the first two positions, the Calculator then argues affirmatively for the third position, namely that intension is measured by the distance from the zero degree of intensity, while remission is measured by the proximity to that zero degree. He compares the concepts of intension and remission with ideas of " more " and " less " (38).

In the same way that something more distant from zero quantity *(a non quanto)* is said by that fact to be " greater " so something more distant from the zero degree of its intension is said to be " more intense "... Remission is the opposite of intension. Hence, if intension is said to be measured by the distance from zero degree, it follows that remission is to be measured by the proximity to zero degree.

Before proceeding further with the Calculator's treatment of the criticism of the third position, it would be well to suggest what the Calculator believes concerning the measurement of intension and remission, and then see how our suggestions are born out by his arguments. It would seem that the following formulations hold :

(1) The numerical measurement of intensity is equivalent to the expression $x = f(1/y)$ where x is the intensity, and y is the magnitude of the line drawn to that xth degree of intensity *from* zero.
(2) The numerical measurement of *remission* is equivalent to the expression $z = f(1/y)$, where z is the remission, and y is the magnitude of the line drawn from that xth degree of intension *to* zero.
(3) Note well that I have used the expressions *from* zero and *to* zero. Thus there is conceived a directional difference between intension and remission. So even in the case where the scalar value of y is one, intension and remission cannot be considered to be the same. This will be clear in the arguments outlined below.

Repeating these statements more succinctly we can say that

---

(38) A, f. IV : Ideo ponatur tertia positio pro qua sic arguitur. Sicut aliquid plus distat a non quanto sic ipsum dicitur esse maius, ergo per idem sic aliquid plus distat a non gradu sue intensionis sic ipsum dicitur esse intensius. Patet prima pars positionis. Secunda arguitur sic : Remissio opponitur intensioni; ergo si intensio habeat attendi penes distantiam a non gradu, sequitur quod remissio habeat attendi penes appropinquationem non gradui.

intension and remission are numerical reciprocals and from a vector or directional standpoint are reversed. (i.e., 180° out).

One of the principal arguments which SWINESHEAD has outlined against this position is that if we accept it then a given maximum degree of intension must be a certain amount remiss. According to the opponents of the third position this could not be. These opponents would hold that at the maximum degree of intension, remission must be zero, a situation that the third position provides for only when the maximum degree of intension is infinite (e.i., with $z = f(1/y)$, the z will equal zero only when y is infinite). Hence the argument seeks to show that since the maximum degree is not at all remiss, a given remission must be instantaneously lost as it passes to maximum intension. This is impossible because (39)

to lose remission is nothing else than to acquire intension as is obvious from the fact that remission is related privatively to intension. And a *privative* being acquired is nothing else than a *positive* being lost nor is a *privative* being lost anything else than a *positive* being acquired. Since, therefore, no latitude of intension is suddenly acquired, it follows that no latitude of remission is suddenly lost...

If then the remission is not suddenly lost, then according to the third position the maximum degree of intension must be a certain amount remiss, which was thought to be impossible.

The Calculator answers the argument by freely admitting that according to the third position any maximum degree short of infinity is a certain amount remiss. But he finds this to be an easily imagined possibility (40).

The third principal position must be supported in light of what has been said and the conclusion that the maximum degree is remiss is conceded. But when

---

(39) A, f. 1v : Sed deperdere remissionem non est aliud quam acquirere intensionem, ut patet eo quod remissio se habet privative respectu intensionis et privatum acquiri non est aliud quam positivum deperdi, necque privativum deperdi est aliud quam positivum acquiri. Cum ergo nulla latitudo subito acquiratur, scilicet intensionis, sequitur nullam latitudinem remissionis subito deperdi respectu *a*.

(40) A, f. 1v-2r : Pro dictis sustinenda est tertia positio principalis et conceditur conclusio quod gradus summus est remissus. Et tunc quando arguitur quod nulla caliditas est caliditate summa intensior, ergo illa non est remissa, negatur consequentia, quia etsi nulla caliditas sic foret intensior de facto, non repugnat tamen illi caliditati, quod aliqua foret illa intensior et imaginando caliditatem intensiorem illa foret remissa sicut nunc est.

152

it is argued that no calidity is more intense than a (given) maximum calidity and hence that maximum calidity is not at all remiss, the *consequentia* is denied. Because although no calidity would be *de facto* more intense, it is not impossible with respect to that calidity that there would be some calidity more intense than the maximum. And by imagining a more intense calidity, then as it now is it would be a certain amount remiss.

Assuming that the third position is correct, when we proceed to investigate the implications of it, three further positions or opinions are advanced, the first two of which are not correct, while the third is (41).

The first posits that every degree is just as intense as it is remiss. The second supposes that there is some degree just as intense as it is remiss, and conversely, although not every degree is just as intense as it is remiss. The third posits that no degree is as intense as it is remiss.

He holds that these same positions can be supposed for " largeness and smallness and every other latitude which is considered positively and privatively with respect to something."

The arguments which the Calculator advanced for the first supplementary position (and which he later refutes) are aimed at proving that intension and remission are identical and thus any degree is just as intense as it is remiss. The first argument along this line tells us that (42)

Intension of a degree is measured by distance from the zero degree and remission by its nearness to that degree, but every degree is *distant from* the zero degree by the same amount it is *close* to the zero degree because the mutual propinquity and distance of things do not differ with respect to something.

In terms of the representations we have made of the Calculator's arguments above, this type of argument will not hold up. It

---

(41) A, f. 2r : Circa tertiam positionem tres versantur positiones. Prima enim ponit quod omnis gradus est ita intensus sicut remissus. Secunda ponit quod aliquis est gradus ita intensus sicut remissus et econverso, sed non quilibet. Tertia ponit quod nullus gradus est ita intensus sicut remissus. Et sicut iste positiones ponunt de intensione et remissione ita etiam ponunt de magnitudine et parvitate et de omni alia latitudine que privative vel positive respectu alicuius consideratur (1) vel dupliciter scilicet positive et privative consideratur. Variant : (1) A, C have *contrariatur.*

(42) A, f. 2r : Intensio gradus attenditur penes distantiam a non gradu et remissio penes accessum ad non gradum, sed omnis gradus equaliter distat a non gradu sicut ipse est propinquus non gradui, quia non differunt respectu alicuius propinquitas aliquorum et distantiam inter illa.

is merely pointing out that the numerical value of y is the same for both intension and remission, but the intension and remission are not the same, but rather reciprocals with contrary directions.

While omitting the details of the next arguments in favor of the first supplementary position, which attempts to identify intension and remission, I wish to point out that in the course of these arguments he clearly states as a commonly accepted part of the basic position of measuring intension by distance from zero and remission by propinquity to the zero degree that a " degree of infinite remission is nothing else but a zero degree of intension " (43). This fits the formulation $z = f(1/y)$ mentioned above. Similarly the reciprocal nature of intension and remission is admitted even though the argument using it is considered unsound by the Calculator. He tells us " that a double degree of remission is nothing but a one-half degree of intension and a one-half degree of remission is nothing but a double degree of intension " (44). It is true that the Calculator later modifies these identifications saying that in addition to the fact that intension and remission are reciprocals they differ also in the way we consider them directionally. Hence these statements I have just quoted will be true only for the numerical relationships of intension and remission.

SWINESHEAD after advancing the supporting arguments for the

---

(43) A, f. 2v : Item sequitur eandem realiter esse latitudinem intensionis et latitudinem remissionis, quia intensio habet attendi penes recessum a suo non gradu et remissio penes accesum versus non gradum sue intensionis, quia remissio non potest attendi penes distantiam a non gradu remissionis, quia *non gradus remissionis est gradus infinite intensionis.* Sed quilibet gradus per infinitum distat (1) a gradu infinito intensionis. Ergo si penes huius distantiam haberet remissio attendi sequitur omnem gradum infinite remissum existere. Et sic notandum est quod ab omni gradu remissionis usque ad non gradum remissionis est latitudo infinita et usque ad gradum infinitum remissionis solum latitudo finita consistit, *quia non est aliud gradus infinite remissionis quam non gradus intensionis* et ab omni gradu usque ad non gradum intensionis est latitudo solum finita. Ergo et cetera. Variants : (1) B, C, ES add *a non gradu remissionis quia per infinitum distat* (Note : I have italicized the pertinent lines in the body of the note.)

(44) A, f. 2v : Ex his etiam est notandum quod ab omni gradu remissionis usque ad suum duplum remissionis privative accipiendo (1) remissionem, est in duplo brevior latitudo quam inter ipsum et suum subduplum remissionis, *quia non est aliud gradus duplus remissionis quam gradus subduplus intensionis et gradus subduplus remissionis quam gradus duplus intensionis.* Variant : (1) A has *considerando de remissione.*

first supplementary position, then argues against that position (45). He shows that if you accept this theory and take any two degrees *a* and *b*, with *b* being more intense than *a*, then not only must the proportion of the intension of *b* to *a* be the same as the proportion of the intension of *b* to the remission of *a*, but also it must be the same as the proportion of the remission of *b* to the remission of *a*. But this is not so, for the proportion of the intension of *b* to *a* will be a proportion of greater inequality (i.e., *b*/*a* is greater than one), while the proportion of the remission of *b* to *a* will be a proportion of lesser inequality (i.e., where say *b*/*a* is less than one). Or to put it even more simply, these proportions of the two intensions and the two remissions will not be equals, but rather reciprocals.

After dismissing the first of these supplementary positions, SWINESHEAD advances the second position, namely that there is some degree which is just as intense as it is remiss. As in the case of the first position he supports it first (46) and then refutes it.

In favor of the second [supplementary] position it is argued as follows : Some degree *a* is taken. If *a* is more intense than it is remiss, it is posited that *a* be remitted towards zero degree. Then because intension is remitted to zero and remission will be intended to infinity, it follows that sometime the remission will be just the same as the intension since now the remission is less than the intension.

Also take the mean degree between the intension and remission. Because the remission will be increased equally fast as the intention is decreased—since to increase remission is nothing else than to decrease intension, then in that instant in which the intension will be at the mean degree, the remission will be at the same degree.

Now it is true that intension and remission could have the same numerical value (in our modern formulation when *y* is equal to one), but since they are directional contraries, they are not

(45) A, f. 2v.

(46) A, f. 2v : Pro secunda positione arguitur sic. Capiatur aliquis gradus qui sit *a*. Tunc si *a* intensior quam ipsemet sit remissus, ponatur *a* remitti ad non gradum. Tunc ex quo intensio remittetur ad non gradum et remissio intendetur in infinitum, sequitur quod aliquando erit remissio tanta sicut pro tunc erit illa intensio eo quod non nunc est remissio brevior quam intensio. Item si capiatur gradus medius inter intensionem et remissionem, sequitur quod ex quo eque velociter maiorabitur remissio sicut minoratur intensio, quia non est aliud maiorari remissionem quam intensionem breviari. Ergo in illo intanti in quo illa intensio erit sub illo gradu medio erit remissio sub eodem gradu.

the same, i.e., they are not comparable. But SWINESHEAD does not yet adopt this line of criticism I have suggested, because he has not yet outlined the basic differences between intension and remission as far as being contraries of direction. This he does in supporting the third supplementary position. But at the very end of the essay when discussing certain doubts, he notes what is substantially the criticism which I have outlined. Supposing we have a remission with a value of 4 degrees and also an intension of 4 degrees. Are not these two equal? No, the Calculator tells us, they are not equals, because they are not comparable terms (47).

Omitting the less important criticism that the Calculator now employs against the second position, we can pass immediately to the third supplementary position which he accepts. In brief his argument is : No degree is as intense as it is remiss because intension and remission are not comparable things. Although they can be numerically equal, they are directionally different. But he points out initially that remission can be thought of in two ways. In one way it really is the intension. In the other way it is the reciprocal or privative of intension. The first way of speaking is not often employed and so he will adopt the second way of speaking (48) :

---

(47) A, f. 4v : Contra hoc obicitur sic : stat quod remissio alicuius gradus ut quatuor et sit intensio alicuius gradus ut quatuor. Ergo videtur quod ille sint equales. Negatur consequentia, quia non sunt comparabiles adinvicem ut sic, sicut ante planius est argutum.

(48) A, f. 3v : Sequitur ergo tertia positio ponens quod nullus gradus est ita intensus sicut remissus... Unde illa est positio quam inter ceteras reputo magis veram. Pro cuius intellectu primo est notandum quod remissio potest dupliciter considerari. Uno modo ut est ille gradus intensus vel ut est illa intensio eo quod idem est realiter intensio cum remissione ut est argutum. Alio modo etiam consideratur remissio ut scilicet est privativum respectu intensionis. Primo modo sunt illa eadem, hoc est sic remissum, hoc est sic intensum. Et in illo modo loquendi de remissione potest concedi quod remissio habet attendi penes id penes quid habet intensio attendi, et ita potest dici omnem gradum ita intensum sicut remissum existere. Contra quem modum loquendi rationes adducte contra primam positionem non procedunt. Ille tamen modus non est multum usitatus. Ideo loquendum est de remissione secundo modo, dicta scilicet prout est quid privativum respectu intensionis. Et dicendum est secundum illum modum nullius gradus intensionem sue remissioni correspondere, sicut argumenta contra illas positiones probant liquide.

(Since the second position has been rejected), the third position which posits that no degree is as intense as it is remiss follows... Whence this is the position I judge to be truer. For its understanding it should be noted in the first place that remission can be considered in two ways. In one way it is the intense degree or is like that intension, since in this way intension really is the same as remission, as was argued. In the other way, remission is considered as a privative *(privativum)* with respect to intension. In the first way they are the same. This thing is so intense or is so remiss. In this way of speaking about remission, it can be conceded that remission is to be measured by the same thing as intension and in this fashion it can be said that every degree is just as intense as it is remiss. The arguments adduced against the first position are not valid against this method of speaking. However, this method of speaking is not used much. Hence one ought to speak of remission in the second way as that which is privative with respect to intension. And it ought to be said according to this method that the intension of no degree corresponds to its remission, as the arguments against the two preceding positions clearly prove.

The Calculator now answers the second supplementary position in the light of remission as a reciprocal or privative of intension. Notice in this passage that he uses two analogies to explain the difference between intension and remission.

The first of these analogies is with proportions of greater and lesser inequality. A proportion of greater inequality for the scholastics resulted when the numerator is more than the denominator $a/b > 1$. Now the Calculator's analogy is simply this : Suppose a and b are the same in both kinds of proportions. Then the proportion a/b differs from b/a even though the numbers a and b are the same, that is, even though the proportions are the same according to thing *(res)*. The proportions are said by him to differ according to " reason " (ratio). This means that the proportions differ as to how we consider the members, whether we are considering the proportion of a to b or of b to a. Similarly, intension and remission which are reciprocal so differ, e.g., y and $1/y$ differ in " reason " (i.e., they both vary as a function of y but one as a direct and the other as an inverse ratio). SWINESHEAD goes on to tell us that this rational distinction in proportions had been made in Master THOMAS BRADWARDINE's *Tractatus de Proportionibus*. BRADWARDINE made the distinction primarily because he applied proportions to movement. This application of proportions being made, it was axiomatic for him that movement could only arise when the motive power was greater than the resistance, or if a were the motive power and b were the resistance, when $a/b > 1$, i.e., was a proportion of greater inequality. And

RICHARD SWINESHEAD AND LATE MEDIEVAL PHYSICS   157

movement could never arise when that proportion of motive power to resistance was a proportion of equality, $a/b = 1$, or a proportion of lesser inequality $a/b < 1$. BRADWARDINE then adopts the view that velocity follows a geometric proportion of the original proportion a/b, with a the motive power and b the resistance (49). This formulation permits him to distinguish clearly between proportions of greater inequality, proportions of equality, and proportions of lesser inequality when applied to movement and thus permits him to conclude that where motion is concerned, or geometric proportionality, those three kinds of proportions : greater inequality, equality, and lesser inequality are not comparable (50). And so SWINESHEAD follows him and similarly

(49) For a history of the law of movement in terms of the motivating force and resistance usually associated with the name of ARISTOTLE, consult CLAGETT, *Giovanni Marliani*, etc. Chapter 6. However, this latter treatment leaves out the influence of PHILOPONUS' criticism of ARISTOTLE's formulation, which criticism influenced the Middle Ages through the work of AVEMPACE as reported by AVERROES. The traditional Aristotelean formulation of a dynamic expression of movement in terms of the force and resistance can be represented in modern symbolization :

$$\frac{V_2}{V_1} = \frac{P_2 \, R_1}{P_1 \, R_2} \qquad \text{where } V_2 \text{ and } V_1 \text{ are velocities,}$$

where V_2 and V_1 are velocities, P_1 and P_2 are motive powers, and R_1 and R_2 are resistances.

Or this formulization may be shortened to :
$$V \propto P/R.$$

Now BRADWARDINE and his successors altered this form of the law because of an apparent inconsistency therein, namely that if the force equals the resistance the formula gives a positive value for the velocity, even when experience shows that if force and resistance are equal no movement arises. Thus BRADWARDINE in his *Tractatus proportionum motuum* stated categorically that no velocity could arise unless P were greater than R, i.e., unless the ratio were a proportion of greater inequality. He then altered the basic Aristotelean law which we can rather artificially represent in modern terminology as follows :

$$\frac{P_2}{R_2} = \left(\frac{P_1}{R_1}\right)^n \qquad \text{where } \frac{P_1}{R_1} > 1 \text{ and } n = \frac{V_2}{V_1}$$

Or rewriting this formula :
$$\frac{V_2}{V_1} = \log a \, (P_2/R_2), \text{ where } a = \frac{P_1}{R_1} > 1.$$

Consult THOMAS BRADWARDINE, *Proportiones* etc., Venice, 1505, ff. 14r-v; Paris BN Fonds latin 6559, f. 54v.

(50) He says that they are not comparables because in his formulization, if P_1/R_1 is a proportion of equality, then any other proportion P_2/R_2 which should be greater than the original proportion must be equal to one, for no matter how many times P_1/R_1 is taken (i.e., regardless of what integer n is) P_2/R_2 will equal

158

distinguishes intension and remission, as not being comparable.

The second analogy used by SWINESHEAD to clarify the " rational " (we might say directional) difference which distinguishes intension and remission and proportions of greater and lesser inequality is a kind of vector analogy. When there is a distance represented by line *ab*, the magnitude of the line is the same regardless of whether we proceed from *a* to *b* or from *b* to *a*, but the movements are not the same because the directions of movement are different, i.e., the spaces from *a* to *b* and from *b* to *a* are the same according to thing, but are different rationally since we are considering them from different directional points of view. This analogy is quite apt, for the Calculator told in the introductory lines of his treatise that in one way of speaking intension and remission are movements.

So much for a brief introduction to the significant passage where he makes the analogy. Here follows the main parts of the passage (51) :

to one when $P_1/R_1 = 1$. Or if P_1/R_1 is less than one, then P_2/R_2 must always be less than one, so long as *n* is an integer. Now we have assumed for movement that P_2/R_2 must be greater than one, hence when we are implying geometric comparisons or taking *n* as a series of integers, we can never have P_1/R_1 equal to or less than one, or to put it another way, we can never employ proportions of equality and lesser inequality. So in this special " geometric " way these latter proportions are not " comparable " to proportions of greater inequality.

(51) A, f. 3v : Ideo pro illo argumento dicitur, posito quod *a* sit intensum, quod sua intensio non est maior nec brevior sua remissione, neque etiam sibi equalis accipiendo remissionem privative respectu intensionis... ymo sicut dicitur de proportione maioris inequalitatis et brevioris inequalitatis que proportiones realiter non differunt nisi tantum secundum rationem sicut via ab *a* ad *b* (1) econtra a *b* ad *a* (2), eadem est secundum rem, differunt tamen secundum rationem. Consimiliter habitudo maioris ad minus et minoris ad maius sunt eedem secundem rem, sola ratione differentes, et habitudo maioris ad minus est proportio maioris inequalitatis et habitudo minoris ad minus est proportio minoris inequalitatis. Ideo ille proportiones secundum rationem tantum differunt et proportio maioris inequalitatis non est maior neque brevior proportione minoris inequalitatis, distinguendo proportionem maioris inequalitatis contra proportionem brevioris inequalitatis, ut venerabilis Magister Thomas Braduardini in suo tractatu de proportionibus liquide declaravit... Et consimiliter sicut dictum est omnino de proportione maioris inequalitatis, sic dicendum est de intensione et remissione distinguendo intensionem contra remissionem accipiendo scilicet remissionem respectu intensionis privative... et sic dicendum est quod nullius gradus intensio remissione eiusdem gradus est maior neque brevior neque ei equalis... Et sic salvuntur omnia argumenta que sunt facta contra illam positionem, quia omnia fundantur super hoc, quod illud penes quid attenditur intensio et illud penes

The answer to the argument of the second position is as follows : When it has been posited that *a* is intense, it is remarked that its intension is neither more nor less than its remission, nor even is it equal to it when we accept remission privatively with respect to intension... We speak in the same way of a proportion of greater inequality and one of lesser inequality. These proportions really do not differ except according to reason, (i.e., the manner in which we consider them), just as the way *(via)* from *a* to *b* and the converse way from *b* to *a* are really the same according to thing (*rem*, magnitude), but different according to reason (i.e., direction). Similarly, the ratio of a greater to a lesser quantity and that of a lesser to a greater are the same according to thing (i.e., numbers are the same), but only differ according to reason (i.e., the way we consider the numbers), and the ratio of the greater to the lesser is a proportion of greater inequality while the ratio of the lesser to the greater is a proportion of lesser inequality. These proportions differ according to reason and a proportion of greater inequality is no more nor less than a proportion of lesser inequality, distinguishing a proportion of greater inequality from one of lesser inequality as the venerable Master Thomas Bradwardine had made completely clear in his *Tractatus de proportionibus*... And similarly just as it has been stated everywhere concerning the proportion of greater (and lesser) inequality so it ought to be remarked concerning intension and remission, distinguishing intension from remission and accepting remission privatively with respect to intension... And so it ought to be said that the intension of no degree is more than, less than, or equal to, the remission of the same degree.

In concluding his discussion of the third supplementary position he reiterates that all of the arguments used against it can be solved by considering the " rational " difference of intension and remission when we think of remission as a privative of intension.

The tractate is completed by the advancement of three " doubts " of statements that appear to be fundamental in the position of measuring intension and remission adopted by the Calculator (52) :

(1) Whether uniform acquisition of intention follows from uniform loss of remission.
(2) Whether remission is increased equally proportionally and with equal velocity as intension is decreased.

---

quid attenditur remissio secundum equalitatem et inequalitatem comparantur. Quod tamen (3) remissio penes illud attenditur universaliter (4) est negandum. Variants : (1) A has *a ad d et enconverso;* (2) A omits *ab b ad a ;* (3) ES adds *ut;* (4) ES has *uniformiter.*

(52) A, f. 4r : His ergo completis in illa materia restat ulterius esse dubitandum : primo numquid ex uniformi deperditione intensionis sequitur uniformis acquisitio remissionis. Secundo numquid eque proportionaliter et eque velociter maioratur remissio sicut intensio minoratur. Tertio numquid si a non gradu remissionis continue eque velociter incipiant aliqua duo acquirere de remissione continue manebunt eque remissa.

(3) Whether two things which begin from zero degree of remission to acquire remission equally fast continue to remain equally remiss.

He advances arguments against all three of these statements, but the negative arguments are for the most part based on a misunderstanding of the use of proportions or on impossible cases. And so he shows that the doubts are not justified and the affirmative statements hold, except in that part of the second one which declares that remission is increased with the same velocity with which intension is decreased. He admits that intension is decreased and remission increased equally proportionally, e.g., when intension is *twice* as less remiss, remission is *twice* as great, etc. But since intension and remission are not comparable because they are rationally different according to direction, so then no actual movement of intension (positive or negative), i.e., of increasing or decreasing intension, can be compared with any positive or negative movement of remission. Hence, we are not justified in saying that they move with equal velocity simply because one decreases proportionally as the other increases. This criticism can be pointed up by a specific passage from among the concluding paragraphs of the treatise (53) :

Therefore, it is said, as before, that if intension and remission are not mutually comparable according to equality and inequality, neither should the intensive and remissive movements be compared. For if the movements are comparable, then the things which have been acquired by the movements are comparable, and so it is denied that remission is increased equally swiftly as intention is decreased. However, it is concluded that they take place equally proportionally, etc.

SWINESHEAD has, then, in this treatise been able to examine the various ways of measuring intension and remission, firmly convinced as he was that qualities can be treated quantitatively. He examined and showed the logical difficulties inherent in measuring intension by reference to the decreasing distance to

---

(53) A, f. 4v : Ideo dicitur quod sicut intensio et remissio ut predicitur non sunt adinvicem comparabiles secundum equalitatem et inequalitatem, sic neque motus ad intensionem et remissionem debent adinvicem comparari. Si enim sunt motus comparabiles (1), acquisita per illos motus sunt comparabilia. Et sic negatur quod eque velociter maioratur remissio sicut minoratur intensio conceditur tamen quod eque proportionaliter et cetera. Variant : (1) A has *equales*, other copies as in text.

the maximum degree of intension (and remission by the distance from that maximum). Equally untenable was the position which, although it measured intension correctly, still attempted to measure remission by the distance from the maximum. He was led therefore to advance and support what he believed to be the correct position, namely that intension of qualities must be measured by the distance from the zero degree of intension and remission by the distance to that degree. After advancing that position he further clarified it by showing that no degree at all was equally intense as it was remiss, because intension and remission are fundamentally different in the way we consider them with respect to direction. It is true that they are so related that gain in remission is accompanied by loss in intension, and that this gain and simultaneous loss take place equally proportionally. But the motions of loss of intension and gain of remission cannot be said to take place with equal velocity because this would imply that these movements are comparable and this would not be true.

(University of Wisconsin.)

IV

The Pre-Galilean Configuration Doctrine: «The Good Treatise on Uniform and Difform [Surfaces]».

I. - Introduction.

The importance of the medieval configuration doctrine for Galileo's geometric treatment of uniform acceleration has been recognized and, with some limitation, accepted [1]. Galileo's use of the vestiges of the doctrine is most fruitfully present in his proof of the fundamental uniform acceleration theorem (Theorem I, Proposition I) in the « Third Day » of his *Discorsi e dimostrazioni matematiche intorno a due nuove scienze* [2], where

---

[1] P. DUHEM: *Études sur Léonard de Vinci*, Paris, 1913, III, pp. 375-398, 562-566, 574-583. Cf. P. DUHEM: *Le Système du monde*, Paris, 1956, VII, chap. 6, and particularly pp. 550-561 H. WIELEITNER: *Ueber den Funktionsbegriff und die graphische Darstellung*, in *Bibliotheca mathematica*, 3. Folge, XIV (1913-1914), p. 242; A. KOYRÉ: *Études galiléennes*, Paris, 1939, II; M. CLAGETT: *The science of mechanics in the Middle Ages*, Madison, 1959, chap. 6, and particularly p. 414. A. MAIER sees the medieval doctrine as fundamentally altered, suggesting that all that remains from the doctrine as expounded by Oresme is a « schlichtes Mittel der graphischen Erläuterung und Veranschaulichung von Begriffen, Voraussetzungen, Beweisen und Resultaten ». See A. MAIER: *An der Grenze von Scholastik und Naturwissenschaft*, Rome, 1952, 2nd ed., p. 384. But even if this is correct, the remnant of the doctrine was not without historical significance for Galileo.

[2] *G. G.*, VIII, pp. 208-212, where the first two theorems and the first corollary to the second theorem are given. For an English translation, see M. CLAGETT: [*cit. n.* [1]], pp. 409-418. A discussion of the Merton theorem with the appropriate literature has been given in the same work, chapters 5 and 6, and pp. 630-631 *note*, 646-647, 649, 654-655. Note that the medieval form of the theorem usually emphasized that the *spaces* traversed in equal times were equal, while Galileo's theorem states that the *times* for the traversal of the equal spaces are equal, a change in wording without great mathematical significance. I have pointed out that, while Galileo was almost certainly not acquainted with either of Oresme's treatments of the configuration doctrine, the Merton theorem was published in printed texts at least seventeen times prior to Galileo, and that in four of these accounts the familiar triangle and rectangle appear (M. CLAGETT: [*cit. n.* [1]], p. 414).

Galileo states the Merton College acceleration theorem in slightly different
form and gives a geometric proof employing a right triangle to represent
uniform acceleration from rest and a rectangle to represent uniform motion
at the speed of the middle instant of the period of acceleration. Theorem I
states:

« The time in which a certain space is traversed by a moving body uniformly accelerated from rest is equal to the time in which the same space
would be traversed by the same body traveling with a uniform speed whose
degree of velocity is one-half of the maximum, final degree of velocity of the
original uniformly accelerated motion ».

This theorem is the basis of the proof of his celebrated second theorem;

« If any body descends from rest with a uniformly accelerated motion, the
spaces traversed in any times at all by that body are related to each other
in the duplicate ratio of these same times, that is to say, as the squares of
these times ».

Incidentally, the use of a right triangle for uniform acceleration and a
rectangle for uniform motion also plays an important part in Salviati's
explanation (*spiegatura*) of the first corollary to that theorem which held
that in uniform acceleration from rest the spaces traversed in any number
of equal and consecutive time periods starting from the first instant of
motion « will be related to each other as the odd numbers beginning with
unity, *i.e.*, 1, 3, 5, 7 ... ».

The origins and spread of the configuration doctrine in the fourteenth century from the initial accounts of it by Nicole Oresme [3] in his
Quaestiones super Geometriam Euclidis and his *De configurationibus qualitatum et motuum* through its use by other authors in the fourteenth, fifteenth, and sixteenth century have been the object of considerable research.
However, much work still remains to be done both in the area of the
doctrine's initial acceptance and alteration and in that of its employment

---

[3] In addition to the works cited in *footnote* [1], we should also note that H. L. L. Busard
has published a text of the *Quaestiones super Geometriam Euclidis* (N. ORESME: *Quaestiones super Geometriam Euclidis*, ed. H. L. L. BUSARD, Leiden, 1961, 2 fasc.), and that
I am publishing the first complete edition of ORESME: *De configurationibus qualitatum
et motuum* together with an English translation and commentary (M. CLAGETT: *Nicole
Oresme and the medieval geometry of qualities and motions*, Madison, 1968). This latter
work will give further examples of the use of the configuration doctrine. Incidentally, I
have recently discovered in the newly found N. ORESME: *Questiones in libros physicorum
Aristotelis*, book VI, quest. 8 (MS Seville, Bibl. Colomb. 7.6.30, f. 71r, c. 2), an analogical
use of the configuration doctrine to illustrate the geometric summation of an infinite
series. I give the complete passage in the aforementioned work. I also argue there
the virtual certainty of an earlier dating of his *Questions on Euclid* in comparison with
the *De configurationibus*. I have no doubt also that the *Questions on the Physics* is even
earlier, and dates from about 1350.

in early modern times. This paper is concerned with one of the earliest and most distinctive tracts treating the configuration doctrine, the tract included in Erfurt codex, Amplon. Q. 325, ff. 43^r-45^r. W. Schum in his catalogue of the Amplonian manuscripts suggests that the cursive hand in which this and the succeeding two pieces is written dates from the middle of the fourteenth century [4]. And indeed this hand is very much like other German hands of this period published elsewhere by Schum [5]. Since the scribe does not give his name, nor that of the author, nor even the title, we have to reason from circumstantial evidence for the date, title, and provenience of this work. The key question concerns the time when this codex was put together. We know from the Amplonian catalogue of about 1412 that the codex existed in its present form, at least by that date [6], and, of course, this date must be a *terminus ante quem* for all of the works of the codex. However, certain of the tracts preceding the treatise which we are considering are dated by Schum as being of later hands [7]. So presumably, if Schum is correct, the various items were bound together some time after the composition of our treatise but prior to the preparation of the catalogue of 1412. However, if the pieces were written successively in an already bound codex, as they well might have been, we than have more specific evidence of the date of our tract. For item number 9 of the codex, which immediately follows two small tracts (items number 7 and number 8) in the same hand as our tract (item number 6), is a *Tractatus Iohannis de Wasia de proportionibus*, whose colophon informs us that it was compiled at Paris by Iohannes de Wasia in 1369. Furthermore, we are told by Schum that it is in the hand of Iohannes de Wasia. If this is correct, and if the codex was already bound at this time, then the *De uniformi et difformi* predates 1369. And even though it is in a different hand than Iohannes de Wasia's tract it is at least possible that it stems from his pen ultimately [8], as I suggested in an earlier work [9].

Unfortunately we cannot be precise as to a *terminus post quem* of the tract. The similarity of the views of the author of this tract with those

---

[4] W. SCHUM: *Beschreibendes Verzeichniss der Amplonianischen Handschriften-Sammlung zu Erfurt*, Berlin, 1887, p. 559. See items 6-8, and see *footnote* [7] below.

[5] W. SCHUM: *Exempla codicum Amplonianorum Erfurtensium*, Berlin, 1882, pp. 24-25. See particularly the specimen plate XXII, no. 48, dated 1369.

[6] W. SCHUM: [*cit n.* [4]], p. 812, item 31.

[7] W. SCHUM: [*cit. n.* [4]], p. 559. Items 1-3 are dated by Schum as « spätes 14. Jh. », item 5 as « etwas älter », and presumably items 6-9) (item 8 being the *De uniformi*), since they are all in the same hand, are being referred to under item 8 when he says « In feiner kl. Cursive, die mehr der Mitte des 14. Jh. anzugehören scheint ».

[8] See *note* [41] of the text below.

[9] M. CLAGETT: [*cit. n.* [1]], p. 637.

found in the *De velocitate motus alterationis* of Johannes of Casali and in the *Quaestiones super Geometriam Euclidis* of Nicole Oresme suggests that it was probably written after 1350 [10]. Finally, in view of the German hand and the fact that the codex includes a number of French works (such as commentaries of Buridan and a tract of Johannes de Lineriis), we can tentatively suppose that the work was composed at Paris and written down by a German student or master.

In selecting the title, I have employed the basic title given in the Amplonian catalogue of about 1412, *Tractatus bonus de uniformi et difformi*. One should perhaps add the term *superficiebus* to it in order to delineate clearly the chief emphasis that this tract puts on surfaces [11].

I have said above that the author of this treatise depended for the basic technique of graphical representation of qualities and motions either upon the work of Casali or the tract of Oresme on Euclid, or even on both. For the general distinctions of « uniform », « uniformly difform » and the like, it is clear that our author's ultimate source lies in the works of the English schoolmen of Merton College, and most particularly in the *Regule solvendi sophismata* of William Heytesbury [12], which in a number of instances he seems not to have understood properly. Still, in spite of its dependence on previous works and the mathematical fuzziness and ineptness of its author, the *De uniformi et difformi* includes several novel twists. In the first place, we should note that the author frames his propositions entirely in the sphere of geometry. The propositions, then, relate to uniform and uniformly difform surfaces. It is only after the propositions are given and proved in geometrical terms that the author then suggests, as corollaries, that they apply to uniform and uniformly difform qualities and motions. In framing the propositions entirely in geometrical terms the authors borrows some of the language of the philosophical treatment of uniform and difform qualities and motions and applies it to surfaces. In this respect, the author is perhaps developing what was merely an off-hand description in Oresme's *Quaestiones*, when the latter spoke of the altitude of a rectangle as being uniform and that of a right triangle as being uniformly difform [13].

Another point of individuality in our tract is the generality achieved for uniform and uniformly difform figures. Thus instead of restricting

---

[10] For similarities with views of Oresme and Casali, see *notes* [21-22] and [25] of the text below. On the date of Casali's work, see M. CLAGETT: [*cit. n.* [1]], p. 332. Cf. also A. MAIER: *Die « Quaestio de velocitate » des Johannes von Casale O. F. M.*, in *Archivum Franciscanum historicum*, 1960, pp. 276-306.

[11] See *note* [18] of text below.

[12] See *notes* [29,31,34,36,39,40,43] of the text below.

[13] N. ORESME: [*cit. n.* [3]], quaest. 10, pp. 25-26, lines 32-33, 1-23.

himself to rectangles for uniform surfaces and right triangles for the first kind of uniformly difform surfaces, he posits that the figures need only be parallelograms to be uniform surfaces and triangles to be the initial kind of uniformly difform surfaces [14], although to be sure in the diagrams of some of his propositions the parallelograms are drawn as rectangles and the triangles as isosceles triangles (see Figg. 9, 10, 11, 12). If indeed the author wrote his tract after that of Casali, he seems to have borrowed the horizontal orientation of his latitude lines from the Italian author, apparently rejecting the common vertical orientation given these lines by Oresme and his followers [15]. Finally, we ought to suggest that our author ignores (if indeed he knew of) Oresme's suggestion that the configuration doctrine might be extended to three dimensions to represent a surface quality. Rather he proposes the superposition of two two-dimensional figures, one upon the other [16]. Incidentally, in erroneously framing proposition III, the author shows a woeful lack of knowledge of elementary geometry [17].

* * *

In my text of this work, I have for the most part kept the medieval orthography, such as *sicud, relicum* (for *reliquum*), *-e* for *-ae*. However, where there is a variation in form like *equidistans* and *equedistans*, *ti* and *ci* before vowels (as in *proportio* and *proporcio*), I have tended to retain the now more conventional first forms. Note also that the first two times (lines *9, 12*) the scribe has used the phrase *uniforme difformis* for « uniformly difform » while everywhere else he has used the more common *uniformiter difformis*. I have accordingly employed the latter throughout. I have also kept *equevelociter* as one word, as our author gives it, since this was usually the way it was written in the fourteenth century.

My line numbers indicate the lines of the manuscript, except that when a words was divided in the manuscript so that half of it appeared on one line and half on the other line, I have kept the whole word on the preceding line of my text. This has occured in the following instances:

43^v: line *1-2* (*petran/seuntur*), *25-26* (*extre/mum*), *26-27* (*dica/tis*), *38-39* (*D/I*), *45-46* (*exten/se*);

44^r: *20-21* (*inci/piens*), *22-23* (*E/F*), *36-37* (*quadri/laterum*), *44-45* (*as/ sumptum*);

44^v: *1-2* (*DBC/E*), *38-39* (*mo/ventium*), *52-53* (*pertran/suerunt*);

45^r: *3-4* (*parallelo/grammo*), *5-6* (*doc/tores*), *7-8* (*assig/nant*).

The figures in the manuscript are very crudely drawn. Such alterations as I have made in them are indicated on each drawing. In the variant readings the Erfurt manuscript is referred to by the siglum *H*.

---

[14] See *notes* [19-21] and [27] of the text.

[15] See *note* [21] of text below.

[16] See *note* [23] of text below.

[17] See *note* [30] of text below.

II. - Text.

[*Tractatus bonus de uniformi et difformi*] [18].

43[r]   Notandum est diligenter qualiter debeamus ymaginare quid sit superficies uniformis et quid sit superficies /2/ uniformiter difformis et quid
superficies difformiter difformis.

Unde sciendum quod superficies uniformis debet ymaginari /3/ tanquam superficies linearum equidistantium ita quod huiusmodi in una
sui parte non sit supereminens et in alia parte depressa sed quod /4/
uniformiter in omni eius parte sit equalis latitudinis et extensionis ita
quod omnes linee tracte parallele a lateribus oppositis sint equales.
/5/ Cuiusmodi est figura quadrangularis: vel quadratum [vel tetragonus
longus] vel figura elmuain vel aliqua alia similis elmuain [19].

43[r]   5 [vel tetragonus longus] *supplevi ex Euclide*; *vide notam* — elmuain [1,2] *corr. ex*
elmouin (*vide notam*)

---

[18] This piece has no title in the manuscript. The title I have used is taken from that
given in the Amplonian catalogue of 1412. As I suggested above, one could perhaps
add *superficiebus* to the title. All five propositions are concerned either with uniform
or uniformly difform surfaces or both toghether. It is true that by means of the corol-
laries these propositions are then related to the conventional statements of the Merton
College authors concerning uniform and uniformly difform qualities and motions. One
might therefore argue that a more completely descriptive title would be: « On uniform
and uniformly difform surfaces, qualities and motions ». But the most characteristic
twist of this tract in relationship to other works of the time concerned with the configuration doctrine is the transfer to the geometrical figures themselves of some of the
philosophical vocabulary associated with the description of the intension and remission
of qualities and motions. Thus we find the surface being described as a « uniform
surface » or a « uniformly difform surface » and the like. And furthermore statements
about these surfaces and their relations are made the point of departure for statements
about qualities and motions, given as corollaries. Hence the addition to the title of
superficiebus emphasizes the central and fundamental role of the surfaces.

[19] The interesting feature here is that the author does not limit himself to a rectangle
as an example of a uniform surface, specifying as he does that any parallelogram is a
uniform surface. In fact he mentions specifically a square, a rhombus, and a rhomboid.
In listing the « quadrangular » figures his obvious source is the *Elements* of Euclid
book I, def. 22 (see the Adelard II-Campanus version given in the Ratdolt edition,
Venice, 1482): « Figurarum autem quadrilaterarum alia est quadratum, quod sit equilaterum atque rectangulum; alia est tetragonus longus, que est figura rectangula sed
equilatera non est alia est helmuaym que est equilatera sed rectangula non est; alia
est similis helmuaym, que opposita latera habet equalia atque oppositos angulos equales, idem tamen nec rectis angulis nec equis lateribus continetur ». Note: I have altered the punctuation. Incidentally this passage from Euclid also provides justification for the addition of *tetragonus longus*, whose omission was no doubt an oversight

De hiis [20] dubitaret tamen /6/ aliquis forsan an sit superficies uniformis, et ita dico quod sic: Et propter hoc declarandum suppono quod sit unum latus /7/ huiusmodi figure AB et relicum latus sibi oppositum CD et tertium sit AC [et] quartum sibi oppositum sit BD. Dico tunc /8/ quod ista non debet ymaginari dividi per lineam GH nec per lineam FL sed per lineam GI [vel per lineam] KL, ut patet in /9/ hac figura [Fig. 1]. Isto modo debet ymaginari superficies uniformis.

Superficies vero uniformiter difformis debet ymaginari ut /10/ superficies cadens inter lineas rectas non equidistantes sed ex una parte et uno puncto concurrentes et ex alia parte discurrentes /11/ ita quod una illarum linearum copuletur extremitati unius linee stantis erecte et in aliqua superficie

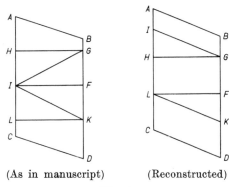

(As in manuscript)   (Reconstructed)

Fig. 1.

---

7 AC corr. ex HG — [et] supplevi licet — [vel ... lineam] supplevi

8 FL corr. ex FI — sed corr. ex scilicet —
9,12 uniformiter corr. ex uniforme

---

on the part of the author. For he certainly would not have excluded an oblong from the possible figures, particularly as he used an oblong, at least in proposition III, to represent a uniform surface. It ought to be remarked that the generality specifically expressed here is implied in the qualifications that Oresme makes in establishing his configuration system in the *De configurationibus*, I.i, when he says that the line of intensity « could be extended in any direction whatever except that it is more fitting to imagine it standing up perpendicularly ». M. CLAGETT: [*cit. n.* [1]], p. 369, lines 55-59). In short, the generality expressed by the unknown author of this tract could have been deduced from a consideration of what I have called Oresme's « suitability doctrine », as applied to uniformity, namely, that the only essential point in choosing the figure to represent uniform intensity is that the intensity or latitude lines be equal.

It should be apparent that in achieving the generality one looses (1) the geometrical convenience of dealing with rectangular figures, and (2) the feeling that the figure approximates the physical situation. In connection with the latter Oresme in both of his treatments thought it fitting to have the intensity or latitude line be perpendicular, for one gets the sense that the intensity lines, while imagined as extending out of the subject in the direction of height, were at least within the « extent » of the subject. On the other hand, if one had the intensity lines at an obtuse angle to the subject base line one would have the feeling that the quality extends beyond the subject in both directions.

Incidentally, it will be noticed from the text and variant reading that I have changed the author's *elmouin* to othe form common in the medieval Euclid manuscripts of the Arabic traditions, *elmuain*, standing as it does for the Arabic *al-mu'ayyin* (= *rhombus*).

[20] I assume that the doubt is only about the propriety of using a rhombus or rhomboid

et alia copuletur alteri |12| extremitati [21], cuiusmodi est figura trian-
gularis et talis superficies vocatur « uniformiter difformis
incipiens a non gradu », ut patet in |13| hac figura
[Fig. 2]. Sed alia est « superficies uniformiter difformis
incipiens a gradu » et illa ymaginatur tanquam super-
ficies |14| cadens inter lineas rectas non equidistantes
que in nulla parte concurrunt in aliquo uno puncto, et
ex una parte ille linee |15| terminantur, scilicet una
super extremo unius linee ex eadem parte et alia super
alio extremo illius linee ex eadem parte, |16| et ex alia
parte consimiliter super extremitatibus unius linee, et illa linea que in
una parte terminatur que est brevior terminat illam superficiem |17|

Fig. 2.

to represent a uniform surface. At least the example he gives is that of a rhomboid.
To make sense out of the example, I have had to correct the figure somewhat and to
substitute AC for the scribe's HG in line 7 and FL for his FI in line 8. The point which
is obviously being made by the author is that, to show that the rhomboid is a uniform
figure, we must note that all the latitude lines are equal but that in doing so we must
be careful not to take the horizontal lines but rather those which are parallel to the
sides, i.e. not GH nor FL but rather GI or KL. These latitude lines play a crucial part
in his description of uniform and uniformly difform surfaces as the succeeding remarks
show.

[21] As I have pointed out above, the author of this tract has generally followed the tradition of Casali in having his latitude lines horizontal rather than vertical, and this
is incontravertibly shown in every figure and throughout the remainder of the text.
But in this definition the author may have been using a source in the Oresme tradition
of vertical (and perpendicular) latitude lines; at least this seems to be the sense of the
phrase *stantis erecte* applied to the line of maximum degree or latitude terminating
the figure uniformly difform beginning from zero. The example that is apparently in
his mind is an Oresme-like right triangle with the perpendicular lines representing degrees
of latitude. Now one might interpret the phrase as simply meaning « perpendicular »
to one of the including lines and not « vertically perpendicular ». In such a case the
perpendiculars would be horizontal and he would be following his customary orientation
of figures. However, regardless of orientation, if the author is saying that the latitude
lines are perpendicular and the figure is a right triangle, he would seem to be abandoning
the generality achieved earlier in describing any parallelogram as a « uniform surface »
where it was of no concern what angle the latitude lines make with side lines of the parallelogram. But he seems once again to return to that generality when he says in line 11
that the figure representing this type of uniformly difform surface is « triangular »
(« not right-triangular »). And indeed the figure he adds in the margin (Fig. 2) is not
a right triangle. It should be further remarked that in the propositions given later the
triangular figures used for uniformly difform surfaces that begin at zero are never
described as right triangles, and indeed in one case (the erroneous Prop. III) the triangle
is specified (although incorrectly) as an equilateral triangle. Finally, it should be
observed that in the succeeding definition of « uniformly difform beginning from a
degree », the limiting lines representing the first and last degrees are not spoken of as
« standing erectly » and in the figure accompanying this definition (Fig. 3) these lines
are given in horizontal orientation and they are not perpendicular to either of the
including lines. And so again we have both horizontal orientation and complete generality of figure.

quantum ad extremum eius remissius et alia terminans illas lineas ex altera parte est longior quam linea sibi opposita et terminat /18/ illam superficiem quantum ad extremum eius intensius, sicud est in ista figura [Fig. 3].

Fig. 3.

Et hic debetis considerare quod in superficie uniformiter difformi /19/ inter lineas non equidistantes plures linee traherentur a punctis in istis lineis non equidistantibus inter que puncta /20/ si esset equalis distantia in una linea signata omnes tales linee essent adinvicem proportionales proportionalitate arismetica [22], sicud in ista /21/ figura [Fig. 4].

Fig. 4. – (Note: I have added letters in brackets). The parallel lines have been made equidistant although in the manuscript no care was taken to do so. The two figures appear on the opposite margins.

Et notanter dixi quod superficies uniformiter difformis est inter lineas rectas non equidistantes ad denotandum /22/ quod superficies que est ad modum vinearum non est uniformiter difformis, sicud in hac figura [Fig. 5].

Superficies vero difformiter /23/ difformis est ubi nulla respicitur uniformiter et quando linee sint tracte, sicud dictum est, non sunt adinvicem proportionales proportionalitate /24/ arismetica, ita quod superficies difformiter difformis debet ymaginari tanquam superficies ubi nulla respicitur uniformiter, sicud in /25/ ista figura [Fig. 6].

Hoc ergo applicando ad qualitates et ad motus considerandum est quod in qualitatibus ymaginantur duo, scilicet intensio /26/ et extensio secundum gradus et subiectum, scilicet intensio secundum gradus et

22 vinearum *corr. ex* venearum *vel* nenearum

[22] This simply means that in a uniformly difform figure (see Fig. 4) where latitude lines $a, b, c, d, \ldots$ are equally distant, $a—b = b—c = c—d = \ldots$. This is an analogous description to Oresme's description applied to a right triangle (*De configurationibus*, I.xi): « if any three points [or the subject line] are taken, the ratio of the distance between the first and the second to the distance between the second and the third is as the ratio of the excess in intensity of the first point over that of the second point to the excess of that of the second over that of the third point, calling the first of these points the one of greater intensity » (M. CLAGETT: [*cit. n.* [1]], pp. 372-373, lines 180-184). As a matter of fact, Oresme had earlier in his *Questiones super geometriam Euclidis* (N. ORESME: [*cit. n.* [3]], quest. 12, p. 34, lines 7-9) given a definition that is almost precisely that found in this tract. Compare also JACOBUS DE SANCTO MARTINO: *De latitudinibus formarum*, prop. 30 (M. CLAGETT: [*cit. n.* [1]], pp. 394, 400) where in distinguishing between uniformly difform and uniformly difformly difform, the author notes that in the former « the increments of degrees equidistant among themselves are equal to each other ».

extensio per subiectum. Et sic possible est aliquam qualitatem esse uniformem /27/ secundum intensionem graduum [et] esse difformem tamen secundum extensionem subiecti. Exemplum sicud si una albedo sit uniformis in gradibus /28/ et extendatur per superficiem uniformiter difformem, scilicet per superficiem alicuius trianguli.

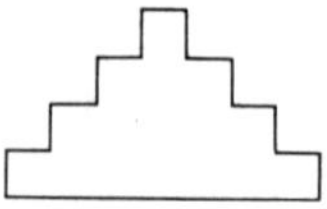

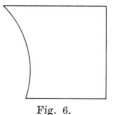

Sed qualitas potest alio modo uniformis /29/ secundum extensionem subiecti et uniformiter difformis secundum intensionem gradum. Exemplum ut si una albedo uniformiter difformis /30/ in gradibus, que esset que ut unum, duo, tres, quatuor, et cetera, extenderetur per unam superficiem quadrati vel per aliquam superficiem uniformem.

Fig. 5.

/31/ Sed qualitas potest tertio modo esse uniformiter difformis tam secundum intensionem graduum quam secundum extensionem subiecti. Exemplum ut si /32/ aliqua albedo uniformiter difformis per superficiem uniformiter difformem extenderetur, puta per superficiem alicuius trianguli. Hoc autem /33/ potest aliquotiens per ymaginationem diversificari: Uno modo sic quod gradus intensior qualitatis sit in extremo superficiei latiori /34/ et gradus remissior in extremo minus lato.

Fig. 6.

Alio modo sic quod gradus intensior qualitatis sit citra extremum superficiciei /35/ minus latum et gradus remissior qualitatis sit in extremo eiusdem superficiei magis lato. Postea ymaginatur de qualitate /36/ uniformiter difformi et eius subiecto acsi unus triangulus alteri esset superpositus, basis ⟨unius⟩ basi alterius, conus /37/ unius cono alterius, sicud posset apparere in tali figura [Fig. 7] [23].

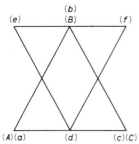

Fig. 7. – (Note: I have added all the letters).

De qualitate uniformiter difformi extensa per subiectum uniformiter /38/ difforme cuius

27 [et] supplevi   29 post si del. H sit   36 ante alterius scr. et del. H albus

[23] We should suppose in connection with this statement that △ ABC representing uniformly difform intensity is to be superimposed on the equivalent triangle abc representing the uniformly difform subject. In case (2) a similar type of superposition in the same plane is employed but in this case the vertex of △ ABC representing uniformly difform intensity, is thought of as lying in the base of triangle def which represents the uniformly difform subject. There is then here no attempt to represent such variations in three dimensions as Oresme suggested in the De configurationibus (I.xvii; see H. WIELEITNER: [cit. n. [1]], pp. 214-215) and earlier in his Questions on Euclid (N. ORESME: [cit. n. [3]], quaest. 10, p. 27, lines 11-26). Incidentally, I know of no other author to use the two-dimensional superpositions used here.

gradus remissior est in extremo magis lato ymaginor recte sicud de duobus triangulis quorum unus /39/ alteri superponeretur et basis unius trianguli cono alterius trianguli et similiter conus unius basi alterius, sicud patet in /40/ figura superius habenda [Fig. 7].

Et notandum idem suo modo quantum ad motum. Potest dici quod quidam est motus uniformis et quidam uniformiter difformis [24].

/41/ Item motuum uniformiter difformium quidam est uniformiter difformis ex parte subiecti, uniformis cum respectu /42/ temporum, et talis est motus semidyameter qui movetur ad motum circuli quia illa semidyameter movetur uniformiter quo /43/ ad tempus; et ideo possibile est quod equalibus partibus temporum vel temporis equale spacium pertranseat quando tamen talis semidyameter /44/ movetur uniformiter difformiter quo ad partes subiecti, quia interim quod movetur uniformiter quo ad tempus eius partes a centro /45/ remotiores moventur velocius quam partes centro propinquiores, eo quod maiorem circumferentiam in eodem tempore et proportionaliter arismetice /46/ describunt secundum quod partes que remotiores sunt a centro maiores circumferentias describunt.

Aliquis vero motus potest esse /47/ uniformis tam quo ad tempus quam quo ad partes subiecti vel mobilis, quod ad presens reputo unum exemplum ut si aliquis /48/ descenderet et tunc in equalibus partibus temporis equale spacium pertransiret et tunc quelibet pars eius moveretur ita velociter, siqui /49/ ita esset, esset motus uniformis quo ad tempus et quo ad subiectum.

43^v Similiter aliquis motus potest esse difformiter / difformis. Exemplum ut esset motus quo in equalibus partibus temporis inequalia et inproportionalia proportionalitate arismetica pertranseuntur /2/ vel aliquo alio modo difformiter se habentia et consimili; vel est dicendum de intensione motus, de qua tamen inferius ulterius /3/ dicam.

Et si aliquis querat quo modo superficies uniformiter difformis diffinitur dicatis quod isto modo superficies uniformiter difformis /4/ est cuius quarumlibet duarum partium sibi invicem immediatarum linearum tractarum ab uno latere non equidistantium ad aliud laterum /5/ longissima que non est in una parte est brevissima que non est in alia, sicud linea que est longissima que non est in illa parte versus /6/ extremum latius est brevissima que non est in illa parte versus extremum strictius.

Et circa istud debetis considerare quod non nulli /7/ magistri hoc hoc vel illi aliquod consimile solent sic exponere: « Linea longissima que non est in parte latiori » est illa linea que non est ibi, /8/ nec eque

---

45 partes *corr. ex* partis   *47* unum: idem ? *H*

43^v *3* difformis [1] *corr. ex* difformiter   *4 mg.* diffinitiones terminorum   *5* longissima [1] *corr. ex* longissimam

---

[24] For the common distinctions between uniform and uniformly difform given here, see M. CLAGETT: [*cit. n.* [1]], pp. 248-249, 326-327, 376, 460-462, 630-634.

longa secum est ibi, nec aliqua minus longa est ibi, et quelibet longior ea est ibi; et «illa est brevissima que non est in parte /9/ strictiori», *i.e.*, illa non est ibi, nec aliqua secum ita brevis vel eque brevis cum ea est ibi, neque aliqua longior ea est ibi, sed quelibet /10/ brevior in parte strictiori est ibi.

Salva tamen istorum gratia istud non est bene dictum; et hoc quo ad illas dicens partes: « linea /11/ longissima que non est in parte latiori est illa que non ibi, » hoc enim potest concedi; et similiter quando dicitur « et quelibet longior illa est ibi » /12/ hoc est falsum, quia ex hoc sequitur quod linea longior basi est ibi, quod tamen est falsum. Et ulterius quando dicunt [post] « et illa est brevissima que /13/ non est in parte strictiori » [quo « illa non est ibi, »] hoc bene est verum. Et quando ulterius dicunt quod « quelibet brevior ea est ibi » verum est quantum ad /14/ superficiem uniformiter difformem incipientem a non gradu; et tamen quantum ad superficiem uniformiter difformem incipientem a gradu /15/ est falsum, nam linea que est brevior quam sit linea terminans partem strictiorem non est ibi. Ergo expositio [linee longissime que non est ibi /16/ et brevissime que non est in alia est insufficiens, ergo] « linee longissime que non est ibi vel in una parte » et « linee brevissime que ⟨non⟩ est in /17/ alia » est insufficiens.

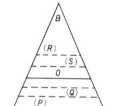

Fig. 8. – (Note: I have added dotted lines and bracketed letters).

Et ergo oportet aliter dicere [25], scilicet quod linea longissima que non est in una parte, scilicet in parte latiori, est /18/ illa que non est ibi, et qualibet longiori illa linea brevior est ibi, et eadem linea nulla brevior est ibi, nec aliqua sibi equalis est ibi; et illa eadem /19/ est brevissima que non est in parte strictiori, et hoc, illa que non est ibi, nec aliqua sibi equalis est ibi, et qualibet breviori illa longior est ibi, /20/ ut patet in hac figura [Fig. 8].

15-16 [linee ... ergo] *in H, sed delendum est?*

[25] The application of the scholastic concept of *maximum quod non* and *minimum quod non* to the definition of uniformly difform qualities or motions was quite popular from about the middle of the fourteenth century. The author's corrected expounding of *linea longissima que non est in una parte* and *linea brevissima que non est in alia parte* follows closely the exposition of Giovanni Casali and that of an anonymous peripatetic author (M. CLAGETT: [*cit. n.*[1]], p. 387; cf. p. 458, lines 175-184). Incidentally the first method of expounding these terms which the author has just rejected is found in the *De proportionibus motuum* of Franciscus de Ferraria of 1352 (M. CLAGETT: [*cit. n.*[1]] p. 503). Referring to Fig. 8, we can see what the author's correction amounts to. When the expounder says in the first part of the exposition that any line longer than *O* is in part *A*, our author objects that a line longer than the base of *A* is not in *A*. And when the expounder says that any line shorter than *O* is in part *B*, this is true for a uniformly difform surface terminating at the apex of *B*, but is not true for a uniformly difform surface that is terminated in a line short of the apex (say, in line *R*), for any line shorter than *R* would then not be in part *B*. So our author (as well as Casali) would substitute

Iterum si aliquis querat alio modo que [est] qualitatis uniformiter difformis, dicatis quod sic: qualitas /21/ uniformiter difformis est cuius quarumlibet duarum partium sibi invicem immediatarum gradus uniformis intensius [26] vel intensissimus /22/ qui non est in una parte eius est remissius [26] qui non est in alia; vel sic: qualitas etc. est cuius gradus intensus est intensior quolibet precedente /23/ et remissior quolibet sequente: Et intensissimus gradus qui non est in *A*, quo quolibet gradu uniformi intensiori gradus /24/ remissior est in *A*, nec aliquis ita remissus est in *A*, ita quod *A* sit extremum intensius; et ille idem est remissimus qui /25/ non est in *B*, quo quolibet gradu uniformi remissiori gradus intensior [est] in *B*, et eodem nullus ita intensus est in *B*, intelligendo per *B* extremum /26/ remissius, ut patet in figura precedenti [Fig. 8].

Iterum si aliquis querat quo modo motus uniformiter difformis in intensione diffinitur, dicatis /27/ quod diffinitur sic: motus uniformiter difformis in intensione est cuius quarumlibet duarum partium sibi invicem immediatarum gradus velocitatis /28/ qui est maxime velocitatis qui non est in extremo intensiori est [ille qui] est remissime velocitatis qui non est in parte remissiori; et intelligatis /29/ hoc iuxta expositionem secundam habitam.

Ex premissis igitur secuntur quedam propositiones.

[I.] Prima est: omnis superficies trianguli correspondet superficiei /30/ quadranguli cuius duo latera opposita sunt equalia linee latera trianguli dividenti per equalia, cuius quadranguli unum de illis lateribus /31/ iacet in basi illius trianguli et relicum iacet per conum [27].

---

21 intensissimus *corr. ex* intensus   *25* [est] *supplevi*   *27* diffinitur *corr.*
ex diffinitioni (*?*)   *29 mg.* propositiones corollarie prima propositio — *ante*
superficies *scr. H et delevi* super

---

the statement in the first part that there always exists in part *A* some line *Q* which is shorter than any line *P* which is itself longer than *O*. Or in the second part he says, there always exists in part *B* a line *S* that is longer than any line *R* which itself is shorter than *O*. The whole treatment of uniform variation by this technique has interesting implications for the analysis of continua.

[26] The use of *intensius* and *remissius* where *intensissimus* and *remissimus* were used by Casali and others was apparently not uncommon (see, for example, the tract of Francischus mentioned in the preceding note: M. CLAGETT: [*cit. n.*[1]], p. 503, asterisk notes). This is no doubt an instance of using the comparative form to stand for the superlative just as one often finds *maior circulus* when what is intended is *maximus circulus* (see M. CLAGETT: *Archimedes in the Middle Ages*, Madison, 1964, I, p. 225 *note*; p. 280, line 19; p. 286, line 90; p. 328, line 2; p. 352, line 56; p. 436, line 3). The justification for this terminology seems to be that the magnitude which is the maximum is in fact greater than any other similar magnitude to which it can be compared in some given set of magnitudes.

[27] Compare this geometric proof with that found in the *De configurationibus*, III.vii see M. CLAGETT: [*cit. n.*[1]], chapt. VI). See also the various proofs of the mean speed theorem given in chapters V and VI of M. CLAGETT: [*cit. n.*[1]]. I have already noted

Exemplum ut si esset triangulus ABC [Fig. 9], latus dividens duo eius /32/ latera per equalia sit DE, et fiat unus quadrangulus cuius latus GF iacet in basi BC trianguli ABC equale linee /33/ DE, et latus sibi oppositum IK transeat per conum trianguli ABC; et sequitur quod superficies quadranguli $IKFG$ correspondet /34/ superficiei trianguli ABC quia superficies quadranguli $FIKG$ excedit superficiem trianguli ex una parte in duobus /35/ triangulis ADI et AEK quos probo esse equilateros [et] equiangulos duobus triangulis DFB et EGC in quibus triangulis /36/ ABC excedit superficiem quadranguli $FIKG$ ex alia parte. Ergo superficies trianguli ABC correspondet superficiei quadranguli $FIKG$.

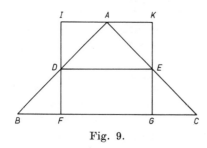

Fig. 9.

/37/ Tenet consequentia unicuique studioso intuenti, quia, si una superficies excedit aliam in una parte in aliquanto et alia eandem ex /38/ alia parte in tanto sibi invicem correspondent.

Ergo assumptum ante probo sic: Et pono quod sint duo latera, scilicet AD et DI /39/ trianguli ADI equalia duobus lateribus DB et DF trianguli DFB, ex eo quod linea eius DE divisit AB in duo equalia; /40/ et angulus D unius angulo D alterius quia contra se sunt positi. Et tunc sequitur per quartam propositionem primi Euclidis relicos angulos /41/ et relica latera esse equalia et totum triangulum toti triangulo esse equalem et consequenter esse probabile triangulum AEK esse /42/ equalem triangulo EGC.

Et ex isto sequitur ulterius quod superficies uniformiter difformis correspondet superficiei uniformi vel /43/ superficies uniformiter difformis correspondet sue medie linee et per mediam lineam debetis intelligere pro nunc [28] illam lineam /44/ que secat duo latera alicuius trianguli in partes equales, et bene vocatur media linea quia equaliter distat ab utroque extremo /45/ superficiei uniformiter difformis, scilicet a cono et

---

34 ante superficiem scr. et del. H super — ante duobus del. H duobus
et DF corr. ex DF et DB   41 ante et [1] scr. H et delevi alia
formi corr. ex uniformiter

39 DB

45 uni-

---

in the introduction that this is perhaps the only one of the medieval proofs not to use a right triangle to represent the uniformly difform surface. An isosceles triangle and a rectangle seem to have been drawn in the figure, and later on f. 45^r, lines 3-4, the author speaks of the parallelogram as a « rectangular parallelogram » (parallelogrammo rectangulo). However the proof is quite general for any triangle with the parallelogram whose base is equal to the bisecting line and whose altitude is equal to that of the triangle, as is clear from a figure like that of triangle ABC and parallelogram $FIKG$ in note [30] below.

[28] See the later discussion of « mean degree », 44^v, lines 39-42.

a basi, i.e., superficies uniformiter difformis correspondet superficiei uniformi extense /46/ uniformiter ad quantitatem illius linee medie duo eius latera secantis ad modum ad quem dictum est.

[II.] Secunda propositio /47/ est quod linea dividens duo latera alicuius trianguli in partes equales est subdubla ad basim eiusdem.

Hoc probo sic: Retenta figura /48/ basis trianguli ABC est linea BC, 44ʳ dupla ad lineam IK que est equalis FG et DE linee secanti latera / per predicta. Ergo etiam linea BC est dupla ad lineam DE.

Consequentia tenet per unam propositionem quinti Euclidis que dicit sic: /2/ « Si alique quantitates equales ad unam tertiam comparentur, tunc illius ad ambas est eadem proportio.» Et assumptum probo, quia linea /3/ FG est equalis linee IK, ut patet ⟨per⟩ diffinitionem quadranguli, et linea BF est equalis linee IA, et etiam linea GC est equalis /4/ linee AK prout in probatione prime propositionis probavi. Ideo est dupla ad lineam IK tota linea BC que est basis.

Et istud oritur /5/ ex eo, quod solet dici quod in omni latitudine uniformiter difformi incipiente a non gradu gradus medius est precise subduplus ad /6/ gradum intensissimum eandem latitudinem determinantem [29].

[III.] Tertia propositio est: cuiuslibet trianguli equilateri [!] quadratum basis est /7/ duplum ad superficiem trianguli hanc propositi [30].

---

46 mg. 2ª propositio   48 IK ... DE *supra scr.* H; (*et scr. et del.* H: DE consequentia tenet per unam propositionem) — *post* linee *scr. et del.* H DE

---

[29] Compare Heytesbury's *Regule solvendi sophismata*, given in M. CLAGETT: [*cit. n.* 1], p. 277, lines 14-16.

[30] As stated this proposition is patently false. It is true not for an equilateral triangle but for one whose altitude is equal to the base. One might argue that a scribe erroneously converted a phrase such as *trianguli equalis basis et altitudinis* to *trianguli equilateri*. But even if this were true, the original proposition would still be rather unsatisfactory, since it would appear to limit the representation of « uniformly difform beginning at no degree » to this specific kind of a triangle. In order to have a proposition general enough to preserve the generality of triangular figure to represent the kind of uniformly difform and to provide a basis for the corollary on motion which the author gives at the end, he should have framed the proposition as follows:
The rectangle arising from the product of the base and altitude of the triangle representing something uniformly difform beginning from no degree is double the triangle. This would be a trite proposition but it would be one from which the corollary would follow and would still preserve the generality of triangular figure implied by the earlier propositions. Incidentally if we assume that the author meant a triangle whose base was equal to its altitude rather than an equilateral triangle, the statement in line *10* « the aforesaid quadrangle is composed of two such squares » would certainly suggest that he thought of the quadrangle which was equivalent to the triangle as a rectangle and

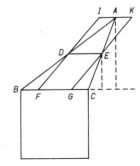

Superficies trianguli est equalis quadrangulo linee dividentis duo latera trianguli, cuius quadranguli /8/ unum extremum est in basi et aliud in cono sed quadratum basis est duplum ad istum quadrangulum. Ergo etiam est duplum /9/ ad superficiem trianguli.

Tenet consequentia de se satis, et assumptum probo quoniam quadratum basis est quadruplum ad quadratum linee /10/ secantis duo latera trianguli, et ex talibus duobus quadratis quadrangulus predictus est compositus, ut patet in tali figura [Fig. 10]. Ergo cum quidquid /11/ est quadruplum ad medietatem sit etiam duplum ad totum, sequitur quod quadratum basis sit duplum ad predictum quadrangulum et per consequens duplum /12/ ad superficiem trianguli que est equalis illi quadrangulo. Et hoc erat probandum.

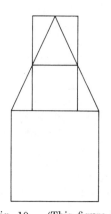

Fig. 10. – (This figure is most crudely drawn in the manuscript. I have redrawn it, employing a triangle having equal base and altitude, rather than one of equal sides).

Sed quod quadratum basis sit quadruplum ad /13/ quadratum linee secantis duo latera trianguli ex hoc patet, quia basis est dupla ad lineam secantem duo latera trianguli /14/ in partes equales, et quelibet est proportio linee ad lineam, talis proportio superficierum huiusmodi linearum proportio duplicata. I rgo si /15/ linee ad lineam est proportio dupla, tunc superficiei ad superficiem illarum linearum est proportio quadrupla, quia proportio quadrupla est duple dupla.

/16/ Et istud oritur ex eo, quod solet dici: si aliquod mobile movetur in aliquo [tempore] vel in aliqua hora continue suum motum intendendo /17/ a non gradu usque ad aliquem certum gradum, et si aliud mobile moveretur in eadem hora continue gradu intensissimo illius /18/ velocitatis uniformiter difformis, secundum pertransiret precise duplum spacium a primo pertransitum.

44^r 9 de *supra scr.* H; (*et scr. et del.* H: per) — quadratum [2] *corr. ex* quadrangulum
        10 quadratis *corr. ex* quadrangulis   12 *post* illi *scr.* H *et delevi* triangulo
        13 quadratum *corr. ex* quandragulum   18 *mg.* 4^a propositio

thus the triangle as an isosceles triangle, as appears to be the case in the figure. However, it can be pointed out that the customary generality of the author's statements could be preserved by supposing merely that the quadrangle is a parallelogram each of whose two halves is equal to the square of the bisecting line, since each half would have its base and altitude equal to one-half of the bisecting line. Thus in a sense one could still say that the parallelogram would be « composed of two such squares ». The accompanying figure illustrates this. For the corollary on motion, M. CLAGETT: [*cit. n.* [1]], p. 278, lines 19-23. Galileo gave a version of this theorem with a geometric proof in his *Dialogue on the two great world systems*, which I have included in M. CLAGETT: [*cit. n.* [1]], pp. 415-416.

[IV.] Quarta propositio est quod /19/ in superficie uniformiter difformi incipiente a gradu, cuiusmodi est superficies quadrilaterum quod abscindit conum trianguli, basis /20/ illius non est dupla ad lineam secantem duo eius latera per equalia.

Probatio: Pono quod si tale quadrilaterum seu superficies incipiens /21/ a gradu uniformiter difformis $ABCD$ [Fig. 11], et linea secans eius duo latera per equalia sit EF, et ulterius /22/ quod basis AB non sit dupla ad lineam EF sed minor quam dupla. Et constituatur unus quadrangulus ad quantitatem linee EF /23/ cuius latera opposita, scilicet unum puta GL iacet sibi invicem basi AB et aliud puta IK super lineam CD que terminat /24/ partem strictiorem illius quadrilateri, qui quadrangulus potest demonstrari esse equalis superficiei illius quadrilateri uniformiter difformis /25/ incipientis a gradu, recte sicud superius demonstratum est

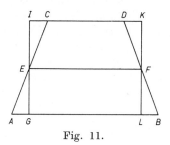

Fig. 11.

illum quadrangulum, scilicet $FGIK$, esse equalem triangulo ABC. Et sic arguo: /26/ sicud linea AB, que est basis quadrilateri $ABCD$ non est dupla ad lineam IK, igitur nec est dupla nec est dupla ad lineam EF /27/ que est equalis linee IK.

Consequentia tenet de se, et assumptum probo quia linea BL est equalis DK et linea AG linee /28/ IC. Et hoc potest probari per equalitatem triangulorum IEC et AEG, ut prius, et linea LG est equalis linee IK. Et /29/ tunc manet in additione CD de tota linea IK. Propter quam linea AB non est dupla ad lineam IK, quia non continet lineam /30/ IK bis sed continet eam semel et cum hoc duas partes, scilicet IC et DK, ex quo ergo continet cum hoc lineam CD /31/ non est dupla ad eam.

Ex quo sequitur ultra quod communiter solet dici inter magistros et doctores: in latitudine uniformiter /32/ difformi incipiente a gradu gradus intensissimus illius latitudinis uniformiter difformis non est duplus ad gradum medium /33/ qui correspondet isti latitudini uniformiter difformi [31], ut patet in hac figura [Fig. 11].

[V.] Quinta propositio est quod omnis superficies trianguli extensa a /34/ cono usque ad lineam duo latera trianguli secantem in partes equales est subtripla ad quadrilaterum quod est reliqua pars totius /35/ trianguli.

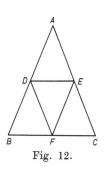

Fig. 12.

Hanc probo sic: Et suppono quod sit unus triangulus ABC [Fig. 12], et linea DE dividens duo latera

22 AB corr. ex A   23 GL corr. ex GB   26 ABCD corr. ex ABC   27 ante Consequentia del. H tenet

[31] See Heytesbury's statement as given in M. CLAGETT: [cit. n. [1]], p. 278, lines 31-33.

illius trianguli, *AB* et *AC*, /*36*/ in duas partes equales, *AB* in puncto *D* et *AC* in puncto *E*. Et sic sequitur quod superficies *ADE* est subtripla ad quadrilaterum /*37*/ *DEBC*.

Probatio: Potraham a puncto de secante qui eque distat a punto *B* [et a puncto] *C* per unum corollarium /*38*/ 31° conclusionis primi libri geometrie Euclidis[32] lineam *DF* equidistantem linee *EC* et lineam *EF* equidistantem linee *DB*, et sic causabuntur /*39*/ ibi due superficies equidistantium laterum, *DFCE* et *EFBD*, et linea *EF* dividit superficiem *DFCE* in partes /*40*/ equales, per unum corollarium 34° conclusionis primi Euclidis[33] quod dicit: « dyametro dividente eam per medium, » et linea *DF* dividit /*41*/ superficiem *EFBD* per medium ut patet per idem corollarium. Ideo triangulus *CFE* est equalis triangulo *DFE* quia uterque /*42*/ eorum est dimidium superficiei *DFCE*, et similiter triangulus *DFB* erit equalis triangulo *DEF*. Ergo omnes isti trianguli tres /*43*/ inter se sunt equales quia omnes tres simul sumpti sunt superficies quadrilateri *DBCE*. Sed unus de istis triangulis, scilicet triangulus /*44*/ *DFE*, est equalis triangulo *DEA*. Ergo totum quadrilaterum est triplum ad ipsum, quod fuit probandum.

Consequentia tenet de se, et assumptum /*45*/ probo quia linea *DF* est equidistans linee *EC* et est sibi equalis, ut patet ⟨per⟩ 34^{am} propositionem primi Euclidis que dicit sic: « Omnis superficies /*46*/ equidistantibus lateribus contenta lineas atque angulos ex adverso collectos habet equales, dyametro dividente eam per medium. » /*47*/ Ex hoc arguo sic: Linea *DF* est equalis et equidistans line *EC*, ut predixi. Ergo similiter erit equalis linee *EA*, que est equalis /*48*/ linee *EC*. Hoc patet per dicta modo: Cum tota linea *AC* sit recta, linea *DF* erit recta et equidistans linee *EA*, et patet hoc /*49*/ per eandem rationem, et per idem patet quod linea *EF* erit equidistans linee *DA*. Ergo dyametro pro linea dyametri *DE* dividente /*50*/ superficiem totam *FDAE* equidistantium laterum per equalia, sequitur triangulum *DEF* esse equalem triangulo *DEA*, quod est demonstrare intentum.

---

37 Probatio: probo *? H* — [et ... puncto] *supplevi*   *38* 31° *corr. ex* 39°   *39* ibi: ibi ibi *? H*   *41 ante* idem *scr.* H *et delevi* est   *42* DFCE *corr. ex* DFC   *44* assumptum *corr. ex* assumpto   *45* 34^{am} *corr. ex* 30^{am}   *49* DA *corr. ex* EA

---

[32] Notice that the text has proposition 39, but since I can make no sense out of that and since some sense can be made out of proposition 31 as the authority, I have accordingly altered the text. We must realize, however, if I am correct, that the corollary referred to is not actually in the text of Euclid. Still it can be proved that the lines which proposition 31 permits to be drawn parallel to *DB* and *EC* respectively would also pass through points *E* and *D* respectively. The Latin is very curious at this point. What I have taken to be *de* in line 37 may be *DE*. But if *DE* were the proper reading, one would find it difficult to emend the text in any simple manner, and in fact several emendments would have to be made. Thus by using Ockham's well known razor I have left the passage as it is.

[33] That is, the last part of the proposition, which the author takes to be a corollary.

44^v / Et ex isto sequitur ulterius, quod totus triangulus ABC sit quadruplus
ad triangulum ADE et sesquitertius ad quadrilaterum $DBCE$.

/2/ Et istud oritur ex communi dicto magistrorum in physica[34],
arismetica, et geometria et astronomia peritorum, scilicet in omni latitudine /3/ motus uniformiter difformi incipiente a non gradu pertransitur
subtriplum in prima medietate temporis ad totum pertransitum in secunda,
et in tota /4/ pertransitur quadruplum ad pertransitum in prima eius medietate et sesquitertium ad pertransitum in secunda.

Et ex hoc sequitur ulterius, quod aliquod /5/ mobile, si moveretur
uniformiter difformiter incipiendo motum suum a non gradu et si in
prima medietate temporis pertransiret unum /6/ pedem, in secunda
medietate temporis non pertransiret nec plus nec minus quam tres quia
in tota hora pertransiret quatuor, ergo et cetera; ⌊et hoc patet⌋ in figura
/7/ immediate facta ⌊Fig. 12⌋.

Circa superius dicta notandum quod primam[35] conclusionem superius
positam posui solum causa exercitii ad intellectus virorum /8/ informan-
dum in speculationibus, nam ipsa est satis dubitabilis et oppositum pos-
sem satis demonstrare per dicta auctorum, quod tamen per presens /9/
prolixitati parcendo omitto.

Iterum et cetera circa predicta ordine premisso debetis advertere de
intensione uniformi et difformi /10/ motuum, quod pro nunc est talis differentia inter istos modos loquendi proportionem maiorationum vel augmenti
et inter maiorationem et /11/ augmentationem proportionis. Nam aliquotiens proportio aliquorum continue maiorabitur quando tamen maioratio
proportionis diminuitur. Verbi gratia, /12/ ut si proportio duorum ad
unum maiorabitur ad unam proportionem que est dupla ad ipsam, scilicet
quadrupla, per additionem duorum /13/ ad duo, et secundo si ipsis quatuor addantur iterum duo, illa proportio augetur et tamen illa augmentatio diminuitur, quia augetur primo proportio /14/ duorum ad unum
per additionem duorum ad duo usque ad proportionem quatuor ad unum,
que est proportio quadrupla, et postea ad proportionem duorum /15/ ad
unum et si quatuor adduntur, proportio maiorabitur quia proportio

---

44^v *1* sesquitertius *corr. ex* sesquitertium   *7, 9, 19* Circa: contra *? H*   *7* primam: illam *? H*

---

[34] For Heytesbury's statement, see M. CLAGETT: [*cit. n.*[1]], p. 280, lines 84-87; 281,
lines 95-100.

[35] The reading may be *illam* instead of *primam*. But since this passage is the first of
a series of comments or clarifications pertinent to the « first » conclusion, it is clear
that whether *illam* or *primam* is the actual reading, it is the first conclusion that is meant.
For those who held some other view than the mean speed theorem, see M. CLAGETT:
[*cit. n.*[1]], pp. 264, *note*[10]. It is here that the author stresses that his purpose is to train
understanding in speculations. This is a particularly interesting remark in view of
Oresme's proemium in the *De configurationibus* telling us that his purpose is to present
a treatise that is not only useful as an exercise but also as *disciplina* (see M. CLAGETT:
[*cit. n.*[1]], p. 367, lines 5-6).

sextupla est maior quam quadrupla et tamen maioratio /16/ proportionis diminuitur ex eo quod prima proportio maiorabitur ad duplum et secunda minus quam ad duplum [36]. Et sic intelligo pro nunc quod /17/ maioratio proportionis continue diminuitur quando proportio continue maioratur; sed hoc est difformiter quia non ad tantum in secunda parte temporis /18/ sicud in prima nec ad tantum in tertia sicud in secunda et sic ultra; cum hoc tamen stat quod proportio continue maioratur eodem modo et illo /19/ stante consimiliter loquendo possibile est quod aliqua proportio continue diminuatur, cuius tamen diminutio continue cum hoc intenditur.

Et circa hoc /20/ est ulterius advertendum quod quando inter aliquos terminos est aliqua proportio et termino maiori in prima parte proportionali addatur aliquantum, /21/ proportio maiorabitur vel saltem aliquantum proportio maiorabitur; et si in secunda ⟨addatur⟩ tantum, iterum maiorabitur proportio sed ⟨non⟩ in /22/ tantum sicud in prima parte proportionali; et si in tertia addatur tantum, et sic iterum maiorabitur proportio et non in tantum sicud in secunda.

Iterum et cetera /23/ si minor terminus diminuatur maiori termino non augmentato nec diminuto proportio maiorabitur. Similiter si residuum diminuatur /24/ deperdendo tantum quantum prius, tunc plus quam prius maiorabitur proportio.

Iterum et cetera si maior terminus minori termino stante /25/ nec aucto nec diminuto in prima proportionali decrescat, proportio diminuitur; et si in secunda parte temporis tantum decrescat, /26/ adhuc diminuitur plus.

Iterum et cetera si minor terminus crescat maiori termino stante, proportio diminuatur; et si in secunda parte /27/ temporis tantum crescat

---

[36] *i.e.*, $6:1 > 2:1$; but $4:2 = 2$ and $6:4 = 3:2 < 2$. It is obvious that the author is not distinguishing between the changing value of a ratio and the changing values of an increment of a ratio. Rather he is distinguishing between arithmetical increase and proportional increase of a ratio, *i.e.*, between the arithmetical increase of a given ratio and the decreasing value of the further ratio between any given value of the given ratio and its predecessor in the series. Thus all he indicates by his example is that if in two successive time periods a given ratio increases from 2 to 4 to 6, then the ratio itself is increasing (*i.e.* by 2 each time), while the further ratio of consequent to antecedent term in the series of changing ratios is decreasing (*i.e.* $4:2$ is greater than $6:4$). That is: $P_1 < P_2 < P_3$ because (where $P_1 + \Delta P = P_2$, and $P_2 + \Delta P = P_3$) $P_1 < (P_1 + \Delta P) < (P_2 + \Delta P)$; at the same time (where ΔP is a constant $P_2/P_1 > P_3/P_2$, or

$$\frac{P_1 + \Delta P}{P_1} > \frac{P_2 + \Delta P}{P_2}.$$

This distinction which he will later apply to local motion (see *note* [37]) he takes from Heytesbury's treatment of the « velocity of augmentation ». Heytesbury had held that such « velocity varies with ratio of the total quantity, composed of the quantity previously existing ». (See C. WILSON: *William Heytesbury: medieval logic and the rise of mathematical physics*, Madison, 1956, p. 129). This distinction between arithmetical and proportional increase is applicable to his next five examples. Incidentally, Swineshead disputed this view of the velocity of augmentation (*ibid.*).

quantum in prima, proportio non diminuitur in tantum sicud in prima parte temporis. Et ista ex predictis satis clare /28/ ut videtur mihi elicio per modum problematum disputabilium.

Consequenter advertendum est quod consimiliter differt « motus intenditur » et « intensio /29/ motus intenditur, » quoniam possible est quod aliquis motus intendatur et quod eius intensio continue remittatur [37]. Exemplum pono, primo quod aliquis /30/ motus intendatur ad duplum et quod postea intendatur ad sesquialterum et postea ad sesquitertium, tunc motus continue /31/ intenditur et tamen intensio eius continue remittitur.

Iterum et cetera possibile est quod aliquis motus continue remittatur et tamen quod remissio /32/ motus continue intendatur. Exemplum ut si aliquis motus primo remitteretur ad duplum ita quod fieret in duplo tardior quam iam /33/ prius erat vel quam iam esset et postea ad quadruplum et iterum postea ad sextuplum, tunc velocitas istius motus continue remittitur et /34/ tamen remissio eius continue intenditur, ideo et cetera.

Iterum debetis advertere quod velocitas intensionis attenditur penes acquisitionem latitudinis /35/ motus recte sicud velocitas motus localis localis attenditur penes spacium linealem maximum descriptum et cetera [38].

Item scitote /36/ quod motus dicitur intendi uniformiter quando in aliqua parte temporis intenditur ad aliquem certum gradum velocitatis et intenditur in tempore sibi /37/ equali ad tantum et cetera [39].

---

[37] Using the conventional meanings of *motus* and *intensio motus* one would expect this passage to mean that there is a difference between « increasing velocity » and « increasing acceleration », At least this is what was intended in the cases discussed by the English authors. [The whole question of the rate of deceleration or acceleration accompanying velocity changes is most brilliantly treated by Richard Swineshead at Merton College prior to 1350 (M. CLAGETT: [*cit. n.*[1]], pp. 292-294, 296-297, 302-304)]. But, in fact, the author here has only distinguished between arithmetical increase and proportional increase. Once more he is applying the concept of Heystesbury concerning the velocity of augmentation but this time to local motion. His first example is one where the velocity arithmetically increases by a constant amount, and thus proportionally decreases (*i.e.*, if the velocity grows from 2 to 4 to 6 to 8, the ratio of a consequent velocity to its antecent is decreasing: *i.e.*, $8:6 < 6:4 < 4:2$.

[38] Here the author appears to have abandoned the velocity of augmentation concept. Note that a similar statement is found in Richard Swineshead's short tract *De motu* (M. CLAGETT: [*cit. n.*[1]], p. 245, lines 28-30): « Sciendum est etiam quod consimiliter habet intensio motus ad motum sicut se habet motus ad spatium ». As I noted there in connection with the Swineshead passage, this can be rendered $A/V = V/S$, and it is clear that there the concept of acceleration (*intensio motus*) has been constructed from the analogy to speed itself, but in the place of the acquisition of space we have the acquisition of speed. I also noted there (pp. 241, 251) the similarity of this with the analysis of uniform acceleration by Galileo.

[39] This definition is ambiguous. The phrases « ad certum gradum » and « ad tantum » obscure the fact that it is the equality of *increments* acquired in any equal parts of time that is crucial for the definition of uniform intension. Much more precise is Heytesbury's

Item notandum quod tunc motus intenditur difformiter quando in equalibus partibus temporis non ad equales /38/ gradus intenditur [40].

Item suppono ad presens quod proportio velocitatum in motibus sequitur proportionem proportionum potentiarum moventium /39/ ad suas resistentias [41]. Et hoc de illo.

Propter summe dicta est advertendum nunc quid sit medius gradus alicuius /40/ latitudinis uniformiter difformis incipientis a gradu vel a non gradu; est enim omnis ille gradus qui ab utroque extremorum distat /41/ recte ad modum ad quem dixi superius quod linea que secaret duo latera superficiei uniformiter difformis equaliter distaret /42/ ab utroque extremorum [42].

Et secundum hoc ponunt quidam talem regulam in hac materia [43],

---

definition (M. CLAGETT: [*cit. n.* [1]], p. 241, lines 91-93): «Uniformiter enim intenditur motus quicunque, cum in quacunque equali parte temporis equalem acquirit latitudinem velocitatis». Compare the *De motu* of John of Holland (M. CLAGETT: [*cit. n.* [1]], p. 249, lines 37-38).

[40] Once again our author has obscured the fact that it is the inequality of the *increments* in equal times that defines the difformly difform intension. He does this by using the rather ambiguous expression « ad equales degrees ». Compare this with the precision of Heytesbury (M. CLAGETT: [*cit. n.* [1]], p. 242, lines 95-97): «Difformiter vero intenditur aliquis motus, vel remittitur, cum maiorem latitudinem velocitatis acquirit vel deperdit in una parte temporis quam in alia sibi equali».

[41] The author gives here the common form of what is now called Bradwardine's function namely $(F_2/R_2) = (F_1/R_1)^{V_2/V_1}$, *i.e.*, the velocity increases arithmetically as the ratio of force to resistance increases geometrically (M. CLAGETT: [*cit. n.* [1]], p. 348). On the use of the expression *proportio proportionum* in expressing this exponential function, see p. 441, *note* 39. Cf. N. ORESME: «*De proportionibus proportionum*» and «*Ad pauca respicientes*», ed. E. GRANT, Madison, 1966.

Notice that here I have translated the phrase *proportio proportionum* by the more precise « ratio of ratios » rather than by the conventional rendering of « proportion of proportions », a change which I urged upon Mr. Grant and which he adopted for his work. Incidentally the use of the phrase *ad presens* in connection with the author's supposition of Bradwardine's function may imply that he intends to discuss it or prove it later. Now in actuality a treatise *De proportionibus* of Johannes de Wasia does begin on 46ʳ (47ʳ!) although to be sure in a different hand, and perhaps it is this tract that our author has in mind. If so, this would substantiate the suggestion made above that this tract was composed by Johannes de Wasia. However, in rebutal one might say that « the ratios of velocities in motions » is such a common subject that the author may have simply intended to leave the subject alone as not being particularly germane to the present argument.

[42] Unlike the last four theorems, the first theorem was restricted simply to the equality of the uniform and uniformly surfaces. And so now, the author intends to relate that theorem to motions and qualities, starting out with an observation on « mean degree ».

[43] The rule seems first to have been stated by William of Heystesbury in 1335 (M. CLAGETT: [*cit. n.* [1]], pp. 262-263), and was many times stated and proved in the course of the fourteenth, fifteenth, and sixteenth centuries, as I have shown in chapters 5, 6 and 11

quod omnis latitudo motus vel qualitatis uniformiter /43/ difformis sive incipiat a gradu sive a non gradu cum ad aliquem gradum sit terminata in extremo eius intensiori suo medio /44/ gradui corresponderet, i.e., tanta est precise quanta est latitudo uniformis sub medio gradu latitudinis uniformiter difformis; /45/ vel sicut si aliquod mobile moveretur uniformiter difformiter intendendo motum suum et aliud in eadem hora moveretur /46/ uniformiter medio gradu illius latitudinis uniformiter difformis, talia mobilia equalia spacia pertransirent et equevelociter /47/ moverentur.

Hanc probo sic [44]: Si *A* mobile a *C* gradu intenderet motum suum usque ad gradum duplum ad *C*, et [si] *B* /48/ ab eodem *C* gradu remitteret motum suum uniformiter [45] usque ad non gradum latitudinis motus, tunc *A* et /49/ *B* tantum perransiret in hac hora describendo spacia linealia quantum continue describerent sibi continue [in] ista hora *C* /50/ gradu, [i.e.,] quantum si continue *C* gradu fuissent mota—A in prima medietate hore *C* gradu [et] *B* similiter in secunda medietate *C* gradu—et non plus /51/ nec minus.

Probatio: quantumcunque *A* magis pertransit intendendo motum suum a *C* gradu usque ad gradum duplum ad *C* in prima medietate /52/ hore tanto *B* minus per sui motus remissionem a *C* gradu usque ad non gradum pertransitum. Ergo precise tantum ambo pertransiverunt /53/ quantum continue *C* gradu uniformiter fuisset pertransitum ab eis in ista tota hora. Ergo motus ipsorum correspondet *C* /54/ gradui qui est medius inter ambas latitudines *A* et *B*, quia si *B* in prima medietate intensisset [!] motum suum a ⟨non⟩ gradu usque /55/ ad gradum *C* et *A* in secunda parte hore ab eodem *C* gradu intendisset motum suum usque

45ʳ ad gradum duplum / ad *C*, tunc *A* et *B* tantum latitudinem quantum prius acquisivissent nunc acquirunt. Ergo cum /2/ ista latitudo esset uniformiter difformis et *C* medio gradui corresponderet, sequitur propositum.

---

47 duplum *corr. ex* equalium — [si] *supplevi*   *48 post* uniformiter *scr. H et delevi* difformiter (*vide notam* [45])   *49* [in] *supplevi*   *50* [i.e.,] *supplevi* — C² *corr. ex* et — [et] *supplevi* — B *corr. ex* A   *52* minus ... motus *corr. ex* per sui minus

45ʳ   *1 post* acquisivissent *scr. et del. H* motum suum usque ad gradum

---

of M. CLAGETT: [*cit. n.* [1]]. Because of the popularity of the rule it is difficult to say which author or authors provided the wording of the rule adopted by our author. My feeling is that he depended on both Heytesbury and Casali.

[44] This type of proof was developed by the Merton College authors (M. CLAGETT: [*cit. n.* [1]], pp. 287-288, 298-300). In point of actual wording it is closest to the later proof of Blasius of Parma (*ibid.*, pp. 404-405).

[45] Notice that the Latin phrase is actually « remitteret motum suum uniformiter difformiter ». In view of what the author is proving. « uniformiter difformiter » should be simply « uniformiter ». He was no doubt thrown off by the fact that the motion is one of uniform difformity, while the change of motion is simply uniform.

Et est recte de latitudine uniformi /3/ C gradus velocitatis et de latitudine uniformiter difformi cuius C est medius gradus sicud superius dixi de illo parallelogrammo /4/ rectangulo cuius unum latus erat in basi trianguli et relicum in cono et [de] superficie illius trianguli, quod parallelogramum /5/ fuit extensum in latitudine ad quantitatem linee secantis duo latera trianguli in partes equales.

Ex isto inferunt ulterius isti doctores /6/ quod possibile sit aliqua duo mobilia moveri in hora et in qualibet parte illius hore unum illorum velocius movetur alio et /7/ in nulla parte hore moventur equevelociter et tamen in fine hore equalia spacia describerent [46]. Causam huius assignant /8/ quia quamvis in nulla parte temporis equevelociter moveantur tamen in toto tempore vel in hora tota sumendo totam velocitatem [47] /9/ ista mobilia sunt equevelociter mota. Et hoc est verum de talibus mobilibus quorum unum moveretur latitudine motus uniformiter /10/ difformi incipiente a non gradu et relicum moveretur medio gradu eiusdem latitudinis uniformiter difformis, talia enim in fine hore /11/ equalia spacia describerent ex eo quod primum mobile quod uniformiter difformiter movebatur in prima medietate temporis de sua /12/ velocitate amisit in comparatione ad relicum mobile quod uniformiter movebatur medio gradu motus uniformiter difformis /13/ tantum in secunda medietate temporis sibi acquisivit. Et hoc de isto.

---

3 C[1] *corr. ex* cuius   *7* tamen *bis H*   *8* totam *corr. ex* totum   *12 post* velocitate *scr. H et delevi* et

---

[46] Cf. *Tractatus de sex inconvenientibus* (Venice, 1505), sig. h-h2*v*, particularly in the proof of the 5th « inconvenience », where a case is posited of Socrates moving uniformly difformly and Plato moving uniformly at the mean degree, and the motions are said to be equal only at the end of the hour because by that time they will have traversed equal spaces. However, the case differs slightly from that suggested by our author here, since it concentrates on the second half of the hour where Socrates is always moving faster than Plato and yet, when the whole motions (including that of the first half as well) are considered, Socrates by the end of the hour has moved equally fast because he has traversed the same space. Incidentally, the author of the *Six inconveniences* seems to have been at Oxford, or in touch with the Oxford authors, in the 1340's. He quotes Bradwardine and Heytesbury and in turn is quoted by JOHN DUMBLETON (*Summa naturalium*, Vat. lat. 954, 8*r*).

[47] For the concept of « total velocity », see M. CLAGETT: [*cit. n.*[1]], p. 364.

V

THE USE OF POINTS IN MEDIEVAL NATURAL PHILOSOPHY AND MOST PARTICULARLY IN THE „QUESTIONES DE SPERA" OF NICOLE ORESME

When I, as a medievalist interested in the history of physics, confronted the physics of Boscovich with its extensionless atoms, I found myself
reflecting upon the widespread discussion of the nature of points that took
place in the scholastic circles of the fourteenth century. Actually, there
were three interrelated currents of discussion going back to antiquity that
bear on the problem: (1) Discussions of the nature of matter or substance:
whether substance is continuous or atomic; whether, if continuous, there
are natural minima below which substance would not be recognizable;
whether, if atomic, the atoms have magnitude and not qualities; or (as,
for example, was held by certain theologians of Islam[1]) whether the atoms
are without magnitude and yet possessed of the qualities of color, taste,
smell, hot, cold, moistness, and dryness, and so on. (2) Discussions of the
nature of the mathematical continuum and its relation to the so-called physical
continuum; a current culminating in the extraordinary Tractatus de continuo
of Thomas Bradwardine, written after 1328 at Ofxord. (3) The discussion
and continuous use of certain points, such as the center of gravity of the earth
and center of the world, in the context of the physical nature of the cosmos.
There is an overwhelming body of literature, old and new, on the first two
currents, and I shall not try to do anything with these currents except to
mention certain arguments that appear pertinent to my discussion of the
third current as particularly illustrated by the use of points in Nicole Oresme's
Questiones de spera, written in the 1350's at the University of Paris.
   It is well known that Archimedes was the principal author of the
fruitful concept of center of gravity. It was of course a concept, however
it might indirectly have been connected with experience, as Mach has suggested, that came immediately from mathematics. I do not suggest that
Archimedes believed in the physical reality of the point designated as center
of gravity, but it is quite apparent that he believed that in the description
of physical events mathematics was an important and useful technique.

---

[1] See O. Pretzl, „Die frühislamische Atomenlehre", Der Islam, Vol. 19 (1931),
120; H. Wolfson, „Atomism in Saadia", Jewish Quarterly Review, New Series, Vol 37
(1946), 116; cf. S. Pines, Beiträge zur islamischen Atomenlehre (Berlin, 1936).

216

The fruitful relationship between mathematics and mechanics was underlined in the introduction to his treatise *On the Method*,[2] although in that passage Archimedes was primarily concerned with the help that mechanics might give to mathematics rather than the opposite. One perhaps can assume that with Archimedes the concept of center of gravity as a point was a mathematical concept determined by symmetry and related to the mental transformation of physical objects into homogeneous mathematical entities like lines, surfaces, and geometrical solids. We may further assume that for Archimedes center of gravity was at least a grand heuristic device reflecting the importance of mathematics, but without posing anything about the physical reality of a point.

Now we must realize tnat such a concept when joined with Aristotelian ideas about the nature of the finite physical world introduced into natural philosophy two interrelated „points" that seemed to some to have a kind of physical reality: the center of gravity of the earth as a whole and the center of the world. There was a great deal of discussion within the context of natural philosophy of these two points in various commentaries — antique and medieval — on the works of Aristotle. I should like to note in particular the discussions of the eminent fourteenth century schoolman, Nicole Oresme. But before doing so let me call your attention to a summary written by al-Khazīnī, an Arab mathematician of the twelfth century, where concepts of the two points are delineated in a manner quite widely accepted.[3]

> Heaviness is the force with which a body is moved toward the center of the world. A heavy body is one which is moved by an inherent force constantly towards the center of the world. Suffice it to say, I mean that a heavy body is one which has a force moving it towards the central point and constantly in the direction of the center without being moved by that force in any different direction ... the body not resting at any point out of the center .. until it reaches the center of the world .. Any heavy body at the center of the world has the world's center in the middle of it; and all parts of the body incline towards the center of the world; and every plane projected through the center of the world divides the body into two parts which balance each other in gravity, with reference to that plane ...

All of this would have been accepted by Oresme and most of the schoolmen at the universities of the fourteenth century. One of the consequences of accepting this doctrine was the introduction of a widely disputed problem as to whether the earth was in fact at rest in the center of the world. It generally was reduced to two supplementary problems. The first was whether the earth rotated at the center of the world on its axis, and while this is an extremely interesting question, which has been much examined by Duhem and others, it does not bear directly on the question I am posing concerning the centers of gravity and of the world. The second sub-problem is of more direct interest; it is one which concerns whether or not the earth is constantly moving in an oscillating motion of translation through the center of the world. Nicole Oresme, following his master John Buridan, takes up this problem

---

[2] Archimedes, *Methodus*, edit. J. L. Heiberg in *Archimedis opera omnia*, Vol. 3 (Leipzig, 1913), 428.

[3] M. Clagett, *The Science of Mechanics in the Middle Ages*, (Madison, 1959) 58—59.

in a work entitled *Questiones de spera*, a work which is extant only in manuscripts.[4]
      It is the third question, „whether the earth is naturally at rest in the
center of the world,“ which first interests us. Initially he distinguishes three
different centers: 1) center of the whole world — „a point equidistant from
all parts (i. e. points) of the convex surface of the whole world;“ 2) center
of magnitude of the earth — a point such that if a plane were passed through
it to divide the earth into two parts „there would be no greater volume in
one part than in the other;“ and 3) center of gravity of the earth — the point
through which a similar dividing plane passes such that „there is no more
gravity in the one part than in the other.“ After a discussion of these distinct
points, Oresme presents a number of conclusions: (1) The earth can be
said to be properly located when its center of gravity is at the center
of the world, and the two are imagined to be the same point. (2) The
whole earth is not uniformly heavy in all of its parts since there are climatic-
geologic changes which bring about uneven features on the earth's surface.
If it were perfectly uniform, the whole earth would be covered with water.
(3) Hence, with the unevenness and nonuniformity of the earth accepted,
it is evident that the center of gravity and the center of magnitude of the
earth are not the same point. And so (4) if the center of magnitude were
at the center of the world, the earth would not be properly located. (5) Moreover, if there were no air surrounding the earth to act as a resistance to its
motion of translation, i. e., if the earth were in a vacuum, it would be infinitely easy to move the earth. Thus even if Socrates walked on the earth, he
would change the center of gravity of the earth and thus produce a slight
motion. The same thing would take place if a piece of wood burned. But
actually (6) as the result of the resisting air, the earth does not move in response to such small changes, and thus one would say that the earth is not
in fact — or at least it could be that it is not — in its natural place at the
center of the world. Still it is possible that enough gravity can be added
to one part that the resistance of the air is overcome. In this case the earth
moves, the new center of gravity seeking the center of the world. He adds
that it is possible for the earth as a whole to be properly located and yet
to have no part of the earth at its center. The case he imagines is that of
a hollow stone at the center of the earth, a stone so located that its center
of gravity is at the center of gravity of the whole earth. In such a case there
would be no matter at all at the center of the world. These are some of the
possibilities enumerated by Oresme. In the course of completing the question,
he does note that a „center“ is not properly a place (*locus*) but is rather a
way of locating (*ratio locandi*) something which has weight. This becomes
evident in the definition of the true place (*verus locus*) of the earth: „That
surface is the true place of the earth which surrounds the earth when the

---

      [4] An edition of the Oresme tract is under preparation by a student of mine, Garrett
Droppers. It will be based on there manuscripts: Florence, Bibl. Ricc. 117, 125r—135r;
Vatican lat. 2185, 71r—77v; Venice, Bibl. Naz. Marc. VIII, 74, 1r—84. The third question
which I paraphrase below begins on 126r in the Florence MS and 72r in the Vatican MS.
For Buridan's discussion see his *Quaestiones super libris quattuor de caelo et mundo*, ed. of
E. A. Moody (Cambridge, Mass., 1942) 158—59, 231—32.

center of gravity of the earth is at the center of the world." This is an Aristotelian definition of place in the full cosmos, namely the innermost surface of that which contains the thing placed. Thus these centers have position but they must not be identified with „place".

Oresme also brings the discussion of these two points, the center of gravity of the earth and the center of the world, into another question hotly disputed by the schoolmen, namely that of the possible plurality of worlds. Aristotle had of course assumed and declared that there was only one cosmos, ordered into the heavenly spheres and the sublunar order of the four elements with the earth at the center. But in 1277 it was made clear in the doctrines condemned at Paris by Étienne Tempier that, however reasonable in philosophy the doctrine of a single cosmos might be, one could not, without falling into heresy, say that it is impossible for God to make a plurality of worlds, for this would place a limitation on the plenitude of God's power. Henceforth schoolmen began to admit the possibility of a plurality, although in general not believing in it. Oresme, in his French commentary on the *De caelo* of Aristotle, speculates on how the possibility might be a reality in the following interesting passage,[5] a passage that obscures the carefull distinction he is wont to make regarding „centers" and „places."

> All heavy things of this world tend to be conjoined in one mass *(masse)* such that the center of gravity *(centre de pesanteur)* of this mass is the place or center of this world, and the whole constitutes a single body in number. And consequently they all have one [natural] place according to number. And if a part of the [element] earth of another world was in this world, it would tend towards the center of this world and be conjoined to its mass ... But it does not accordingly follow that the parts of the [element] earth or heavy things of the other world (if it exists) tend to the center of this world, for in their world they would make a mass which would be a single body according to number, and which would have a single place according to number, and which would be ordered according to high and low [in respect to its own center] just as is the mass of heavy things in this world ... I conclude then that God can and would be able by his omnipotence *(par toute sa puissance)* to make another world other than this one, or several of them either similar or dissimilar, and Aristotle offers no sufficient proof to the contrary. But as it was said before, in fact *(de fait)* there never was nor will there be any but a single corporeal world ...

We have to this point illustrated in some detail Oresme's use of these points in a physical discussion, but we have not really cast sufficient light on his views as to the ultimate nature of these points except to affirm that they are not properly „places" (in the Aristotelian usage of that term) but rather are positions. We do know that Oresme was conscious of the necessity of discussing the nature of points in general, a necessity arising from his usage of these terms, for he introduces as a first problem in his *Questiones de spera*, „whether the definition of a point [given by Euclid] which holds that'a point is that which has no part' is a sound definition."[6] Before sug-

---

[5] Clagett, *Science of Mechanics*, 592—93. For the original text, see Nicole Oresme, *Le Livre du ciel et du monde*, ed. of A. D. Menut and A. J. Denomy in *Mediaeval Studies*, Vol. 3 (1941), 243—44.

[6] See the manuscripts noted in footnote 4: Florence MS, 71r, c. 2; Vat. MS 125r.

gesting Oresme's answer we ought to note that Aristotle and thus most of the schoolmen held that a point is a position, limit, or terminus in a continuum. In fact, for them, the character of a continuum is determined by (a) its continuous divisibility *ad infinitum*, and (b) the mutual connexity of its parts so that upon division the point terminating the one part is also the point beginning the other and thus is possessed in common by both parts. Needless to say, in this view the point is without magnitude and the continuum is not *composed* of points but only *contains* points. As a mathematical entity the point has only *potential* existence in nature — as do all mathematical entities — and it is only because of this potential existence in nature that we are able by the process of abstraction to give it a kind of *actual* existence in our minds. Fundamentally, it is because mathematical entities are indeed abstractions from nature that we can apply mathematics to nature, according to the Aristotelian.

Now Oresme jumps into the discussion immediately, presenting three ways to conceive of a point, rejecting the first and noting that the other two are both probable ways. He says:[7]

> For the solution of this question, there is one way which posits that a point is nothing other than a body of the genus of quantity or substance, if quantity is substance. But the supporters of this view have to concede that a point is divisible and that it is moved and make many other similar concessions ... But all of these conclusions (or concessions) do not accord with the ancient expositors and authors. Accordingly I shall not treat this way. Now in addition to this there are two other, probable ways. The one is that a point does not exist in the nature of things but is only feigned by imagination. The other is that a point is an indivisible accident which is reductively posited in the genus of quantity.

While Oresme speaks of both of these as „probable ways," in this first question he produces arguments only for and against the second of the two ways. That he does not argue determinatively for this position is shown by the whole tentative form of the argument. The substance of the argument is that it could be supported that the point is an „indivisible accident in the soul" in spite of the many arguments that would appear to be against it. It is not clearly stated but this way probably accepts a kind of potential existence in nature of the sort that Aristotle attributes to mathematical entities. As the result of his argument he would modify the Euclidian definition as follows: a point is that which has no part but does have position in a continuum.

Now in the second question which is entitled „whether the definition of a line holding that a 'line is length without breadth whose extremities are two points' is a sound one" Oresme argues in some detail for the first of the

---

[7] *Ibid.* (in the text established by Mr. Droppers): „Sciendum quod de solutione istius questionis est una via, que ponit quod punctus non est aliud quam corpus de genere quantitatis vel substantie si quantitas sit substantia. Et isti habent concedere quod punctus est divisibilis, et quod movetur, et sic de aliis ... et quia non bene concordant cum dictis antiquorum expositorum et auctorum, ideo non pertracto eam. Et preter istam sunt alie due vie probabiles. Una est quod punctus non est aliquid in rerum natura sed solum fingitur per ymaginationem. Alia est quod punctus est unum accidens indivisibile quod ponitur in genere quantitatis reductive."

two probable ways mentioned in the first question, namely that points, lines, and surfaces are fictitious entities of the imagination. He introduces the argument with the following general remarks:[8]

> It was seen above [in the first question] how it can be supported that points, lines, and surfaces are certain accidents made distinct from the body and in the genus of quantity. Now it remains to be seen how one can support another way which posits that lines, points, etc. are nothing *(nihil)* but are only imagined to exist.

As in the first question he argues the position by suggesting counter-arguments which he then refutes. To the argument that because a point is imagined to exist it is therefore something that exists in nature or otherwise the imagination is false, Oresme answers that one must distinguish between „imagining" something and „believing" or „opining" something. If mathematicians „believe" that points exist, Oresme argues that they believe so falsely. For imagination properly is neither true or false. Or to the argument that there is a „science" of points and lines and therefore they must exist since one cannot have a science of non-being, Oresme answers indeed such a science can exist and it is called a „hypothetical science" of the sort that includes such propositions as „if a line exists, it is length without breadth." To the argument that the center, axis, and poles of the world, as well as place and surface, are existing things since the astronomers assert propositions about these entities such as the proposition that the place of the earth is that toward which the earth is moved and that the place has a conservative power *(virtus conservativa)*, and accordingly points and lines are real entities, he argues that they are speaking loosely, and all such statements (if they have any validity) can be explicated in terms of bodies, leaving points like the center of the world and its poles as merely imagined entities. In the course of the argument, he refutes in the manner I have already indicated, the assertion that the center of the world is a „place" in the Aristotelian sense of that word. He also refutes mathematical arguments transferred to the physical world, such as that a sphere touches a plane in only one point. The up-shot of the whole argument is that like the position supported in the first question to the effect that these mathematical entities are accidents in the soul, it can also be argued with some probability that they are simply imagined enti-ties. While Oresme does not argue certainly for either of the two probable positions, he quite clearly refutes the view holding that they are *separate* and *real* physical entities.

Turning away from the *Questiones*, we can find some still tentative support for the view that points are fictitious entities in the first chapter his famous *De configurationibus qualitatum et motuum*, wherein he proposes a two dimensional coordinate system to represent the distribution and change of

---

[8] *Ibid.* (in Mr. Droppers' text): „Visum est supra quomodo potest substineri quod puncta, linee, et superficies sunt quedam accidentia distincta a corpore de genere quanti-tatis. Nunc restat videre qualiter alia via potest substineri que ponit quod linee, puncta, etc nihil sunt, sed solum ymaginantur esse".

qualities in a subject and various kinds of local motions. The first paragraph tells us:[9]

> Every measurable thing except numbers is conceived in the manner of continuous quantity. Hence it is necessary for the measure of such a thing to imagine points, lines, and surfaces and their properties in which, as the philosopher wishes, measure or ratio is immediately found ... And even if indivisible points or lines do not exist, still it is necessary to conceive of them mathematically for the measures of things and the recognition of their ratios.

Later in the same treatise he recognizes that the procedure of using mathematical entities as if they were real occasionally leads to paradoxes and difficulties which are not actually in the nature of physical things but are rather resident in the mathematical devices used for their measure.[10] I think then, that, if pressed, Oresme would probably have held that the points such as center of gravity and center of the world are also imaginative devices that help us understand the world rather than physical realities.

But even of this is so, his constant reiteration of these concepts in a physical context gives them a kind of quasi-physical reality as limits of some kind. It is perhaps this kind of usage that leads later to the concepts of points as centers of force or some kind of activity, and thus leads to Leibnitz and even ultimately to Boscovich.

---

[9] Bruges, Stadsbibl. 486, 159 r: „Omnis res mensurabilis exceptis numeris ymaginatur ad modum quantitatis continue. Ideo oportet pro eius mensuratione ymaginari puncta, lineas, et superficies aut istorum proprietates in quibus, ut vult Philosophus, mensura seu proportio per prius reperitur ... Et si nichil sunt puncta indivisibilia aut linee, tamen oportet ea mathematice fingere pro rerum mensuris et earum proportionibus cognoscendis."

[10] *Ibid.* (Part III, Chap. 4), 171v, c. 2. „Item cum prior casus de alteratione subiecti AB non videatur naturaliter impossibilis et tamen naturaliter impossibile est aliquid subito fieri de summo frigido summe calidum et ita de aliis,inde potest sumi argumentum ad probandum quod punctus non est aliquid realiter indivisiblie nec linea est aliquid nec superficies, quamvis eorum ymaginatio sit conveniens ad rerum mensuras melius cognoscendas, ut tactum fuit in primo capitulo prime partis. Multa alia possunt faciliter inferri iuxta ymaginationem predictam que tamen ipsa ymaginatione non prohibita viderentur quibusdam aut nimis difficilia, aut forsan impossibilia."

VI

FRANCESCO OF FERRARA'S
" QUESTIO DE PROPORTIONIBUS MOTUUM "

The editor presents the complete text of the *Questio de proportionibus motuum* composed by Francesco of Ferrara at Padua in 1352. The text has been prepared from its unique copy in Oxford, Bodleian Library, MS Can. Misc. 226, 58r-63r. An introduction by the editor indicates briefly the potential importance of the tract for the evaluation of the influence in Italy of Thomas of Bradwardine's *Tractatus de proportionibus*, written at Oxford in 1328 and constituting the point of departure for Francesco's work. The editor also establishes with virtual certainty that the unique copy of Francesco's treatise is in the hand of the author. The text is supplemented by an extensive *Index verborum* that includes philosophical and scientific terms, as well as the names of the authors on whom Francesco depended.

The development of the quantitative treatment of qualities and motions in the fourteenth century has received considerable scholarly treatment in the last generation. One of the problems connected with describing this development is to trace the spread of English treatises on natural philosophy to France and Italy. Important among such treatises was Thomas Bradwardine's *Tractatus de proportionibus* written in 1328 [1]. Its earliest influence on an Italian author is evident in

---

[1] Edited and translated by H. L. CROSBY, Jr: *Thomas of Bradwardine, His Tractatus de proportionibus* (Madison, Wisc., 1955). Cf. also the *Tractatus brevis proportionum*

Johannes de Casali's *Questio de velocitate motus alterationis* composed in 1346[2]. That influence is not surprising in view of Casali's residence in England[3]. Less well known than Casali's treatment is the *Questio de proportionibus motuum* composed by Francesco of Ferrara in 1352 and extant in an unique copy: Oxford, Bodleian Library MS Can. Misc. 226, 58r-63r. Its main purpose is to explain and discuss the conclusions of Bradwardine's tract. And indeed Francesco refers to Bradwardine as *Magister proportionum* (see 58r, 13) or simply as *Magister* (59r, 19). I first called attention to Francesco's treatise when I published excerpts from it in my *The Science of Mechanics in the Middle Ages*[4]. Now almost twenty years later it seems appropriate to publish the full text in view of the increased interest in Bradwardine and his so-called " exponential function " relating velocity and the ratio of force to resistence[5]. To keep the text within publishable bounds, I shall reserve detailed analysis of the text to some future publication. However, the reader may find some preliminary observations concerning Francesco's dependence on Bradwardine's work and his possible dependence on the works of Bradwardine's successors at Merton College as well as on Casali's tract in my earlier exposition[6]. I should also note that the reader can easily find out how Francesco used earlier works if he consults the *Index verborum* under the following entries: Aristoteles, Phylosophus or Philosophus, Averroes, Commentator, Euclides (where there is also a reference to the *Liber de curvis superficiebus Archimenidis* here falsely attributed to Euclid but in all likelihood written, or translated from an unknown author, by Johannes de

---

abbreviatus ex libro de proportionibus D. Thome Braguardini Anglici, which I reprinted and translated in my *The Science of Mechanics in the Middle Ages* (Madison, Wisc., 1959, 2nd. pr., 1961), pp. 465-94. I had twenty years earlier discussed Bradwardine's reinterpretation of Aristotle's views on the proportion of motions in my *Giovanni Marliani and Late Medieval Physics* (New York, 1941), pp. 125-44, stressing Marliani's misunderstanding of that reinterpretation in the Italian author's *Questio de proportione motuum*.

[2] M. CLAGETT, *The Science of Mechanics*, p. 332, and *Nicole Oresme and the Medieval Geometry of Qualities and Motions* (Madison, Wisc., 1968), pp. 66-70. See also E. SYLLA, *Medieval Quantification of Qualities: The " Merton School "*, " Archive for History of Exact Sciences ", Vol. 8 (1971), pp. 9-39 (and particularly pp. 34-35).

[3] Cf. M. CLAGETT, *Nicole Oresme*, p. 70, n. 27.

[4] M. CLAGETT, *The Science of Mechanics*, pp. 495-503.

[5] See A. G. MOLLAND, *The Geometrical Background to the " Merton School "*, " The British Journal for the History of Science ", Vol. 4 (1968), pp. 108-25. Cf. his essay review *Ancestors of Physics*, " History of Science ", Vol. 14 (1976), pp. 54-75 (and particularly pp. 57, 67-70). In both articles Molland discusses and refers to the earlier literature on Bradwardine's proportionality theory and shows the anachronistic dangers of expressing his " law " as a logarithmic relationship. For the very rich development of Bradwardine's " law " in the hands of Nicole Oresme, see E. GRANT's introduction to his *Nicole Oresme: " De proportionibus proportionum " and " Ad pauca respicientes "* (Madison, Wisc., 1966).

[6] M. CLAGETT, *The Science of Mechanics*, pp. 498-99.

Tinemue) [7], Jordanus, Commentator de ponderibus. And, of course, the reader will find particularly instructive a direct comparison of the treatises of Bradwardine and Francesco now that they are both readily available. Incidentally, it should be noted that the references to Aristotle in the *Index* include roman numerals that represent the particular books of the *Physics* and the *De caelo* which are being cited.

The little that is known about Francesco of Ferrara comes from the Bodleian manuscript itself. Folios 58r-69v comprise a section written in a hand that is different from those that precede and follow it. It includes not only the treatise here edited but several other shorter philosophical pieces. All of these pieces (except perhaps the fragments on verso of folio 69) were almost certainly written down by Francesco himself, as various notes on these sheets reveal. Thus on folio 58r the opening title of our tract reads " Questio determinata per me Francischum de Ferraria Padue ad preces quorundam scolarium anno domini M ĊĊĊ LII die X^a mensis Decembris. Deo gratias. " Then on folio 63r, a few lines below the end of the treatise, the scribe has written " Francischus. " Further, on folio 63v we read (upside down) on the bottom of the page: " M̊ iii xlvii. Indictio xv. " Beneath this (with the page still turned upside down) are two lines. The first reads: " Est questio utrum totum sit sue partes aut res distincta a suis partibus. " The second states: " Iste quaternus est mei francisci de ferraria, filii condam domini Bartholomey. " At the top of the same sheet (now turned right-side up) is written: " Questio de proportionibus determinata " and below it " (Francischus de ferraria), " with the parentheses included. Then on folio 64r is written the title of his tract in a form somewhat similar to that given at the beginning of the tract: " Questio de proportionibus motuum determinata per magistrum Francischum de ferraria in studio paduano. Anno domini M iii lii die x^a mensis decenbris. Amen. " From all of this we can conclude the following. (1) Francesco spelled his name in two ways: Franciscus and Francischus. (2) This section including the tract and additional short pieces was written out by Francesco and was part of a booklet (*quaternus*) owned by Francesco. (3) His father was named Bartholomeus and was probably dead by 1347 (or at least was dead by the time that the *ex libris* on folio 63v was written). (4) Francesco taught at the University of Padua and " determined " this *questio de proportionibus* there on 10th of December, 1352, at the request of certain students. One might be tempted

---

[7] M. Clagett, *Archimedes in the Middle Ages*, Vol. 1 (Madison, Wisc., 1964), pp. 450-507 for the Latin text and English translation. See also " *Addenda* ", p. 720.

6

to identify our Francesco with a Francesco of Ferrara present in the
residence of the *podestà* of Padua in January, 1378, except that the
latter's father is named Benvenuto [8].

I indicated above that certain other questions were included in this
section of the manuscript written by Francesco. But only two of them
are particularly relevant to the subject of the treatise that I have edited.
The first (66r-66v) begins: " Questio prima. Utrum motus sequatur
proportionem agentium ad passa vel utrum proportio velocitatum in
motibus sit secundum proportionem agentium ad passa seu potentiarum
moventium ad potentias resistentivas. " The second begins: " Questio
secunda. Utrum in motibus velocitas sit at[t]endenda penes latitudinem
aquirendam an penes latitudinem et extensionem simul. " The other
short pieces and indeed the whole manuscript deserve further study,
particularly because the description of the manuscript in Coxe's catalogue is quite inadequate [9].

In my text below I have made a literal transcription of Francesco's
copy, staying close to his orthography. For example, I have everywhere
followed Francesco in dropping one of the double consonants that
appear conventionally in Latin words. Thus I have followed Francesco
by giving *adere* for *addere, atrahere* for *attrahere* and *inteligere* for
intelligere. For these and many other examples consult the *Index
verborum* below. I have, however, always rendered Francesco's rather
frequent " ci " before a vowel as " ti " (e.g. *circumferentia* instead of
Francesco's *circumferencia*). The enclitic *-que* was often written separately by Francesco but I have joined it to the preceding word in
the conventional manner. I have uniformly capitalized proper names,
though Francesco's practice is inconsistent. The whole tract includes
only two simple figures, which I have given in their proper places in
the Variant Readings following the text. I have added the folio numbers
of the Bodleian manuscript to the margins of the text below. Furthermore, I have added line numbers within the body of the text [e.g., (1),
(2), etc.]. These in fact represent the actual lines of the folio pages in
the Bodleian manuscript. The reader will also see that I have used

---

[8] M. CLAGETT, *The Science of Mechanics*, p. 498.

[9] H. O. COXE, *Catalogi codicum manuscriptorum bibliothecae Bodleianae pars tertia*
(Oxford, 1854), misc. 226. For example, Coxe assigns the *Practicella ex dictis Mesue
abbreviata* (folios 21v-24r) to Dinus de Garbo de Florentia, though in fact, it was done
by John of Parma. The piece on folios 28r-31bis r is described as *Questiones meta-
physicae* of Jean of Jandun. It is rather questions that he disputed on Aristotle's
Physics. In fact, corrections (some small, some more significant) can be made in Coxe's
description of many of the succeeding tracts in the codex. In describing Francesco de
Ferrara's *Questio de proportionibus motuum*, he implies (by omission) that it occupies
the whole section 58r-69v, though, as I have said above, Francesco's principal tract
ends on folio 63r and thereafter are included other short questions.

these folio and line numbers in my *Index*. The *Index* is limited to philosophical and scientific terms. Words that appear on almost every page like *maior*, *minor*, etc., have not been included. The addition of an asterisk to a number in the *Index* indicates that the word in that line has been added by the editor. This has not been done when the word has also been used by the author in the same line.

[Questio Francischi de Ferraria de proportionibus motuum]

[Proemium]

58r   (1) Questio determinata per me Francischum de Ferraria Padue ad preces quorundam scolarium (2) anno domini M C̊C̊C̊ LI̊I̊ die Xᵃ mensis Decenbris. Deo gratias.

(3) Quoniam materia de proportionibus agentium ad passa et moventium et resistentiarum adinvicem in velocitate et tarditate motu(4)um eorundem, de earundemque causis, que multum naturali phylosophie existit utilissima, multum latitat hodie phylosophos natura(5)les modernos, sitque scientia multum ardua et dificilis quod modernorum doctorum et naturalium sapient[i]um insinuat disco(6)lia que signum dificultatis in rebus existit eficax ac perspicuum, fueritque latitanter ac obscure aliqualiter in forma tractatuum a quampluri(7)bus valentissimis pertractata, tamen ut intellectus studentium in ea clarius elucescant ad sepissimas preces quorundam commotus proposui in forma (8) questionis rescribere, diminutis adere, aditis quoque superflua iuxta posse melius resecare, quam sub tituli questionis forma et titulo proponemus ut sic: (9) Utrum velocitas et tarditas in motu sit atendenda penes proportionem potentiarum moventium ad potentias resistentes; quod (10) nichil est aliud nisi querere que est causa quod unum agens movet suam resistentiam in quocunque motu sic tanto gradu velocitatis et non maiori nec mi(11)nori et quare unum agens certo gradu velocitatis movet velocius vel tardius suam resistentiam quam aliud suam; utrum videlicet hec sint quia unum agens (12) tantam proportionem habet ad suam resistentiam et non maiorem nec minorem et quia unum agens habet maiorem proportionem ad suum passum quam aliud (13) ad suum, ideo unum velocius movet alio.

[Rationes Quod Non]

Et arguitur primo quod non, primo rationibus quas Magister proportionum aducit. Primo sic, aliqui sunt (14) duo motores quorum unus continue in maiori proportione excedit suam resistentiam quam alius suam et tamen unus non movet suam resistentiam velocius (15) quam alius suam. Igitur velocitas in motu non est atendenda penes proportionem potentie motive ad resistentiam. Consequentia tenet pro tanto, quia si atenderetur penes (16) rationalem proportionem ubi esset maior proportio agentis ad passum, illud velocius moveretur. Sed antecedens declaratur, posito casu quod sint duo luminosa (17) exempli gratia, *a* et *b*, equalia intensive et extensive, et iuxtaponantur *a* et *b* duo opaca, scilicet *c* et *d*, equalia similiter inter se, minora tamen *a* et *b*, sic quod (18) ad eis creentur unbre equales. Deinde accipio duo mobilia, *e* et *f*, que sint mobilia ad intrinseco et ponam *e* mobile in cono un(19)bre *c* et *f* mobile in cono unbre *d*. Deinde pono quod *a*, quod est corpus luminosum, continue crescat, *b* semper manente in eadem dispositione, quia eo scilicet (20) non aucto nec diminuto et quod *c* et *d* opaca obstacula decrescant et diminuantur continue uniformiter quousque coru[m]pantur et non moveantur alio (21) motu nisi motu coruptionis et quod *e* et *f* continue sint in cono unbrarum quas sequuntur, quem casum clarum videas in figura. Tunc (22) arguo sic: *e* mobile continue se habet in maiori proportione ad suam resistentiam, scilicet ad *c*, quam *f* ad suam, scilicet ad *d*, et tamen *e* non continue movetur velocius (23) *f*. Igitur asumptum verum, scilicet quod sunt aliqui duo motores et cetera. Antecedens probatur quo ad utranque partem, posito quod *e* continue se habet in maiori proportione ad suam (24) resistentiam, scilicet ad *c*, quam *f* ad suam, scilicet ad *d*, quia propter augmentationem *a* luminosi resistentia *e*, scilicet *c*, continue erit minor et minor magis quam resistentia *f*. Similiter (25) *a* continue augmentabitur; ergo resistentia *e*, scilicet *c*, continue erit minor quam resistentia *f*, scilicet *d*; ergo si continue erit minor resistentia, *e* continue se habebit in maiori (26) proportione ad *c* quam *f* ad *d*. Sed quo ad aliam partem probatur, scilicet, quod *e* non continue movebitur velocius *f* quia in eodem tempore *c* et *d* corumpantur (27) per positum et in eodem tempore equalia spacia pertransibunt. Ergo in eodem tempore *e* non continue movebitur velocius *f*.

Secundo ad idem principaliter sic: (28) Si questio esset vera, tunc sequeretur [quod] *a* [haberet] equalem proportionem ad *c* sicut *b* ad *d* et tamen *a* moveret velocius quam *b*; igitur questio falsa. Consequentia clara est, (29) sed antecedens declaro, sumpto quod *a* sit una potentia sicut 16 et *c* resistentia sua sicut 8, *b* sit potentia sicut 4, *d* resistentia sua sicut duo, sic quod *a* (30) et *b* debeant pertransire illa duo spacia, scilicet *c* et *d*. Tunc arguo sic: *a* equalem proportionem habet ad suam resistentiam, scilicet *c*, sicut *b* ad suam, scilicet (31) ad *d*, quia utraque est proportio dupla. Modo una proportio dupla non est maior alia dupla et tamen *a* movet velocius quam *b*, quia si non velocius igitur eque (32) velociter; erunt igitur isti duo motus eque veloces. Sed contra, illi motus sunt eque veloces per quos in equali tempore spacium pertransitur equale. (33) Sed per *a* in equali tempore pertransitur quadruplum spacium a spacio pertransito a *b*. Igitur motus *a* non est equalis motui *b*. Maior istius prosillionis est (34) Phylosophi sexto physicorum, sed minor patet, quia cum *a* equalem proportionem habeat ad *c* sicut *b* ad *d* in eodem tempore in quo *a* pertransibit totum *c*, *b* (35) pertransibit totum *d*. Sed constat quod *c* est quadruplum ad *d*, quia ad ipsum se habet sicut 8 ad 2; quare et cetera.

Tertio, sunt aliqua duo mobilia quo(36)rum unum semper movet velocius alio et tamen equalia spacia pertransibunt in eodem tempore. Igitur questio falsa. Consequentia tenet quia pertransitus equalium spaciorum in equa(37)li tempore non procedit nisi ex equali proportione et hoc supono. Antecedens declaratur licet debiliter, ut ipse arguitur, sumpto quod *a* moveatur velocius super su(38)um spacium in duplo quam *b* super suum que spacia sint equalia, sed quidem cum *a* erit in fine sui spacii iterum redeat ad principium. Tunc sic *a* continue (39) movet in duplo velocius quam *b* ut casus suponit et tamen *a* et *b* equalia spacia pertransibunt, ut sumptum patet. Igitur [et cetera.]

Quarto, aliqua est potentia movens (40) aliquod mobile per aliquod spacium in aliquo tempore, que potentia si dupletur movebit illud mobile per idem spacium in triplo velocius. Igitur questio falsa. Consequentia tenet ut in priori, (41) sed antecedens deduco, posito quod *a* agens se habeat ad *b* passum sicut 6 ad 4 et moveat *a*, *b* per *d* spacium in una hora, tunc *a* movet *b* per aliquod spacium in aliquo (42) tempore, quod si dupletur movebit in triplo velocius. Hoc asumptum probatur, quia dupletur *a*, tunc duplum ad *a* est 12. Modo 12 movebit 4 in triplo velocius (43) quia ad ipsum se habet in proportione tripla, ut patet; quare et cetera.

Quinto, sunt duo motores non equalem proportionem habentes ad idem mobile qui movebunt idem mobile (44) per idem spacium equali velocitate; igitur et cetera. Consequentia clara, sed antecedens probatur sic: 9 et 8 sunt duo motores non equalem proportionem habentes ad 1 et tamen ipsum movebunt equali (45) velocitate. Quod non equalem proportionem habeant patet, quia una est proportio 9^{la}, alia 8^{la}. Sed quod movebunt equali velocitate sic ostenditur, sumpto quod dualitas moveat unitatem (46) per spacium pedale aliqua velocitate, tunc sic aliud potest movere unitatem per spacium pedale et dimidium equali velocitate. Hoc autem esse non potest nisi ternarius. Tunc sic (47) ternarius movet per spacium et cetera. Sed proportio 9 ad 1 est tripla ad proportionem 3 ad 1. Ergo movebit in triplo velocius vel per triplum spacium equali velocitate et similiter 8 (48) per triplum spacium equali velocitate, quia si duo movet unitatem per spacium pedale, ergo 4 per bipedale, quia dupla erit proportio; ergo 8 per tripedale, quia triplicaretur (49) proportio; quare et cetera.

Sexto ad idem, aliqua potentia movet aliquod mobile aliquod tempus; cuius medietas movebit medietatem mobilis equali velocitate in eodem tempore et tamen non equalem pro(50)portionem habent totum ad totum sicut medietas ad medietatem. Quare questio falsa. Consequentia est plana, sed assumptum deduco, sumpta una supositione posita secundo de proportionibus: (51) omnes proportiones sunt equales quarum denominationes sunt equales. Sunt autem denominationes proportionum nomina earum numeralia, ut dupla, tripla, et cetera. Deinde cum illo (52) suposito quod *a* sit una potentia se habens ad suam resistentiam, scilicet ad *b*, sicut 10 ad 6, tunc arguitur, *a* movet *b* per aliquod spacium aliqua velocitate, ergo eius medietas (53) per idem spacium in equali tempore movet medietatem *b* per regulam Aristotelis septimo physicorum in fine et tamen eorum proportiones non sunt equales quia proportio *a* totius ad totum *b* est (54) sicut proportio 10 ad 6, sed proportio medietatis *a* ad medietatem *b* est sicut proportio 5 ad 3, sed ille proportiones non sunt equales; quod probatur quia earum denominationes (55) non sunt equales; quare nec proportiones per supositum. Quod denominationes non sint equales patet, quia proportio *a* ad *b* est proportio superbitertia (!); proportio [medietatis *a*] autem ad [medietatem] *b* est proportio (56) superbipartiens; quare et cetera. Quod autem ille proportiones sic denominentur inferius ostendetur in nominum declaratione.

Septimo, sunt aliqua duo gravia quorum unum (57) est gravius alio mota per idem medium in equali tempore que equali velocitate moventur;

ergo et cetera. Asumptum breviter probo accipiendo duo gravia *a* et *b* descendentia per *c* medium (58) uniforme, et sit *a* piramidalis figure descendens per conum et *b* quadrangularis et late sed sit multum tenue, sed tamen sit in duplo gravius *a* et cetera. (59) Manifestum est quod *b* si esset figure piramidalis sicut *a* moveretur velocius quam movebitur, quia figura impedit maxime, ut patet quarto physicorum. Pono igitur quod figura latitudinis (60) et tenuitas tantum impediant eius descensum quantum eum adiuvat gravitas per quam *b* excedit suam levitatem. Et quod hoc sit possibile ad experyentiam patet; (61) ad evidentem experyentiam patet de lintiamine (!) magno et parvo plumbo positis in aere plumbum velocius

descendit, posito quod lintiamen sit gravius eo. (1) Hoc autem non est nisi propter dispositionem figure. Potest igitur ut in exemplo illo patet tantum figura impedire quantum gravitas adiuvare. Tunc igitur *a* et *b* sunt duo mobilia gravia quorum (2) unum est gravius alio in duplo, scilicet *b* quam *a*, mota per *c* medium in eodem tempore que equali velocitate moverentur; quod patet quia si non equali velocitate moverentur, hoc (3) esset quia *b* moveretur velocius et tunc non moveretur velocius nisi propter iuvamentum gravitatis. Sed hoc non, quia tantum est impedimentum figure ut ostensum est quantum iuvamentum quod prebet gravitas. (4) Quare velocius non movetur.

Octavo, aliquod mobile movetur continue per aliquod tempus in duplo velocius alio et tamen in eodem tempore quadruplum spacium pertransit; quare et cetera. Consequentia tenet semper ut prius, (5) sed antecedens declaratur, sumpto quod sint duo concentrica *a* et *b* et sit *a* in duplo maior *b*. Deinde supono unam supositionem Euclidis de curvis superficiebus et est (6) ut allegat iste conclusio quinta, licet secundum aliam cotationem sit tertia, et est: Omnium angulorum et circulorum circumferentie suis dyametris sunt proportionabiles. Igitur qualis est (7) proportio dyametri *a* ad suam circumferentiam talis est proportio dyametri *b* ad suam circumferentiam. Igitur a permutata proportione que fieri potest ut patet per supositionem (8) Euclidis: Si fuerint quatuor quantitates proportionabiles permutatim erunt proportionabiles. Igitur qualis est proportio dyametri *a* ad dyametrum *b* talis est proportio circumferentie *a* ad (9) circumferentiam *b*. Sed dyameter *a* est precise dupla ad dyametrum *b*. Igitur circumferentia *a* erit precise dupla ad circumferentiam *b*. Igitur quilibet punctus in circumferentia *a* precise in duplo velocius (10) movetur quam aliquis punctus in circumferentia *b*, et tamen quilibet punctus in circumferentia *a* quadruplum spacium pertransibit ad spacium pertransitum in circumferentia *b*. Quod probo quia sicut probatur per tertiam quarti (11) proportionum: Omnium duorum circulorum est proportio alterius ad alterum tanquam proportio sui dyametri ad dyametrum alterius proportio duplicata Et cum proportio *a* ad (12) *b* sicut simul proportio dupla et duplicata et proportio duplicata sit quadrupla, erit proportio *a* ad *b* quadrupla et cum quilibet punctus in circumferentia *a* movetur ad motum *a*, quilibet punctus in cir(13)cumferentia *a* quadruplum spacium pertransibit ad spacium pertransitum a puncto in circumferentia *b* et tamen quilibet punctus in *a* precise in duplo velocius movebitur puncto in circumferentia *b*.

(14) Nono ad idem arguo sic: Aliqua sunt eque gravia equalem proportionem habentia ad suas resistentias quorum unum movetur velocius alio, igitur questio falsa. Consequentia manifesta est, sed assumptum (15) declaro. Et accipio equilibram cuius canonum sit precise duplum alterius et

vocetur maius *a* minus vero *b*. Et hiis sint apensa duo pondera, scilicet *c* et *d*, ita quod *c* (16) sit duplum ad *d* et ponatur *c* in brachio *b*, *d* vero in brachio *a*. Tunc ut habetur octava de ponderibus: Si brachia libre fuerint proportionabilia (17) ponderibus apensis ita quod breviori gravius apendatur, tunc eque gravia secundum situm erunt apensa. Sed sic est in proposito; quare *c* et *d* sunt eque gravia et equalem proportionem (18) habent ad suas resistentias et tamen quando *d* movetur, movetur velocius *c* quia in eodem tempore maiorem portionem circuli describit. Igitur maius spacium pertransit in eodem tempore. Igitur (19) velocius movetur; quare et cetera.

Decimo: Sunt aliqui duo motores moventes duo mobilia per aliquod spacium in aliquo tempore qui coniuncti movebunt illa duo mobilia in medietate temporis (20) per idem spacium. Igitur antecedens probo, posito quod *a* sit una potentia motiva movens *b* ferrum et sit potentia *a* ad resistentiam *b* dupla et *c* sit potentia movens *d* ada(21)mantem alicuius certi ponderis et sit etiam *c* duplum ad *d*. Tunc patet antecedens, quia *a* et *c* sunt duo motores moventes duo mobilia, scilicet *b* et *d*, per aliquod spacium (22) in aliquo tempore; et si coniungantur *c* et *a*, movebunt *b* et *d* in duplo velocius quia adamans trahit ferrum et ad motum adamantis movetur ferrum ex se. Nec (23) maiorem dificultatem habent *a* et *c* in movendo adamantem cum ferro sicut sine ferro, sed si ambo coniuncta moverent adamantem sine ferro, moverent (24) in duplo quam ipsa separata. Ergo et nunc movent velocius. Et confirmatur quia tunc adamans velocius traheret parvum ferrum quam magnum. Consequens est falsum, quia utrumque (25) movetur equali velocitate ad motum seu secundum quod magnes movetur. Consequentia tenet, quia maiorem proportionem ad ferrum parvum quam ad magnum.

Deinde postea (26) arguitur aliis rationibus de quibus magis deduco quam de istis.

Prima est: Si motus in velocitate et tarditate sequeretur proportionem, tunc sequeretur quod ignis intensus citius (27) calefaciet calidum remissum quam calidum intensius. Consequens est falsum, quia citius calefaciet et assimilabit sibi similius et quod minima indiget calefactione et quod minus ei (28) resistit. Tale autem est calidum intensum. Sed consequentia patet, quia maiorem proportionem habet ad calidum remissum cum sit minus quam ad calidum intensum, ut de se patet. Et probatur (29) per regulam proportionum: si due quantitates [in]equales ad tertiam comparentur, maior quidem maiorem, minor vero minorem, obtinebit proportionem, illius quidem ad ambas, ad (30) maiorem quidem proportio minor, ad minorem vero proportio maior erit. Igitur per secundam partem huius regule cum ignis summus comparetur ad magis calidum et (31) ad minus calidum, ad maius proportionem minorem, ad minus vero proportionem maiorem habebit.

Secunda: Sequeretur quod unum movens velocius moveret unum (32) lapidem minorem versus deorsum quam lapidem maiorem. Consequens est falsum quia lapis minor magis resistit moventi quam lapis maior, igitur tardius movetur. Consequentia (33) tenet, quia ubi maior resistentia ibi motus tardior. Quod autem magis resistat sic deduco, nam aliquid dicitur resistere alicui vel quia (34) inclinatur ad motum seu locum vel qualitatem oppositam ei quam agens intendit producere vel quia naturaliter inclinatur ad non paciendum. Modo corpus magis grave (35) utroque modo resistit minus, primo quia minus inclinatur ad locum sursum quam [minus] grave et ad quiescendum similiter sursum quam minus grave. Ydeo plus inclinatur ad deorsum. Ergo (36) minus resistit. Ergo velocius movebitur. Sed quod tardius patet, quia movens minorem proportionem habet ad ipsum cum sit maior. Ergo proportionem minorem ad eum habet ut (37) evidenter patet sequi per regulam precedentem et deductionem consimilem.

Tertia ratio est: Nullum agens alterans alterat ab aliqua proportione; igitur consequentia manifesta est. Antecedens probatur (38) quia ponitur quod *a* agens calidum agat in *b* inducendo in ipsum caliditatem. Tunc *a* agit in *b* et a nulla proportione agit; quia, si ab aliqua agat, ponitur igitur quod a dupla; (39) tunc sic *a* immediate post hoc inducet aliquam caliditatem in *b*. Igitur immediate post hoc habebit aliquod iuvamentum a *b*. Igitur immediate post hoc aget a maiori proportione (40) quam a dupla. Igitur non a dupla, quod est propositum.

Quarto tunc sequeretur quod *a* ageret in *b* ab infinita proportione et per consequens velocitate infinita. Consequens est satis absurdum, sed consequentia (41) patet. Nam agit in *b* ab aliqua proportione et non a proportione finita; igitur ab infinita. Quod non agat secundum proportionem finitam probo, quia ponitur igitur si sic, sit quod a dupla (42) agat. Tunc *a* habet ad *b* proportionem duplam; igitur ad medietatem quadruplam. Sed plus aget in medietatem *b* quam in totum *b*. Igitur plus aget a quadrupla quam a dupla et similiter probatur (43) quod plus a octupla et cetera.

Quinto sequeretur quod illud quod haberet infinitam proportionem ad aliud in virtute esset virtutis infinite. Consequens est falsum. Consequentia tenet, quia si velocitas et (44) tarditas atenduntur penes proportionem et velocitas et tarditas dependent a potentia et virtute eiusdem, ergo virtus alicuius et similiter quantitas atendetur. Et probatur ex proportione, sed falsitas (45) consequentis aparet de homine in perfectione respectu asini, nam homo ad asinum infinitam proportionem habet quia in perfectione in infinitum excedit ipsum. Quod patet, quia si non (46) in infinitum, igitur per finitum. Ponitur igitur exempli gratia quod in centuplo; tunc cum sit aliquod animal brutum in centuplo perfectius asino, sequitur quod ille et homo erunt equalis (47) perfectionis, quod est absurdum. Similiter corpus infinitam proportionem habet seu in infinitum excedit, quod in proposito est, superficiem quia plusquam in duplo, plusquam in triplo et cetera. Cum (48) igitur superficies sit quanta, detur quod corpus sit quantitatis infinite. Similiter corpus partes proportionales habet quarum quelibet est maior superficie. Igitur corpus in infinitum est maius (49) superficie; quare quodlibet corpus erit infinitum.

Sexto *a* et *b* sunt duo agentia equalem proportionem habentia ad *c* et *d* resistentias et tamen *a* velocius agit in *c* quam (50) *b* in *d*. Ergo consequentia clara est ut precedentes. Sed antecedens sic deduco, sumpto quod *c* et *d* sint due terre simplices equales, quibus aproximata sint *e* et *f* (51) agentia equalia et sit *e* aer et *f* ignis. Et alterent *e* et *f*, *c* et *d* ad quemdam gradum caliditatis certum equale in utroque *ec* et *fd*. Deinde removeatur *e* et *f* (52) a *c* et a *d* et aproximentur eis duo ignes in summo, scilicet *a* et *b*, qui alterent ea ad summum gradum caliditatis deducendo. Tunc *a* et *b* sunt duo agentia equalem (53) proportionem habentia ad *c* et *d*, tamen *a* velocius aget in *d* quam *b* in *c*. Nam *d* est similior *a* quam *c*, *b*; quod patet quia in *d* est caliditas ignis et forma substantialis ignis, (54) sed in *c* est caliditas aeris et forma substantialis aeris; quare *d* est similior *a* quam *c*, *b*.

Septimo tunc sequeretur quod ignis agens in aquam infinitam caliditatem induceret in (55) aquam. Et per consequens ipse esset vigoris infiniti intensive. Consequens est contra Aristotelem in quampluribus locis, et ultimo physicorum et primo celi. Consequentiam ego declaro, posito quod (56) *a* ignis summus agat in *b* aquam summam remittendo frigiditatem quousque in ea generetur forma ignis et hoc uniformiter corumpendo eius frigiditatem et uniformiter inducendo (57) latitudinem caliditatis et hoc secundum partes proportionales huius temporis, exempli gratia hore, sic quod in prima parte proportionali hore corumpatur prima pars proportionalis frigiditatis et (58) inducatur prima pars proportionalis caliditatis, scilicet medietas latitudinis caliditatis; et in secunda, secunda; et in tertia, tertia; et sic secundum proportionem et assumptionem partium proportionalium. Tunc sic *a* in (59) secunda parte proportionali hore transmutabit *b* in duplo velocius quam in prima et in tertia in triplo velocius et sic in infinitum. Et cum sint infinite partes proportionales hore, igitur infinita (1) velocitate agit; tunc sic, si aliqua velocitate agit, aliquam caliditatem inducet; si dupla velocitate agit, duplam caliditatem; dico intensive. Ergo si infinita velocitate agit, infinitam (2) caliditatem inducet, quod est probandum. Quod autem in secunda parte proportionali *a* aget in duplo velocius quam in prima probatur, quia in secunda parte proportionali *a* habebit ad *b* in duplo maiorem (3) proportionem, quia resistentia *b* erit in duplo minor. Ergo iuxta hanc opinionem in duplo velocius aget. Et si dicatur non sequitur quia agens debilitatur, primo quia est potentie finite, secundo quia re(4)patitur ab aliqua, contra, quia maius est iuvamentum quod recipit *a* a caliditate inducta in *b* in prima parte proportionali quam eius debilitatio. Tantum igitur vel plus faciet iuvamentum seu (5) auxilium quam eius debilitatio; propter igitur debilitationem non stabit quin in duplo velocius agat. Item hoc est regula Aristotelis expressa in septimo physicorum in fine, quod si aliqua potentia movet aliquod (6) mobile per aliquod spacium in aliquo tempore, eadem potentia movebit medietatem mobilis per idem spacium in duplo minori tempore. Igitur similiter in proposito: si *a* inducit in *b* in prima parte proportionali (7) medietatem latitudinis caliditatis seu primam partem proportionalem caliditatis, in secunda inducet secundam in duplo minori tempore et per consequens in duplo velocius. Item ex hac responsione sequitur quod nullum (8) agens possit agere uniformiter per aliquod tempus. Consequens est falsum. Consequentia de se patet, quia ipsum in actione continue debilitatur. Falsitas consequentis probatur, nam eo concesso sequitur hec conclusio, (9) quod aliquod agens continue per horam difformiter aget aliquam qualitatem et tamen in fine hore erit illa qualitas uniformis per totum. Consequens est absurdum, nam quero a quo producitur talis uni(10)formitas, non ab agente, quia agens, ut suponitur, continue difformiter agit. Consequentia probatur, sumpto quod ignis summus agat in aliquam resistentiam inducendo in ea caliditatem summam, (11) sic quod in fine hore sit caliditas summa inducta in tali resistentia. Tunc ignis talis continue difformiter aget ut ponit caliditatem que caliditas in fine hore erit uniformis. (12) Quod patet, quia si non, ergo difformis; sed cum pars qualitatis difformis sit intensior quam tota, ut probat suficienter una conclusio de latitudinibus, igitur pars istius qualitatis erit intensior (13) caliditate tota, sed caliditas tota est caliditas summa. Igitur

59r

erit aliqua caliditas intensior caliditate summa, quod implicat. Secundo non
(! *del?*) est notandum quod ignis sicut nec aliquod agens repa(14)tiatur; si
enim reactio esset possibilis, tunc idem simul et semel moveretur motibus
contrariis, quia scilicet simul et semel calefieret et frigefieret. Consequens
est absurdum. Sed consequentia declaratur per communem (15) modum
deducendi eam, quia sit *a* agens principale calidum, *b* reagens frigidum.
Deinde accipio partem repassam in principali agente *a* que vocetur *d* et
accipio partem non re(16)passam que vocetur *c*. Tunc arguo sic: Non
minoris virtutis, ydeo maioris forte est *c* ad calefaciendum partem repassam, scilicet *d*, quam *b* ad infrigerandum per reactionem; (17) sed *b* ipsam
d partem frigefaciet. Ergo *c* ipsam calefaciet. Quare *d* simul calefiet et
infrigerabitur.

Secundo tunc a proportione equalitatis et minoris inequalitatis
pro(18)veniret motus, cuius oppositum inferius declarabo.

Tertio sequitur quod minus calidum velocius ageret quam magnum in
idem calidum. Consequens est inconveniens. Consequentiam ego declaro,
sumpta regula quam ponit (19) Magister in tractatu de proportionibus
motuum, capitulo quarto, et est conclusio prima, quod si fuerit proportio
primi ad secundum et secundi ad tertium, erit proportio primi ad tertium
dupla ad proportionem (20) primi ad secundum et secundi ad tertium.
Deinde ponitur quod sint 3 corpora apta agere et reagere: *a*, *b*, *c*: *a* sit
sicut 16, *b* sicut 8, *c* sicut 4. Tunc sic talis est proportio et (21) habitudo,
si *c* reagat in *b* et *b* reagat in *a*, *c* ad *b* qualis *b* ad *a*. Ergo proportio *c* ad *a*
in reagendo est dupla ad proportionem *c* ad *b* et *b* ad *a*, sed a pro(22)por-
tione dupla provenit motus in duplo velocior. Ergo *c* quod est minus
velocius aget in *a* quam *b* quod est magis calidum.

Quarto tunc sequitur quod aliqua actio foret perpetua. Et arguo modo
(23) de reactione inter equalia. Consequentia est contra Philosophum
octavo physicorum ostendente nullum motum eternum preter motum circularem. Consequentiam probo, posito quod *a* ignis summus agat in *b* aquam
(24) summam. Tunc *a* ita fortiter agit in *b* sicut *b* in *a*. Vel igitur est dare
finem istius actionis vel non. Si non, propositum sit sic. Ponatur igitur
quod in *d* instanti huius temporis (25) illa actio finiatur. Tunc in *d* instanti
erit *a* calidum sicut *b* frigidum aut igitur utrum est summum. Et tunc ita
erunt [ambo] suficientia ad agenda sicut prius, cum sint equalis potentie
[et] equalis (26) contrarietatis. Aut erunt remissa; ponatur igitur quod
usque ad medium gradum, sic quod utrum habeat medium gradum totius
sue latitudinis. Tunc cum sint aproximata et sufi(27)cienter ydeo magis
contraria, cum sint calida solum [ad] medium, igitur ob defectum contrarietatis non deficient agere nec ob defectum proportionis, quia eandem
pro(28)portionem habent sicut in principio; quare et cetera.

Quinto sequitur quod idem simul calefieret et frigefieret, posito quod
inter *a* calidum summum et *b* frigidum remissum fit aer (29) medium.
Tunc cum *a* producat frigiditatem in *b*, hanc producere non videtur nisi
producat per totum medium, et cum simul *b* reagat in *a*, hoc esse non
potest (30) nisi reagat prius frigiditatem in medium; quare aer simul
calefiet et frigefiet.

Sexto si reactio esset possibilis, (31) sequeretur idem inconveniens
quod prius, videlicet quod idem simul calefieret et frigefieret; quod probo,
sumpto quod sit unum corpus uniformiter difforme non calidum secundum
se (32) totum. Tunc cum partes istius sint contrarie, quia partes extremi

intensioris contrarie sunt partibus extremi remissioris, igitur maxime ille agent et reagent. Si sic, tunc arguo: nulla est (33) pars huius corporis que secundum se totam manebit non remissa per partem remissiorem. Igitur quelibet manebit remissa, similiter nulla est pars eiusdem *a* que secundum se totam manebit non (34) intensa per aliquam partem intensiorem. Igitur quelibet manebit intensa. Igitur quelibet manebit intensa et remissa. Consequentia satis clare tenet. Sed antecedens declaratur, quia si aliqua pars manebit (35) non remissa, sit illa *b*. Tunc *b* est diformiter calida. Igitur una pars *b* contrariatur alteri. Igitur agent adinvicem. Igitur si pars intensior agat in remissiorem, pars intensior (36) per reactionem remitteretur a parte remissiori; et similiter probatur quod non quelibet manebit non intensa. [Septimo] et idem argui similiter potest de corpore uniformiter calido in quo caliditas excedit (37) frigiditatem; quare et cetera.

Octavo principaliter arguitur quod motus non sequatur proportionem, quia tunc sequeretur quod ignis in summo agens in calidum aliquod remissum infinita velocitate ageret. (38) Consequens est absurdum, quia tunc in instanti produceret motum. Consequentia probatur, quia talis ignis summus infinitam proportionem habet ad talem calidum remissum; quod probo, quia ad unum remisse (39) frigidum talis ignis aliquam proportionem habet. Sed in infinitum maiorem proportionem habet ad tale calidum remissum quam ad frigidum remissum. Igitur ad calidum remissum infinitam proportionem (40) habet. Tunc sic si ad ipsum aliquam proportionem habet, aliqua velocitate aget; si duplam proportionem habet, dupla velocitate aget; et si triplam, tripla, et sic deinceps. Igitur si infinitam proportionem (41) habet, infinita velocitate aget. Sed quod talis ignis in infinitum maiorem proportionem habeat ad calidum remissum quam ad frigidum remissum probo, quia ad frigidum remissum aliquam proportionem (42) habet et ad frigidum in duplo remissius duplam proportionem; ad [frigidum in] triplo remissius, triplam [proportionem], et sic deinceps. Cum igitur maiorem proportionem habet ad calidum remissum dictum quam ad (43) quodcunque frigidum quantumcunque remissum, ergo et cetera. Quod autem maior sit proportio ignis ad calidum remissum, exempli gratia, duorum graduum quam ad quodcunque frigidum remissum probatur, quia ipsum (44) sibi minus resistit quam quodcunque frigidum; quod patet, quia ipsum est sibi similius; quare et cetera.

Nona ratio: Tunc corpus motum motu violento velocius moveretur in fine quam in (45) principio, cuius oppositum insistat Aristoteles secundo celi et quarto physicorum, ponens quod in hoc differt motus naturalis a violento, quia motus naturalis in sui fine velocitatur, violentus autem tardatur. (46) Consequentiam declaro, quia corpus violenter motum maiorem proportionem habet ad resistentiam suam, scilicet ad medium, in fine quam in principio, quia medium est minus. Ergo velocius tunc movebitur.

(47) Decimo tunc sequeretur quod una guta aque posset scindere unum lignum forte. Consequens est falsum. Consequentiam ego deduco, sumpto vel accepto aliquo dividente quod possit (48) dividere et dividat lignum datum in una hora. Deinde accipio tantam distantiam in aere quod istud dividens precise eam dividat in una hora et sit (49) dividens *a* et lignum *b* et distantia in aere data *c*. Tunc *a* equalem proportionem habet ad *b* et c. Igitur *b* et *c* sunt equalis resistentie. Consequentia tenet per illam regulam Euclidis: (50) Si fuerit duarum quantitatum ad unam quantitatem proportio una, ipsas esse simpliciter equales necesse est; et si

unius ad duas eadem proportio, ipsas equales existere (51) constat. Tunc sic quecunque sunt equalis resistentie, quicquid potest [agere] in unum et in reliquum, sed una guta aque potest dividere distantiam in aere datam. Ergo similiter et (52) lignum, quod erat probandum.

Undecimo *a* alterans alterat *b* per aliquod tempus et continue alterabit a maiori proportione et maiori et tamen continue alterabit (53) tardius. Igitur questio falsa est. Consequentia nota sed antecedens declaratur, posito casu quod *a* sit agens calidum remissum agens in *b* et assimilans sibi *b* et volo quod *a* in(54)cipiat intendi, sic quod continue intendatur sicut continue alterat. Sed intendatur a minori proportione quam sit illa proportio qua *a* alterat *b*. Tunc patet assumptum, (55) nam *a* alterat *b* et continue alterabit a maiori proportione et maiori, quia caliditas in *a* continue erit maior et maior et resistentia in *b* continue erit minor et minor. Et (56) cum *a* secundum positum alterat *b* secundum proportionem caliditatis *a* ad resistentiam *b*, igitur continue alterabit a maiori proportione et maiori et tamen continue alterabit tardius et tardius, quia post (57) hoc *a* non alterabit velocius quam ipsummet alteretur; aliter ageret ultra gradum quod nullum agens naturale facere potest. Sed *a* continue alterabitur tardius et tardius; igitur et continue similiter alterabit.

[Rationes Quod Sic]

(1) Ad oppositum arguo de mente Phylosophi septimo physicorum ponente motum diversificari in velocitate et tarditate secundum diversam proportionem agentium vel agentis ad passa vel passum, ubi diversas regulas (2) ponit, et de mente Commentatoris ibidem, commento ultimo, dicentis: Hec duo, scilicet velocitas et tarditas, et quantitas temporis sequuntur proportionem inter alteratum et alterans. Si igitur proportio (3) fuerit magna, velocitas erit magna et tempus, breve, et conversim. Et etiam in eodem septimo [physicorum], commento 35, dicebat et arguebat ex duplicitate proportionis potentie moventis ad motum duplam ve(4)locitatem in motu. Unde ait: Cum diverserimus motum, contingit necessario ut dicta proportio potentie motoris ad motum sit dupla proportionis illius, et sic illius velocitas ad illam velocitatem est dupla. (5) Quare expresse patet de intentione Aristotelis et Commentatoris, motum in velocitate et tarditate penes proportionem atendi et diversificari et hoc mellius inferius declarabitur in ponendo hanc opinionem.

(6) Hiis igitur breviter sic argutis, ad solutionem huius questionis que multum dificilis existit propter sapientium in ea discordiam, ut statim patebit, accedo. In cuius quidem de(7)terminatione materie breviori stilo quo potero breviter sic procedam.

Nam primo exponam terminos in questione positos, eorum aliquas distinctiones animadvertendo, concludendo similiter questionis dificulta(8)tem. Secundo circa quesitum ponam opiniones aliorum falsas que varie et diverse sunt, eas suis motivis fortioribus confirmando, deinde iuxta posse melius ea iuxta (9) mentem auctorum reprobando et eorum motiva dissolvendo. Tertio ponam opinionem que videtur in proposito verior et magis consona dictis Phylosophi et veritati, ei aliquas (10) varias et diversas conclusiones animadvertendo. Quarto ex dictis aliqua concludam et eliciam corelaria. Quinto circa dictas conclusiones et corelaria alia dubia et instantias movebo (11) et removebo. Sexto et ultimo breviter solvam rationes principales oppositas.

Primum principale

Accedo ad primum. In questione nostra terminus dubitabilior est ille terminus « proportio ». Propter quod pro eius no(12)tificatione eam difinio et describo. Nam proportio nihil aliud est nisi habitudo et comparatio unius ad aliud. Et hanc accipiendo proportionem communiter, ut ipsam sic difiniam, proportio est (13) duorum comparatorum in aliquo in quo comparantur unius ad alterum habitudo. Propter quod notanter dico « in quo comparantur », quando enim inter unum et aliud proportio aliqua ponitur, non debet similis comparatio ex illa (14) proportione in omnibus, sed precise in quo comparantur. Et hec est diffinitio proportionis communiter accepte, quia proportio proprie et strictius sumpta reperitur in solis quantitatibus que sic di(15)finitur: proportio est duarum quantitatum eiusdem generis unius ad alteram habitudo. Unde isti tres termini--« proportio, » « comparatio, » et « habitudo »-- in proposito sinonima sunt. Ista autem proportio (16) communiter sumpta, de qua noster existit sermo, multiplici modo dividitur: primo, divisione generalissima; secundo, divisione generali; tertio, divisione speciali.

Divisio generalissima est quod (17) proportio est duplex, scilicet rationalis et irrationalis. Proportio rationalis est que immediate denominatur ab aliquo numero; que pro tanto vocatur rationalis, quia variatur et denominationem immediate recipit secundum rationalem proportionem nu(18)meri ad numerum, sicut proportio dupla, tripla, quadrupla, et cetera.

Proportio irrationalis est que non immediate denominatur ab aliquo sed immediate denominatur ab aliqua proportione que immediate denominationem (19) recipit a numero. Et hec proportio per oppositum pro tanto vocatur irrationalis, quia non immediate recipit denominationem secundum rationem numeri ad numerum. Qualis est proportio dyametri ad co(20)stam, dyameter enim ad costam nullam proportionem rationalem habet que sit secundum proportionem numeri ad numerum, nec duplam, nec triplam, nec aliam numeralem, sed immediate denominatur ab (21) aliqua medietate proportionis duple que immediate denominatur a proportione dupla.

Ex quo sequitur differentia inter has duas proportiones, quia proportio rationalis reperitur in quantitatibus et numeris, non (22) tamen in quibuscunque quantitatibus sed in quantitatibus commensurabilibus, quibus est una mensura eas mensurans, sicut linea pedalis, tripedalis, et cetera. Proportio autem irrationalis non reperitur in numeris, ut (23) ostensum est, nec in quantitatibus commensurabilibus, sed in quantitatibus incommensurabilibus, quibus non est aliqua mensura ipsas mensurans, sicut patet in dyametro et costa, nulla enim mensura (24) per reduplationem vel subduplationem posset mensurare dyametrum cum costa. [Ratio huius est quia ut dictum est non se habent adinvicem secundum proportionem numeri ad numerum. Nam accipiatur costa quadrati et signetur per 4 et linea dupla ad illam per 8. Si tunc esset aliquis numerus inter 8 et 4 ad quem 8 se habet sicut ipsum ad 4, tunc illa linea signata per illum numerum medium esset equalis dyametro quadrati. Quia igitur nullus rationalis (medius) est, ideo non est rationalis proportio dyametri ad costam qualis numeri ad numerum. Si enim sic, ideo dyameter foret commensurabilis coste, quia ut ponit Euclides commento 20: omnes quantitates habentes adinvicem proportionem secundum numerum ad numerum sunt commensurabiles. Quod autem nullus sit rationalis numerus medius patet, quia tunc maxime foret 6. Sed hoc non, quia 8 se habet ad 6 in sexquitertia proportione, 6 ad 4 in sexquialtera; quare et cetera.]

De prima est hic sermo, quia de illa sit questio que variari potest secundum gradum numeralem (25) secundum diversificationem agentium et passorum in velocitate et tarditate motus, talis autem est proportio rationalis; quare et cetera.

Secunda divisio generalis est quod proportio illa rationalis dividitur, quia que(26)dam est proportio equalitatis, quedam inequalitatis. Proportio equalitatis est duorum equalium adinvicem habitudo sive illa sint quantitates, ut habitudo linee pedalis ad pedalem, sive sint qualitates, (27) ut habitudo caliditatis 2 graduum ad caliditatem duorum, sive qualitercunque fuerint.

Proportio inequalitatis est duorum inequalium adinvicem habitudo. Et ista subdividitur communiter, (28) nam quedam est proportio inequalitatis maioris, quedam inequalitatis minoris. Proportio inequalitatis maioris est habitudo maioris ad minus, ut quantitatis bipedalis (29) ad pedalem. Proportio inequalitatis minoris est habitudo minoris ad maiorem, ut econtra pedalis ad bipedalem. Proportio autem maioris inequalitatis rursum multiplicem (30) recipit divisionem in species, quia aut est simplex aut composita. Si simplex, tunc dividitur in 3, scilicet in proportionem maioris inequalitatis, multiplicem, superpartientem, et superpar(31)ticularem. Proportio maioris inequalitatis multiplex est habitudo maioris quantitatis ad minorem illam multotiens continentis, ut dupla et tripla et cetera. Proportio maioris (32) inequalitatis superparticularis est habitudo maioris quantitatis ad minorem illam semel et aliquam partem eius (33) aliquotam continentis que si contineat ipsam semel et medietatem eius vocatur proportio emiolia seu sexquialtera, si tertiam partem sexquitertia, si quartam sexquiquarta. Exemplum sex(34)quialtere est proportio 6 ad 4. Exemplum sexquitertie, 4 ad 3. Exemplum sexquiquarte, 5 ad 4, et sic deinceps. Notanter autem dixi « partem aliquotam », quia pos(35)sibile esset aliquid continere aliquid semel et aliquam partem eius que non esset pars aliquota, et tunc non esset proportio multiplex (! superparticularis?) sed aliqua aliarum, sicut patet de 7 ad (36) 5. Est autem pars aliquota pars que multotiens reduplicata redit equaliter suum totum, ut 2 respectu 6.

Proportio autem maioris inequalitatis superpartiens est habitudo (37) maioris quantitatis ad minorem illam semel et aliquas eius partes aliquotas continentis, ex quibus non fit una pars aliquota, sicut 5 ad 3, que multipliciter dividitur (38) in infinitas proportiones singulares quas enumerare et difinire nimis foret prolixum.

Si autem ista proportio maioris inequalitatis sit composita, tunc similiter subdividitur (39) in 2 componendo proportionem multiplicem que est simplex cum proportione superparticulari et cum proportione superpartienti, propter quod vocantur composite. Quarum una vocatur mul(40)tiplex superparticularis et alia multiplex superpartiens. Proportio maioris inequalitatis composita multiplex superpartiens est habitudo maioris quantitatis ad minorem illam multotiens (41) continentis et aliquas eius partes aliquotas ex quibus que non fit una pars aliquota, ut 12 ad 5, que similiter sicut precedentes simplices recipit divisiones varias in (42) species infinitas.

Proportio maioris inequalitatis composita multiplex superparticularis est habitudo maioris quantitatis ad minorem illam multotiens et eius partem aliquotam continentis, (43) ut 9 ad 4, que similiter in infinitum recipit sectionem.

Tertio dividitur proportio divisione specialiori. Nam quedam est proportio geometrica, quedam arismetica, (44) quedam armonica.

Proportio arismetica est aliquorum comparatorum in excessu numerali equalitas ut 3, 2, 1, idem enim est excessus 3 ad 2 (45) et 2 ad <1>, utrumque enim excedit aliud per unitatem.

Proportio geometrica est aliquorum comparatorum in proportione vel proportionibus equalitas, ut 12, 6, 3, eandem enim propor(46)tionem habet 12 ad 6 et 6 ad 3, quia utrumque duplam proportionem.

Ex quo statim patet clara differentia inter istas 2 proportiones. Nam idem est excessus 6 ad (47) 4^{or} et 4 ad 2, nam utrumque excedit per 2, tamen proportio non est eadem, quia proportio 6 ad 4 est sexquialtera et 4 ad 2, dupla.

Proportio armonica [est] (48) aliquorum comparatorum primi ad ultimum proportio equalis proportioni excessus primi et secundi ad excessum secundi et tertii, ut exempli gratia, capiantur tres termini, scilicet 6, 4, (49) 3. Tunc proportio 6 ad 3 est dupla; modo proportio 6 ad 3 in excessu est equalis proportioni excessus 6 ad 4 ad [excessum] 4 ad 3, 6 enim excedit 3 in duplo (50) et binarius qui est excessus 6 ad 4^{or} ad unitatem qui est excessus 4 ad 3 habet similiter proportionem duplam. Ponere autem divisiones inter istas tres propor(51)tiones nimis longum foret nec multum proposito utile; quare non pono.

Ex quibus concludo questionis dificultatem que est an causa motuum (52) in velocitate et tarditate sit equalitas proportionum an equalitas excessuum, videtur enim excessus et proportio maioris inequalitatis idem signare; quod tamen non est, (53) ut in sequentibus ostendam clarissime.

2^m principale

Expedito primo generali nominum videlicet notificatione, venio ad secundum. Circa quesitum enim questionis nostre quinque sunt opiniones (54) famose. Prima opinio est quod velocitas et tarditas in motu atenduntur penes excessum potentie moventis supra potentiam rei mote, ita quod (55) quanto agens excedit suam resistentiam tanto velocius movet, et quanto minus tanto tardius. Et quod causa quare hoc movens movet velocius illo est quia magis ex(56)cedit resistentiam suam quam aliud.

Recipit autem ista opinio fundamentum ex quadam auctoritate Aristotelis primo celi capitulo de infinito ubi sic inquid Phylosophus, proportionaliter (57) oportet secundum excellentiam moveri et excessum. Secundo ex quadam auctoritate Averroys quarto physicorum, commento 3, sic dicente, omnis motus est secundum excessum potentie (58) motoris ad rem motam et eiusdem septimo physicorum, commento 30, sic dicente, velocitas propria uniuscuisque motus sequitur excessum potentie motoris supra potentiam rei (59) mote. Et commento 39 dicit, secundum excessum potentie alterantis supra potentiam alterati fit velocitas, alteratio, diminutio quantitatis et temporis. Idem in locis quampluribus restat.

(1) Est autem illa opinio falsa et erronea, cuius defectus et error in eius reprobatione aparet. Ex hac enim quamplura absurda in phylosophia et inconvenientia sequuntur. Quod ostendo, primo enim (2) sequitur quod potentia existens sicut 6 ita velociter ageret in resistentiam sicut 4 quam potentia sicut 2 in resistentiam sicut 1. Consequens est inconveniens, nam potentia sicut 2 habet maiorem proportionem ad resistentiam (3) que est 1 quam 6 ad 4, nam illa est proportio dupla, ista autem sexquialtera. Sed consequentia tenet, quia maior est excessus 6 ad 4 quam excessus per quem 2 excedunt 1. Nam (4) 6 excedunt 4 per 2, 2 autem excedunt 1 per 1; quare etc.

Secundo sequitur quod si aliquis motor movet aliquod mobile per aliquod spacium in aliquo tempore, medietas potentie (5) non movebit medietatem mobilis per idem spacium in eodem tempore. Consequens est falsum, et contra regulam Aristotelis septimo physicorum. Et patet, nam similiter et secundum eandem proportionem se habet totum movens ad totum motum (6) et medietas moventis ad medietatem moti. Igitur motus erunt eque veloces. Sed consequentia immediate patet, nam maior est excessus per quem totum excedit totum quam excessus per quem (7) medietas excedit medietatem, sicut maior est excessus per quem octonarius excedit quaternarium quam excessus per quem 4 excedunt 2, quia 8 excedunt 4^{or} per 4, 4^{or} exce(8)dunt 2° per 2. Cum igitur ex maiori excessu secundum hanc opinionem velocior sequatur motus, evidenter sequitur quod velocius movebit totum totum quam medietas medietatem, quod erat probandum.

Tertio sequitur (9) quod si duo motores separati movent duo mobilia separata per aliquod spacium in aliquo tempore, illi duo motores coniuncti non movebunt illa duo mobilia coniuncta per idem spacium (10) in equali tempore. Consequens similiter est contra regulam Aristotelis in fine septimi physicorum. Et patet, quia secundum eandem proportionem se habet motor compositus ad motum compositum et motor simplex ad motum (11) simplicem. Sed consequentia declaratur, quia excessus per quem illi duo motores coniuncti movent illa duo mobilia coniuncta est duplus ad excessum per quem quilibet illorum motorum (12) separatorum excedit suam resistentiam; quare in duplo velocius movebunt.

Quarto sequeretur quod si terra parva moveretur in aliquo medio, quod medium in duplo excederet, non posset (13) moveri in duplo velocius in aliquo alio medio. Consequens est falsum. Consequentia patet quia tunc totum movens esset excessus. Sed falsitas consequentis probatur, nam in motu simplicium ad quamcunque (14) velocitatem deveniri potest per medii subtiliationem ut patet quarto physicorum, capitulo de vacuo, ubi ponit quod per subtiliationem medii movente eodem precise erit devenire ad quam(15)cunque velocitatem motus datam.

Ex quo similiter concludo quod idem movetur eque velociter in pleno et vacuo, quod est impossibile, cum in vacuo nullam resistentiam extrinsecam habeat, (16) in pleno autem sic. Sed consequentia tenet, quia si non eque velociter, igitur velocius in vacuo quam in pleno; igitur per finitum vel per infinitum. Non per infinitum, (17) quia tunc infinita velocitate moveretur et per consequens in instanti, quod est impossibile de gravi mixto. Si per finitum, ponitur igitur quod in duplo, tunc pono quod medium subtilietur ad (18) duplum, tunc excedet mobile medium in duplo quam prius. Igitur in duplo velocius movebitur quam prius. Sed cum in vacuo movetur similiter precise in duplo velocius quam prius movebatur in (19) pleno, igitur eque velociter moveretur in vacuo et pleno.

Quinto tunc sequeretur quod motor excedens suam resistentiam per maiorem excessum quam alius suam velocius mo(20)veret illam. Et tunc si motor unus fortior per maiorem excessum excederet suam resistentiam quam motor debilior suam, velocius moveret illam. Hec consequentia est satis (21) clara. Sed falsitas consequentis aparet ad experyentiam. Videmus quod musca, que est motor debilis, velocius portabit unum parvum mobile notando quod excedit parvo (22) excessu quam homo magnum mobile quod maiori excessu excedit. Ymo homo vix ipsum movere poterit.

Ex quibus evidenter patet quod velocitas et tarditas in motu (23) non atenduntur penes excessum potentie motoris supra potentiam rei mote. Multe et varie rationes aduci possent, sed ille suficiant, quia satis eius insuficientiam declarant. (24) Ad motivam eius que super solum auctoritatibus firmamentum recipit, dico quod Aristoteles et Commentator per excessum inteligunt proportionem maioris in(25)equalitatis penes quam atenduntur velocitas et tarditas in motu, ut inferius declarabitur suficienter.

Est igitur secunda opinio que ponit quod velocitas et tarditas in motu atenduntur (26) penes proportionem excessus potentie motoris supra potentiam rei mote. Que simpliciter sola auctoritate movetur Averrois septimo physicorum, commento tertio, dicentis velocitas et tarditas in motu (27) est secundum proportionem excessus potentie motoris supra potentiam rei mote. Est autem illa opinio falsa et multo reprobanda ut precedens eisdam rationibus et motivis, propter quod non (28) aliter eam reprobo.

Et ad auctoritatem motam respondeo similiter ut ad precedentes, quod per proportionem excessus inteligit proportionem maioris inequalitatis geometricam qua motor (29) excedit resistentiam mobilis.

Tertia opinio est quod velocitas et tarditas in motu atenduntur penes proportionem passorum manente eodem agente vel equali et penes proportionem agentium (30) manente eodem passo vel equali. Que opinio rationabilior videtur esse quam precedentes. Que opinio fundamentum suscipit ex aliquibus auctoritatibus. Primo quo ad primam partem veri(31)ficandum suscipit auctoritatem Phylosophi quarto physicorum, capitulo de vacuo, ubi sic inquid Phylosophus, sit b quidem aqua, d vero aer, quanto vero aer subtilior est aqua et incor(32)poralius tanto citius a mobile per d spacium quam per b movebitur. Igitur secundum hanc quidem rationem, i.e., proportionem, per quam quidem distat aer ab aqua et velocitas a velocitate (33) distabit; quare si in duplo subtile est aer quam aqua duplici tempore id quod est a, b pertransibit quam d, que quidem auctoritas expresse insinuare videtur quod secundum variationem (34) aeris et aque passorum seu mobilium in dempsitate et raritate manente eodem agente diversificari motum in velocitate et tarditate.

Secundo confirmatur hec opinio similiter quod ad partem (35) priman per sententiam Phylosophi primo celi, capitulo de infinito, dicentis: ab eodem, scilicet agente, subponitur specialiter magis et minus pati in pluri et minori tempore, quecunque proportionaliter tempora (36) divisa sunt. Similiter ex regula Phylosophi septimo physicorum in fine ponente quod si aliqua potentia movet aliquod mobile per aliquod spacium in aliquo tempore, eadem potentia movebit medietatem mobilis (37) per duplum spacium in eodem tempore vel equali, vel per idem spacium in duplo minori tempore. Ex qua regula expresse patet quod manente eodem agente et movente et variatis resistentiis (38) variatur motus et tempus et spacium.

Ad probandum autem secundam partem huius opinionis, scilicet quod velocitas et cetera sequitur proportionem agentium manente eodem passo, suponit auctoritatem (39) Phylosophi et regulam Phylosophi septimo physicorum que iuxta mentem et expositionem Commentatoris est illa: quod si aliqua potentia movet aliquod mobile per aliquod spacium in aliquo tempore, dupla potentia movebit (40) idem mobile per duplum spacium in eodem tempore.

Secundo accipiunt pro eius confirmatione auctoritatem Phylosophi quarto physicorum, capitulo de vacuo, dicente quod gravia et levia per equale spacium (41) in eodem tempore velocius et tardius moventur secundum proportionem et gravium et levium. Item videtur velle Phylosophus primo celi, capitulo de infinito, ubi ait: analogiam, (42) i.e., proportionem, quam gravitates habent, tempora econtrario habebunt, puta si media gravitas in hoc tempore, dupla in medietate eius. Cui consonat dictum commentatoris de ponderibus conclusione prima, (43) videlicet: Inter quelibet gravia est velocitatis in descendendo et ponderis eodem ordine sumpta proportio. Et multe alie varie et diverse sunt auctoritates idem sonantes. Et (44) confirmatur hec opinio in radice una ratione sic: nam si una potentia sit in duplo maioris virtutis quam alia, in duplo plus potest movere ipsa non variata idem mobile, vel equaliter duplum mobile, (45) quia si non duplum, igitur minus foret, igitur minoris potentie quam precise duple.

Hec sunt dicta et motiva huius opinionis que si subtiliter investigentur potius probant opinionem oppositam quam propositam. (46) Est igitur hec opinio falsa et insuficiens. Insuficiens quidem primo, quia non potest velocitatem in motibus atendi et cetera nisi precise ubi est idem motor et resistentia diversa vel econtra, scilicet ubi (47) tam potentia motiva quam resistentia diversitatem recipiunt non docet. Quare insuficienter ponitur.

Et secundo hec opinio falsa, nam ex ea sequitur hoc inconveniens, quod aliqua potentia motiva quodlibet mobile (48) movere possit, et quod aliquod mobile a qualibet potentia motiva moveri possit. Hoc autem est satis absurdum. Tenet consequentia, nam aliqua potentia motiva potest movere aliquod mobile aliqua tarditate et (49) potest movere aliud mobile dupla tarditate. Igitur duplum mobile secundum hanc positionem quadrupla tarditate; igitur quadruplum mobile octupla tarditate et sic in infinitum. Igitur aliqua potentia motiva quodlibet mobile movere (50) potest. Et similiter arguitur secundum.

Et potest ad huius erroris manifestationem aduci illa ratio de medii subtiliatione aducta in reprobatione prime opinionis.

Item confirmatur, nam (51) si aliqua potentia movet aliquod mobile in aliquo tempore per aliquod spacium, dupla potentia movebit idem mobile per equale spacium in duplo velocius; quare opinio illa falsa. Consequentia clara est; sed antecedens (52) patet ad experyentiam: Primo posito uno homine movente unum mobile ponderosum sic quod vix ipsum movere possit, quod si dupletur potentia sic quod adatur alius [homo] movebunt plus (53) quam in duplo velocius.

Idem patet de pondere in horologiis apenso. Et hoc pro conclusione quadam inferius ostendetur clarissime in reprobando regulas Phylosophi septimo physicorum et sta(54)tim aliqualiter patebit in solutione cuiusdam auctoritatis.

Ad eius igitur motiva dico quod omnes auctoritates dicentes quod velocitas in motibus et tarditas atenduntur penes proportionem (55) passorum manente eodem agente inteligunt proportionem passorum ad sua agentia, ut in tractatu de proportionibus motuum insinuatur evidentissime. Ad primam igitur auctorita(56)tem Phylosophi quarto physicorum, capitulo de vacuo, dico quod Phylosophus inteligit quod quanto proportio aeris est minor propter subtilitatem et incorporalitatem ad dividens quam aque ad idem [dividens] propter eius gro(57)ssitudinem et corporalitatem tanto idem citius movebitur per aerem quam per aquam, et hoc est verum. Et ratio est quia quanto proportio aeris ad dividens est minor quam aque ad idem (58) dividens tanto proportio dividentis ad aerem quam ad aquam et quia velocitas sequitur proportionem, sequitur quod idem citius movebitur per aerem quam per aquam. Et hoc intendit (59) mens Phylosophi et non aliud.

Ad secundam auctoritatem eiusdem primo celi dico quod inteligit, suponatur enim specialiter in pluri et minori tempore magis et minus pati quecunque proportionaliter tempori (60) divisa, i.e., quorumcunque proportiones ad illud idem agens proportionaliter tempori divise sunt. Hoc est quod secundum quod agens habet proportionem magnam vel parvam (1) ad passum, dico, magnum vel parvum sic ipsum movebit in magno et parvo tempore motu velociori vel tardiori.

60v

Ad tertiam auctoritatem que est regula Phylosophi septimo physicorum dico (2) quod non est intentio Phylosophi ponere quod si aliqua potentia movet aliquod mobile in aliquo tempore per aliquod spacium quod eadem potentia moveat medietatem mobilis per duplum spacium in equali tempore vel per idem (3) spacium in duplo minori tempore, accipiendo per medietatem partem quantitativam mediam, quia possibile esset quod moveret plus quam in duplo velocius, ut declarabitur inferius. Sed per medietatem inteligit partem (4) resistentie habentis ad eandem potentiam motivam medietatem totius proportionis quam habet totum movens ad totum motum. Et quod istud sit verisimile declaratur ut declaratum in tractatu (5) de proportionibus motuum per expositionem Averroys super illo passo ubi sic ait: Cum igitur diviserimus motum, i.e., rem motam, contingit necessario ut proportio potentie motoris ad mo(6)tum, i.e., ad rem motam, sit dupla illius, scilicet proportionis, quod tunc non esset nullum nisi inteligeretur modo dicto. Et ratio quia licet totum movens habeat aliquam proportionem ad (7) totum motum, non sequitur necessario ut habeat ad medietatem eius proportionem duplam, ut inferius clare ostendam.

Ad motiva probantia secundam partem: Ad primum dico quod illa regula est (8) falsa, ut inferius declarabo. Aristoteles autem potest exponi multipliciter, ut inferius dicam in conclusionibus. Et ad aliam auctoritatem Phylosophi quarto physicorum dico quod inteligit: secundum proportionem gra(9)vium ad suas resistentias et levium ad suas, gravia et levia velocius et tardius moventur. Et sic exponi debent omnes auctoritates posite que videntur istam opinionem (10) confirmare, cum ipsa sit contra mentem Phylosophi et Commentatoris, ut declarabitur.

Ad auctoritatem autem commentatoris de ponderibus dico quod inteli-

VI

24

git quod inter quelibet gravia est velocitatis in (11) descendendo et proportionis ponderis ad suam scilicet resistentiam eodem ordine sumpta
proportio sic: videlicet quod quodlibet grave velociter moveter deorsum et
tanto velocius movet quanto pon(12)dus gravitatis suam resistentiam plus
excedit et tanto tardius quanto minus et hoc non arguit opinionem illam
sed potius arguit quod sequatur proportionem.

Ex quibus omnibus breviter concludo (13) hanc conclusionem necessa-
riam quod velocitas et tarditas non atenduntur penes proportionem agentium, i.e., secundum quod agentia diversificantur in fortitudine et debilita-
te manente equali passo, nec (14) sequitur diversificationem passorum manente eodem agente.

Quarta opinio generalis est quod velocitas et tarditas in motu non
sequitur excessum nec proportionem potentie (15) motive ad resistentiam
sed sequitur quoddam dominium et habitudinem naturalem motoris ad
motum. Et ratio est quia proportio et excessus sub quantitate debentur.
Hec autem, sicut erronea et (16) vilis, viliter movetur et debili fundamen-
to fundatur, movetur enim propter quandam auctoritatem Averroys octo
physicorum, commento 79, dicentis quod potentia incorporea non dicitur
finita nec infinita, quia finitum et (17) infinitum tantum de corporibus
dicuntur. Nec una potentia potest dici maior vel minor alia, maius enim et
minus solius quantitatis sunt. Nec potentie separate a corpore sunt proportionales nec habent (18) proportionem adinvicem. Quare proportio est
solius magnitudinis ad magnitudinem. Ex istis contingit quod nulla potentia motiva est finita vel infinita, maior vel minor alia, nec aliquo modo
proportiona(19)bilis, quia nulla potentia motiva est corpus sed potentia
existens in corpore vel aliquo modo separata.

Et potest hec opinio sic probari: Proportio est comparatio rerum
unius et eiusdem generis, (20) sed potentia activa, ut patet, non est
eiusdem generis cum passiva.

Tertio tunc si potentia activa et passiva haberent proportionem adinvicem, tunc essent comparabiles; tunc igitur essent (21) eiusdem speciei;
haberent igitur subiectum vel substantiam eiusdem speciei specialissime.
Consequentia illa patet per Phylosophum septimo physicorum ubi vult
quod omnia que debent comparari adinvicem tam subiectum seu (22)
subastantia comparaveris. Quod idem est quam illud vel illa in quo vel in
quibus sit comparatio sint eiusdem speciei specialissime et nullam differentiam habeant. Sed falsitas consequentis probatur quia potentia dividitur
(23) per activum et passivum que sunt differentie repugnantes.

Quarto si inter potentiam motivam et resistentiam esset proportio, illa
esset maxima proportio maioris inequalitatis qua agens ex(24)cederet passum. Et cum omne excedens dividatur in excellentiam et in id quod
exceditur, ut patet quarto physicorum, capitulo de vacuo, sequitur quod
quelibet potentia motiva illo modo dividi posset. (25) Quod est falsum,
quia omnis potentia motiva est indivisibilis et aliqua potentia motiva
corporea est minor secundum quantitatem quam potentia illius rei mote.
Neque valet obici quod Aristoteles loquatur ibi de (26) excessibus secundum quantitatem solum, nam loquitur de excessu subtilitatis ad subtilitatem, ut patet ibidem.

Ista autem opinio sicut abusiva est multo reprobanda, quia si inter
(27) potentias motivas et resistentias non esset proportio quia non sunt
quantitative, tunc nec eodem modo inter voces. Et per consequens totius

musice modulatio deperderetur, nam tonus (28) in sexquioctava proportione consistit, diatessaron in sexquitertia, diapente in sexquialtera. Secundo confirmatur, nam Commentator in septimo physicorum, commento 36 et 38, probat quasdam conclusiones de (29) proportione velocitatum in motibus.

Item primo celi, commento 65, probans illam conclusionem, quod nullum infinitum possit movere finitum capiendo ab adversario quod infinitum possit movere finitum in (30) tempore finito et quod agens finitum potest in eodem tempore movere partem huius passi finiti, tunc capit unum movens finitum quod se habet ad primum movens finitum sicut totum passum (31) finitum ad illam partem. Tunc arguitur permutatim per rationem Euclidis quinto elementorum, sicut se habet magnum movens finitum ad suum totum passum sic minus movens ad partem pas(32)si. Ex quo constat magnum movens finitum movere istud passum in eodem tempore in quo minus movens finitum movet illam partem et quo agens infinitum etiam movet id totum. (33) Item movens excedit potentiam motivam et est maioris potentie ipsa. Igitur si dupletur erit maioris; igitur oportet quod hoc sit secundum aliquam proportionem, quam etiam propositionem Phylosophus et (34) Commentator ponunt septimo physicorum et in locis quampluribus.

Item falsum est quod illa opinio assumit quod excessus solus debeatur quantitati; quia etiam virtuti, cum sit divisibilis sal(35)tim intensive et etiam extensive.

Item quelibet talis potentia est corporea quia virtus extensa ad extensionem corporis. Sic igitur ista opinio est falsa.

Tunc igitur ad eius (36) motiva: ad auctoritatem Commentatoris dicendum quod proportio duplex est, ut supra tactum fuit aliqualiter in difinitione proportionis, scilicet proportio communiter dicta reperta non solum in hiis que sunt (37) eiusdem sed etiam diversorum generum; alia est proportio proprie dicta que in solis quantitatibus et eiusdem generis reperitur. Modo non semper inter potentiam motivam et resistentiam est proportio pro(38)prie dicta, sed communiter; et sic inteligunt Commentatoris et Aristotelis auctoritates. Et ideo proportionabilia proportione dicta communiter non requirunt quod sint eiusdem generis nec per consequens eiusdem (39) speciei, constat enim quod in genere generalissimo fit comparatio dicendo, « forma est magis substantia quam materia, » vel oppositum, « substantia est magis ens quam accidens. » Et per hec solvuntur (40) prima et secunda et tertia.

Ad quartam dico concedendo quod illa proportio que est inter potentiam motivam et resistentiam est proportio maioris inequalitatis qua agens excedit passum. Et cum dicitur omne (41) excedens dividitur in excellentiam et in id quod exceditur, dico quod sicut proportio seu excessus est duplex, sic excedens dividitur dupliciter. Si enim sit excessus seu pro(42)portio proprie quia quantum proprie dividitur, quia talis est vere divisibilis et vere quanta. Si autem sit excessus seu proportio communiter sumpta non dividitur proprie sed commun(43)iter, quod nihil aliud est nisi remitti. Et tunc erit sensus quod omne excedens potest remitti ad extremitatem eius quod exceditur, sic quod virtualiter in excedente continetur tota latitudo qua (44) excellit et que exceditur.

Et ideo si capiatur potentia motiva equalis potentie resistive non dicitur illa potentia dupla ad medietatem resistentie quia in duplo possit movere, sed quia dupla (45) illius est tante virtutis in resistendo sicut illa in movendo. Et hec de illa quarta opinione suficiant.

Quinta est opinio ponens quod velocitas et tarditas in motu (46) atenduntur penes facilitatem et dificultatem agendi agentis in passum. Modo autem hec opinio satis rationabiliter quadam ratione superius aducta probante motum in velocitate (47) et tarditate non sequi proportionem quia tunc a et b sunt duo motores equalem proportionem habentes ad suas resistentias et tamen una movebit in duplo velocius (48) quam alia. Igitur consequentia tenet, quia ei videtur quod penes nihil alius possit atendi velocitas et tarditas nisi vel penes proportionem vel penes facilitatem et dificultatem. Sed antecedens (49) probatur, sumpto quod a sit unum dividens in dupla proportione se habens ad b lignum et c sic medietas dividentis et d medietas ligni. Tunc a et c habent equalem (50) proportionem ad b et c quia utrum duplam habet et tamen a dividit velocius b quam c, d, quia si non velocius, igitur eque velocius. Sed contra per motum et divisionem eque velocem, (51) ut sexto physicorum: passum debet dividi equale: Sed per divisionem a et b passum non dividitur equale, quod patet quia per divisionem a dividitur totum b, per divisionem c (52) dividitur medietas b. Modo totum b est duplum ad eius medietatem. Quare et cetera.

Secondo confirmatur hec opinio, quia si aliqua potentia movet aliquod mobile in aliquo tempore dupla potentia movebit (53) idem per equalia spacia plus quam in duplo velocius ut patuit superius in exemplo de homine et de ponderibus orologiis suspensis. Et magis inferius declarabo. (54) Ad auctoritatem autem Phylosophi septimo physicorum que probat motum sequi proportionem quia si aliqua potentia movet aliquod mobile per aliquod spacium in aliquo tempore, dupla potentia movebit (55) in duplo velocius idem mobile, exponunt adendo « vel plus quam in duplo velocius. » Similiter exponunt aliam quando dicit quod si aliqua potentia movet aliquod mobile (56) per aliquod spacium in aliquo tempore medietas eiusdem potentie movebit medietatem mobilis per idem spacium in eodem tempore adendo « vel minori tempore » et quando Commentator probat causam propter equalem proportionem, dicunt quod per (57) proportionem inteligit facilitatem quia eque facile est agenti toti movere totum passum sicut medietati moventis medietatem passi.

Sed illa opinio quo ad quid est falsa, quo (58) autem ad quid insuficiens, est enim falsa in hoc quod ponit velocitatem et tarditatem non sequi proportionem, quod tamen patebit inferius esse falsum. Est autem insuficiens quia (59) licet verum sit motum diversificari penes facilitatem et dificultatem et verum sit quod potentia dupla aliquando movet plus quam in duplo velocius, bene etiam exponat quo ad quid regulas et (60) auctoritates Phylosophi, tamen hic, facilitas et dificultas non sunt esse suficientes. Ymo non sunt cause, sed quedam consequentie consequentes proportionem.

Item idem dubium restat quare (61) hoc movens facilius movet, aliud vero dificilius, istud enim est eque vel magis dubium quam querere quare hoc movens movet velocius alio, illud autem tardius.

3^m principale

(1) Expedito igitur secundo principali positisque aliorum opinionibus eisque iuxta auctorum sententiam (?) et mentem iuxta posse multius

reprobatis, breviter restat (2) ad tertium accedere principale ad ponendum unam opinionem in nostro proposito veriorem et ipsam cum anexis materiam de proportionibus declarantem sub trium conclusionum (3) numero recoligam breviori.

Prima igitur conclusio est quod velocitas et tarditas in motu causaliter atenduntur penes proportionem agentium ad passa.

(4) Secunda conclusio. Quod illa proportio penes quam atenditur et diversificatur velocitas et tarditas motuum est proportio maioris inequalitatis, sic quod nec a proportione equalitatis nec minoris inequalitatis (5) possit ullatenus fieri motus.

Tertia conclusio quod non quelibet proportio maioris inequalitatis est suficiens ad causandum motum nec actionem.

Primam (6) conclusionem declaro primo auctoritatibus: auctoritate quidem primo Phylosophi et Commentatoris expresse in septimo physicorum, ubi ponit quod si aliqua potentia movet aliquod mobile per aliquod spacium in aliquo tempore, (7) medietas motoris movebit medietatem moti per idem spacium in equali tempore. Nunc autem manifestum est quod illa regula, ut declaratur Commentator ibidem, non est vera nisi quia (8) eandem proportionem habet totus motor ad totum motum et medietas motoris ad medietatem moti. Et idem sonant alie regule et per idem tenent. Et si dicatur quod Phylosophus (9) et Commentator ibidem loquuntur de proportione que atenditur secundum excessum, contra hoc arguo sic geometrice et demonstrative sumendo unam supositionem de proportionibus, quod si fuerint quatuor quan(10)titates proportionales permutatim erint proportionales. Tunc arguo sic: qualis est proportio vel secundum quam proportionem se habet totus motor ad medietatem motoris secundum eandem (11) proportionem se habet totum motum ad medietatem moti. Ergo a proportione permutata, que simul potest per supositionem datam, qualis est proportio totius motoris ad totum motum, talis est proportio (12) medietatis motoris ad medietatem moti, quod erat probandum.

Secundo confirmo hoc idem, quia si de illa proportione inteligeretur illa regula, tunc et de eadem proportione inteligere(13)ntur similiter alique regule. Et per consequens de illa proportione arismetica, scilicet inteligeretur illa regula, quod si duo motores divisi moveant duo mobilia divisa in aliquo tempore per aliquod (14) spacium, ambo coniuncti movebunt illa duo mobilia coniuncta per idem spacium in equali tempore et hoc secundum proportionem arismeticam. Sed hoc est falsum, quia non eodem excessu ex(15)cedit motor simplex motum simplex et motor compositus motum compositum. Ymo excessus per quem motor compositus excedit motum compositum est duplus ad excessum (16) per quem motor simplex motum simplex excedit, ut probatum fuit supra in argumento secundo contra primam opinionem. Quare de illa proportione arismetica Phylosophus nec Commentator inteligunt. Secondo confirmatur (17) hec eadem conclusio de mente Commentatoris expresse quarto physicorum commento 71 ubi sic dicit. Et universaliter verum seu manifestum est quod causa diversitatis vel equalitatis motuum (18) est equalitas et diversitas proportionis motoris ad rem motam. Cum igitur fuerint duo motores et duo mota et proportio alterius motorum ad alterum (19) motum fuerit sicut proportio reliqui motoris ad reliquum motum, tunc duo motores erunt equales in velocitate; et diversificata proportione diversificabitur motus secundum

(20) illam proportionem. Et in fine, eodem commento, ait, diversitas motuum in velocitate et tarditate est secundum hanc proportionem que est inter suas potentias, scilicet motivam et resistivam, (21) et in multis locis idem testantur expresse.

Et potest hoc confirmari ratione: penes illam causam debet atendi velocitas et tarditas in motu. Qua posita ponuntur qua remota (22) removentur. Sed posita proportione ponuntur velocitas et tarditas in tali gradu et ea remota removentur. Quare et cetera.

Ad huius tamen conclusionis evidentiam pleniorem est notandum quod hec conclusio universaliter (23) debet de proportione geometrica, si enim de arismetica inteligeretur, tunc nihil esset aliud dicere nisi ponere velocitatem et tarditatem atendi penes excessum, que fuit prima opinio superius reprobata. Et (24) ideo quando ex variatione proportionis arguitur et concluditur variatio velocitatis et tarditatis, debemus diligenter inspicere an varietur proportio geometrica an arismetica.

Secundo notandum quod ad hoc (25) quod ut huiusmodi conclusio sit vera, presuponere debemus cetera paria ut videlicet figura et omnibus aliis adiuvantibus vel impedientibus motum. Si enim essent duo moventia equalia equalem proportionem habentia (26) ad suas resistentias, scilicet ad medium, quorum unum foret pyramidalis figure et aliud late, tunc movens pyramidalis figure velocius moveret medium, et per consequens velocius descenderet. Quare (27) cetera debent esse paria sicut regula communis comparationis suponit.

Tertio notandum quod sicut velocitas et tarditas atenduntur penes proportionem potentie motive ad resistentiam, sic quod ubi (28) est maior proportio ibi motus velocior, ubi minor proportio ibi motus tardior, ubi proportio equalis ceteris paribus motus equalis sit. Et proportio velocitatum in motibus est atendenda secundum propor(29)tionem proportionum potentiarum moventium ad resistentias; ita quod causa, quare hoc movens, exempli gratia *a*, movet in duplo velocius *b* movente scilicet est quia proportio dupla est proportio qua *a* (30) movet ad proportionem qua *b* movet. Et illa conclusio, ut patet, est principalis, solvens principale quesitum questionis.

Secunda conclusio declaratur: si a proportione equalitatis vel minoris inequalitatis po(31)sset provenire motus, sequeretur quod iste motus non esset velocior nec tardior aliquo motu. Consequens est falsum, quia omni motu dato potest dari motus in (32) duplo velocior et in triplo, ut inferius ostendetur. Consequentia declaratur, quia si esset aliquis motus velocior, aut ille foret motus proveniens a proportione equalitatis, et hoc non quia sicut (33) nec aliqua proportio equalitatis est maior alia sic nec motus velocior; aut foret motus procedens a proportione maioris inequalitatis, et hoc non quia nulla proportio (34) maioris inequalitatis est maior vel minor proportione equalitatis. Ergo nec motus velocior nec tardior provenit. Quod autem nulla proportio maioris inequalitatis sit maior (35) vel minor proportione equalitatis probatur; quia si foret, tunc proportio equalitatis aliquociens sumpta rederet equaliter proportionem maioris inequalitatis, quod tamen est falsum, quia semper red(36)dit minorem. Similiter probatur quod proportio maioris [inequalitatis] non sit minor eadem per consimilem rationem.

Secundo hoc eadem determinatione probo, quia si proportio maioris inequalitatis foret maior (37) proportione equalitatis, seu si proportio

VI

equalitatis foret minor vel excederetur a proportione maioris inequalitatis, tunc igitur secundum aliquam proportionem excederetur.

Ex quo concludo quod sequeretur quod proportio (38) equalitatis que vocetur *b* et proportio maioris inequalitatis nominata *a* forent equales. Consequens est absurdum ut de se clarum est. Sed consequentiam deduco sumendo unam supositionem quinti (39) elementorum Euclidis: Si fuerit aliquarum quantitatum ad unam quantitatem proportio una, ipsas esse simpliciter equales necesse est. Tunc arguo sic: Secundum hanc positionem possibile erit proportionem (40) equalitatis *a* et proportionem aliquam maioris inequalitatis equaliter excedi vel equalem proportionem habere ad aliam proportionem maioris inequalitatis. Igitur dicte due erunt equales, secundum hoc docetur (41) ex supositione dicta. Sed assumptum probo, quia cum proportio equalitatis excedatur secundum aliquam proportionem maioris inequalitatis ab alia proportione maioris inequalitatis, sit igitur quod in duplo. (42) Et, ut clarius videatur, signetur per numeros, i.e., sit proportio equalitatis *b* ·1· ad ·1·, proportio maioris inequalitatis excedens sit 4 ad 1, que sit *a*. Cum igitur *a* sit (43) maior *b*, sit igitur exempli gratia in duplo. Tunc accipio unam proportionem maioris inequalitatis precise subduplam ad proportionem quadruplam, que erit proportio binarii ad 1. (44) Tunc sumo unam supositionem quam inferius declarabo, quod si fuerit proportio maioris inequalitatis primi scilicet extremi ad secundum scilicet medium interpositum et secundi scilicet medii ad tertium aliud (45) scilicet extremum, tunc proportio primi ad tertium est precise dupla ad proportionem primi ad secundum et secundi ad tertium. Sed sic est in proposito, proportio enim maioris inequalitatis est quatuor ad 2 (46) et 2 ad 1. Et ideo proportio 4 ad 1 est precise dupla ad proportionem 4 ad 2 et 2 ad 1. Ergo proportio 2 ad 1 est precise subdupla ad proportionem 4 ad 1. (47) Sed *b* proportio equalitatis posita est precise subdupla ad proportionem maioris inequalitatis *a*. Ergo proportio equalitatis et proportio maioris inequalitatis erunt equales propter habere ean(48)dem proportionem ad tertium, scilicet ad *a*. Et per idem probo quod proportio equalitates non sit maior proportione inequalitatis maioris. Quia si sic, secundum igitur aliquam proportionem (49) maioris inequalitatis foret maior. Pono igitur in duplo sit maior, ita quod proportio equalitatis *b* sit dupla proportionis duple, ita quod proportio equalitatis sit modo sicut 4, proportionis maioris (50) inequalitatis excessa que sit *c* sicut 2 ad 1. Deinde capio proportionem maioris inequalitatis *a* in duplo maiorem proportione *c*, que erit sicut 4 ad 2, et sit *d*. Tunc igitur per eandem (51) supositionem quam prius proportio *b* ad *c* est precise dupla et proportio *d* ad *c* est precise dupla (52), ut patet. Ergo proportio *c* et *d* erunt precise equales per rationem superius positam, quia eandem proportionem habent ad *b*, sed una est proportio equalitatis scilicet *b* et alia proportio ma(53)ioris inequalitatis scilicet *d*; ergo et cetera. Et similiter arguitur de proportione minoris [inequalitatis] si subtiliter investiges; quare et cetera.

61v   (1) Ad huius tamen conclusionis secunde evidentiam est notandum quod hoc maxime veritatem habet in motu locali, in motu autem alterationis similiter pro nunc concedere volo, licet de isto magis dubium videatur. Hanc (2) autem conclusionem de motu alterationis veritatem habere non probo sed eam concedo propter aliqua videre subtilia et hoc gratia collationis, propter illam questionem pridie disputatam an contraria (3) equalia agant et patiantur. Quod autem de motu alterationis hoc dictum veritatem

VI

30

contineat probo de mente Commentatoris secundo celi ponentis mixtum equale ad pondus per alterationem ab (4) intrinseco nunquam corumperetur. Si foret, hoc autem esse non videtur verum nisi propter equalitatem elementorum que motui et transitui repugnat et hoc dico presumendo omnimodam equalitatem. Similiter idem dico (5) et idem arguunt ex inteligente terminos de proportione minoris inequalitatis quod ab ea scilicet nullus motus proveniat.

Tertiam autem conclusionem declaro et hoc inteligo de motu alterationis, (6) quia de motu locali pro nunc, licet dubium habeat, non credo veritatem habere. Hec conclusio patet, sumpto quod *a* sit uniformiter difformiter calidum terminatum in extremo intensiori ad gradum summum (7) exclusive et in extremo remissiori ad non gradum. Deinde accipio duas partes immediatas huius corporis, partes quarum una sit versus extremum intensius terminata in extremo remissiori ad (8) medium gradum totius latitudinis exclusive et aliam similiter, quarum intensior vocetur *a* remissior *b*; *c* autem sit gradus medius. Tunc arguo sic: *a* denominationem seu proportionem maioris (9) inequalitatis et excessus habet supra *b* et tamen *a* non potest agere in *b*. Igitur conclusio dicta vera est. Consequentia tenet clarissime, sed antecedens declaro, quia si *a* ageret in *b* aut igitur [secundum] *c* gradum (10) aut intensiorem aut remissiorem. Non est dicendum quod [secundum] *c* gradum agat in *b*, quia ipsum non habet quia cum istud corpus sit uniformiter difformiter calidum et *a* et *b* sint immediate, (11) *c* est intenssisimus qui non est in *a* et remissisimus qui non est in *b*; sic valet quod *c* non est in *a* et quilibet intensior est in *a*; similiter *c* non est in *b* et quilibet remissior est in (12) *b* et nullus qui non est in *b* est ita remissus sicut iste. Igitur *c* est intenssisimus qui non est in *a* et remissisimus qui non est in *b*, ut patet per istorum terminorum expositionem communem. (13) Si dicatur quod *a* agit [secundum] gradum remissiorem, contra [arguo] per idem, quia nullum remissiorem habet; igitur nullum remissiorem inducit, nec similiter aget [secundum] gradum intensiorem, quia non potest inducere gradum (14) intensiorem quin prius inducat remissiorem; cum non contingat transitus de extremo ad extremum absque medio, physicorum 5°, ita quod alteratio in corpore isto sic disposito fieri (15) non valet, nec mirum si in tali casu reactio non proveniat ex proportione minoris equalitatis cum principalis actio, que per excessum fit, fieri non possit. Hec (16) eadem conclusio similiter vera in corporibus quibuslibet inter que non est distantia qualitativa gradus a gradu et quantitativa similiter, sicut patet de elementis in mixto.

Quod autem hec conclusio de motu loca(17)li non verificetur patet per auctoritatem Iordani de ponderibus dicentis quod quodlibet pondus in equilibra elleveat pondus minus eo, similiter est in aliis motibus et cetera.

4ᵐ principale

Nunc (18) expedito tertio principali venio ad quartum ad eliciendum videlicet ex conclusionibus prefatis aliqua corelaria. Quorum primum est quod si potentia motiva ad potentiam moti sit dupla proportio (19) potentia motiva dupla movebit idem mobile in duplo velocius. Hoc corelarium declaro, quia potentia motiva dupla duplam proportionem habebit ad idem mobile. Ergo in duplo velocius movebit.

(20) Secundum corelarium est quod si potentie moventis ad potentiam moti sit precise dupla proportio, eadem potentia movebit medietatem eiusdem moti dupla velocitate. Patet hoc corelarium per idem, quia eius-(21)dem potentie ad medietatem moti est dupla proportio ad proportionem eiusdem potentie ad totum motum.

Tertium corelarium est quod omni motu dato in infinitum potest motus velocior et tardior (22) dari, quia omni proportione maioris inequalitatis potest aliqua in duplo et in triplo et sic in infinitum maior et minor dari, quia omnis proportio maioris inequalitatis per intensionem potentie et (23) remissionem resistentie in infinitum maiorari potest; quare, et cetera.

Quartum corelarium est quod si potentia motiva aliquam resistentiam moveat ad quam se habeat minus quam in dupla (24) proportione, potentia motiva duplata movebit illam resistentiam plus quam in duplo velocius. Istud corelarium ex predicta conclusione declaro sic: sit *b* potentia motiva se habens sicut (25) 4, *c* resistentia sicut 3; *a* potentia motiva dupla ad *b* sicut 8. Deinde ymaginor unam resistentiam scilicet *d* ad quam *b* se habeat sicut *a* se habet ad *b* et igitur sicut duo. Sequitur (26) quod *a* movebit *d* precise in duplo velocius quam *b* ipsum moveat, quia *a* quod est sicut 8 habet ad *d* quod est sicut 2 precise duplam proportionem ad proportionem quam (27) habet *b* quod est sicut 4 ad idem *d* signatum per 2, quia *a* ad *d* est proportio quadrupla et *b* ad *d* proportio dupla. Modo proportio quadrupla est precise dupla proportionis (28) duple. Cum igitur *c* sit minus *d* sequitur quod proportio *a* ad *c* est maior proportione *a* ad *d* et cum proportio *a* ad *d* sit precise dupla, (29) ergo proportio *a* ad *c* est maior quam dupla. Ergo movebit plus quam in duplo velocius. Sed quia forte aliquis diceret quod falsum est quod *c* sit minus *d*, ymo est maius. (30) Tunc ex hoc etiam quod sit maius habeo intentum ut formalem talem rationem, si proportio *b* ad *c* esset precise dupla, tunc proportio *a* ad *c* esset precise dupla ad proportionem *b* (31) ad [*c*]. Cum igitur proportio *b* ad *c* sit minor quam dupla, sequitur quod proportio *a* ad *c* esset plus quam in dupla ad proportionem *b* ad *c* et per consequens movebit plusquam in duplo velocius, quod erat (32) probandum.

Quintum corelarium est quod si aliqua potentia motiva moveat aliquam resistentiam ad quam se habeat plus quam in dupla proportione, potentia motiva dupla movebit eandem resistentiam minus (33) quam in duplo velocius. Istud corelarium ex dicta conclusione declaro similiter ut precedens variando terminos. Sit *b* potentia motiva se habens sicut 4, *c* resistentia eius sicut 1 et (34) sit potentia motiva dupla ad *b* sicut 8. Tunc [quod] *a* movebit *c* minus quam in duplo velocius ipso *b* patet, quia si proportio *b* ad *c* foret precise dupla, tunc proportio (35) *a* ad *c* esset precise dupla ad proportionem *b* ad *c*, sed quia proportio *b* ad *c* est magis quam dupla, ut patet in exemplo. Idcirco proportio *a* ad *c* est minor quam (36) dupla ad proportionem *b* ad *c* et per consequens minus quam in duplo velocius *a* movebit *c* quam *b*, *c*, quod erat probandum. Ex quibus duobus corelariis concludo quod illa regula (37) Aristotelis septimo physicorum est falsa, si aliqua potentia potest movere aliquod mobile in aliquo tempore, dupla potentia movebit idem mobile in duplo velocius, et hoc clare elicetur esse falsum (38) in duobus casibus: primus, quando prima potentia motiva fuit minus quam dupla; secundus, quando fuit magis quam dupla. In primo enim potentia motiva dupla ad primam movet plus quam in (39) duplo velocius, in secundo minus quam in duplo velocius. Et similiter idem dicitur et probatur de aliis regulis Aristotelis ibidem declaratis. Ad Aristotelem autem pro eius dictorum (40) iustificatione exponemus, primo extorquendo quod in translatione fuit defectus, secundo quod ille regule habent veritatem sumpto quod prima potentia motiva sit precise dupla. Hoc (41) autem non dixit Aristoteles, quia subintelexit, vel arguitur quod non inteligit Aristoteles ut in dicta regula, quod si aliqua potentia movet aliquod mobile in aliquo tempore, dupla potentia movebit (42) idem mobile in duplo velocius, sed movebit tanta velocitate quanta est duple dificultatis ad velocitatem priorem que duplam virtutem et potentiam pro eius completo re(43)quirit. Et sic exponatur: Aliter illa veritatem habere non aprobantur. Quare pro maiori declaratione horum addo duo alia corelaria. Primum et est sextum in ordine (44) est quod si aliqua potentia motiva moveat aliquam resistentiam ad quam se habeat in dupla proportione precise, potentia motiva plus quam dupla ad eam movebit idem mobile minus quam in (45) duplo velocius. Hoc corelarium ut precedentia clarius deduco. Sit b potentia motiva sicut 4, c resistentia sicut 2, a potentia motiva plus quam dupla ad b sicut 12. Tunc (46) si a se haberet ad b sicut b ad c ita quod a esset precise duplum ad b sicut b ad c, tunc a ad c haberet proportionem precise duplam ad (47) proportionem b ad c, sed proportio a ad b est magis quam dupla, igitur proportio a ad c est minor quam dupla ad proportionem b ad c et per consequens a movebit (48) minus quam in duplo velocius. Secundum corelarium et est septimum in ordine est quod si b potentia motiva movet c resistentiam ad quam se habeat in dupla proportione precise, i.e., a potentia (49) minus quam dupla ad b movebit c plus quam in duplo velocius b, quod sic declaro ut precedens si et cetera consimiliter per totum ut precedens. Clarius tamen possunt hec duo (50) corelaria declarari. Probo primum ymaginando unum medium ad quod b se habeat sicut a se habet ad b et sit d, tunc proportio a ad d est precise dupla ad proportionem (51) b ad [d] (?); cum igitur c sit maius d, igitur a ad c minorem proportionem habet quam ad d, sed ad d habet precise duplam; igitur ad c habet minorem quam duplam; movebit (52) igitur minus quam in duplo velocius. Similiter patet secundum, quia cum in casu secundo sit minus d et ad d habet a proportionem precise duplam, igitur ad c, cum sit minus (53) habebit proportionem magis quam duplam, movebit igitur plus quam in duplo velocius.

Ex quibus omnibus concludo hanc conclusionem affirmativam: velocitas et tarditas in motu atenduntur pe(54)nes proportionem potentiarum moventium quocunque motu moveant et quecunque potentie sint dummodo sint potentie naturales ad potentias resistentes. Dico autem « dummodo sint (55) potentie naturales » quia in potentiis voluntariis ut inteligentiis non proprie proportio reperitur, nec proprie dictam veritatem haberent. Dico autem « penes proportionem et cetera » sic (56) quod ubi proportio maior ibi motus velocior arguitur et tempus minus, et ubi minor ibi motus tardior et tempus maior, ubi autem equalis proportio existit ibi motus eque velox et tempus equale. (57) Ex secunda autem conclusione concludo unum talem corelarium quod si contraria equalia aproximarent adinvicem quod ex merito equalitatis quam adinvicem habent nunquam agerent. Hoc satis sequitur (58) clare ex dicta conclusione. Idem similiter de proportione minoris inequalitatis. Ex tertia autem conclusione tale elicio corelarium quod nullum uniformiter difforme potest ab intrinseco alterari (59) quod ratio conclusionem probans ostendit determinative et rationabilis causa est quia partes ille sic immediate adinvicem non distant per latitudinem sed solum per gradum qui gradus ponitur (60) consistere in indivisibili. Similiter secundum aliquod exemplum possibile est alicui agenti aproximare extrinseco quod non patietur ab eo, sumpto excessu supra idem, sicut patet in casu quod *a* sit (61) unum calidum uniforme equaliter habens de caliditate et frigiditate et *b* [sit] unum uniformiter difforme sic quod in extremo intensiori sit quilibet gradus citra illum gradum inclusive quem habet *a*, secundum illa extrema, enim non agent et cetera.

5^m principale

62r

(1) Nunc et quinto principaliter restat aliqua movere dubia et instantias et reservationem (!) circa dictas conclusiones et corelaria pro earum declaratione maiori. Primo enim dubium circa prin(2)cipalem conclusionem que fuit prima conclusio: Quid sit ista proportio penes quam ponuntur velocitas et tarditas atendi? Secundo non videtur quod penes istam proportionem atenditur (3) et velocitas in motu quia tunc sequitur quod medio subtiliato ad duplum motus foret in duplo velocior. Consequens est falsum licet videatur Aristoteles confessisse quarto physicorum capitulo de vacuo (4) ponens quod qualis est proportio medii ad medium in subtilitate et dempsitate talis est proportio motus ad motum in velocitate et tarditate, sed hoc non est verum, nec est eius mens (5) nec intentio. Sed illam propositionem accipit tamquam datam vel exponas modo superius prenotato in solutione tertie opinionis. Falsitas autem huius consequentis oportet, quia tunc ut supra (6) fuit contra opinionem primam argutum sequeretur quod idem moveretur eque velociter in pleno et vacuo deducta ratione ut plene ibidem, sed prima consequentia declaratur quia medio subtiliato (7) ad duplum est in duplo maior proportio mobilis ad medium quia resistentia est in duplo minor, vidilicet grossities et dempsitas medii que per medii subtiliationem ad subduplum diminuta est.

Item (8) si hec opinio sit vera, videtur sequi quod corpus grave velocius descendeat in medio parvo quam magno. Consequens ad experyentiam concluditur esse falsum, sed consequentia patet, quia maior est pro(9)portio mobilis ad minus medium quam ad maius. Item Deus movet et anime inteligentie et non per proportionem; igitur et cetera. Item contra hoc idem vadunt omnes rationes (10) ante oppositum questionis facte.

Secundo principaliter arguitur contra secundam conclusionem multipliciter, primo quia tunc sequeretur quod reactio non foret possibilis quod videtur esse inconveniens propter auctoritatem illam quod omne agens in agen(11)do repatitur et ad sensum videmus de aqua calidissima iuxtaposita aque frigidissime. Tenet consequentia quia reactio non provenit nisi a proportione minoris inequalitatis. Tertio in qua(12)libet proportione equalitatis vel minoris inequalitatis passum dominatur supra aliquam partem agentis ut ipsa inexistit toti; igitur supra illam aget. Quarto accipio canonum et pendulum (13) unius lateris sint maiora canono et pendulo alterius lateris. Et sit maius *a*, minus vero *b*, et sit *a* modicum elevatum supra *b*. Tunc *a* et *b* sunt eque gravia (14) et penduli equales et etiam brachium et nihil movet nisi pondus cum adiutorio penduli vel brachii ex parte una et nihil resistit nisi pondus cum adiutorio (15) penduli vel brachii ex parte altera. Igitur potentia motiva est equalis potentie resistive et tamen ibi erit velocitas et motus quod patet per propositionem secundam de ponderibus: « si in (16) equilibra fuerit positio equalis equalibus ponderibus apensis ab equalitate non discedet et si ab equalitate separetur ad equalitatis situm revertetur. » Hoc autem sine motu (17) esse non potest; quare et cetera. Et idem argutum est sicut, si in equilibra penduli et brachia forent equalia et in uno ponatur unus florinus, statim descendet, alio (18) sursum ascendente. Deinde in alio brachio ponatur alius florinus omnino equalis. Tunc movebuntur ad equalitatem; et hoc non nisi ab equalitate proportionis, quare etc. Item aliter reductio ad (19) temperamentum fieri non posset, quia quero quando fit reductio ad temperamentum. Aut hoc sit ab equali et tunc propositum cum talis reductio absque motu et actione fieri non potest (20) aut fit ab excedenti et tunc non fit reductio ad temperamentum quia agens dominans transformabit ipsum in sui naturam; quare et cetera. Item si aliqua potentia movet aliquod mobile aliqua velocitate, (21) igitur movebit duplum mobile equali velocitate. Sed duplum mobile est equale potentie prime, sumpto quod primum mobile precise in duplo exceditur a potentia data; quare et cetera. Item sequeretur si a (22) proportione equalitatis motus non proveniret quod aliquod corpus grave in aere sursum positum non descenderet. Consequens est satis absurdum, stante medio in eadem raritate quam nunc. Sed (23) consequentiam deduco quia accipio aliquod corpus grave scilicet *a*. Tunc quero: istud excedit hoc medium, exempli gratia, *b*, vel igitur secundum aliquam proportionem vel secundum nullam. Si secundum nullam, igitur conclusio principalis falsa (24) que ponit motum quemlibet fieri secundum proportionem aliquam. Si ergo secundum aliquam proportionem excedit, sit ergo exempli gratia quod per duplam, tunc sic est reperire aliquod corpus grave ad quod istud *a* (25) habet proportionem duplam ut notum est. Sit illud *c*. Tunc *a* mobile equalem proportionem habet ad *b* et *c*. Igitur *b* et *c* sunt equalia; tenet hec consequentia per illam rationem superius positam, si (26) fuerit duarum quantitatum ad unam quantitatem proportio una ipsas esse simpliciter equales necesse est. Si vero unius ad ambas eandem idem continet, tunc sic *b* et *c* sunt equalis (27) virtutis. Ergo propter proportionem equalitatis *b* non poterit movere nec dividere *c* nec per consequens in eo moveri et tamen est corpus grave, quare et cetera. Item magnes equalis virtutis (28) cum ferro ipsum atrahit et tamen ibi est equalitas; igitur [et cetera]. Multe autem et varie alie instantie possent aduci, sed iste suficiant quia istarum solutio satis suficienter ma(29)nifestabit intentionem conclusionis.

Tertio principaliter contra tertiam conclusionem arguitur et hoc auctoritate Phylosophi in quampluribus locis positis: Unumquodque non solum localiter sed etiam qualitative moveri (30) secundum elementi predominantis naturam. Hoc autem non est nisi quia proportio maioris inequalitatis in motu alterationis est suficiens ad causandum actionem. Item si non quilibet excessus (31) et cetera, tunc non quilibet ignis in summo posset generari ex aqua in summo sibi simile. Consequens est falsum. Consequentia tenet quia in qualibet tali continue erit excessus in duplo minor et in triplo minor (32) et sic deinceps. Igitur aliquando erit ignis ex minoratione excessus non suficienter excedens aquam. Igitur tunc non fiet actio. Sed hoc erit ante assimilationem aque ipsi igni. Igitur est (33) tertio si non quilibet excessus et cetera. Igitur conclusio prima posita falsa que ponit quamlibet proportionem esse causam motus.

Sed istis non obstantibus tenende sunt in firmitate conclusiones prefate robo(34)rate auctoritatibus et rationibus fortibus. Ad primam igitur instantiam quando queritur quid sit illa proportio, huius solutio patuit in primo principali in terminorum videlicet notificatione, nam proportio nichil (35) aliud est (proportio dico maioris inequalitatis) quam comparatio quedam virtutis moventis ad virtutem mobilis mediante qua virtus moventis mobilis virtutem excedit.

Ad secundam dico pro eius solutione (36) notandum quod resistentia dicitur esse duplex: quedam intrinseca, altera extrinseca. Modo in casu posito ad motum in duplo velociorem non suficit precisa diminutio resistentie extrinsece, sed (37) requiritur etiam intrinsece, et per consequens talis motus atenditur penes proportionem mobilis ad duplicem resistentiam, intrinsecam scilicet mobilis et extrinsecam medii. Et idcirco quia per divisionem resistentie extrin(38)sece precise non sequitur dupla proportio, ideo motus non erit in duplo velocior. Sed contra: ista solutio non evadit quesitum quia ex eadem responsione sequitur inconveniens datum, quia (39) igitur [per] divisionem utriusque resistentie ad duplum, totum movebitur in duplo velocius. Movebitur igitur eque velociter in pleno et vacuo. Dicendum dupliciter primo quod in isto casu concedendum est idem moveri posse (40) eque velociter in utroque, supositis dictis divisionibus, vel dico quod oportet proportionaliter tantum diminuere de quantitate medii quantum de eius grossitudine demittitur, quantitas enim multa medii (41) quantumcunque subtilis in motu moventi resistit ut de se patet. Sed contra istam responsionem arguo: nam idem inconveniens sequitur in casu isto concedendo quod aliquod eque velociter possit moveri (42) in pleno et vacuo sicut et in aliis casibus, nam aduc corpus cuius resistentia extrinseca et intrinseca diminute sunt. Aliquam habet resistentiam extrinsecam in pleno et in vacuo nullam. Quare eque velociter (43) movebitur cum resistentia sicut absque, quod nullus sane mentis concederet. Dicendum quod non sequitur, et ratio est quia licet in vacuo nullam resistentiam extrinsecam habeat, habet tamen resistentiam intrinsecam ma(44)iorem ut patet in casu posito que plus vel tantum adiuvat ad velocitatem motus quantum resistentia medii impedit. Vel arguitur quod cum positum sit resistentiam extrinsecam diminui: (45) Nota quod in aere ymaginor resistentiam ex parte 3: primo ex parte quantitatis medii que satis resistit, secundam ex parte dempsitatis medii, tertiam ex parte gravitatis (46) et levitatis. Nam cum non dicatur simpliciter gravis nec simpliciter levis sed gravis et levis, ideo resistit gravibus et levibus, levibus in ascendendo et gravibus in (47) descendendo. Si igitur foret ab ipso aere tota ablata levitas, tunc solum levibus resisteret. Si autem foret tota ablata gravitas, tunc solum gravibus resisteret ad motum. Ergo in (48) duplo velociorem causandum oportet resistentiam extrinsecam diminui ad duplum ex parte omnium istorum 3 et tunc consequens dictum non erit inconveniens. Ad aliam instantiam dico quod consequentia non tenet. (49)Cuius ratio est quia licet resistentia medii sit minor extensive non tamen est minor intensive que ambo requiruntur ad causandum proportionem maiorem. Ad aliam dico quod ratio illa (50) non est contra conclusionem positam, questio enim, ut dictum fuit supra, in agentibus naturalibus inteligi debet, cuiusmodi non sunt Deus nec alie inteligentie.

Solutis igitur instantiis prime (51) conclusionis vado ad instantias secunde. Ad primam nego consequentiam. Ad probationem respondeo. Pro eius solutione notandum quod differentia est inter equale agere in equale et a proportione equalitatis pro(52)veniente motum. Nam primum est verum, secundum autem non. Ratio primi est quod quodlibet equale, ut arguebatur secundo, dominatur super aliquam partem alterius et sibi invicem sunt contraria suficienter et suficienter (53) aproximata et ideo agent. Similiter idem aprobo de proportione minoris inequalitatis. Ratio autem secundi est quia a proportione equalitatis provenire est quo merito equalitatis totius ad totum sit (54) apta nata agere, quod non est verum. Tunc igitur nego probationem totam et ideo dico quod omnia equalia motu alterationis possunt ad se invicem transmutare cum sint contraria et unum (55) dominatur supra partem alterius ut ipsa inexistit toti. Et ideo cum sint equalia et similiter et equaliter se transmutant, reducent se vel ad temperamentum vel ad gradum aliquem in quo cessabit (56) actio propter insuficientiam contrarietatis, ut inferius declarabo. Reactio igitur non provenit ratione dominii supra totum sed suficit dominium supra partem. Sed contra istam responsionem arguitur: omne agens (57) potens agere in partem *a* potest agere in *a*, sed nihil potest agere in *a* nisi quod habet proportionem maioris inequalitatis respectu *a*. Igitur et cetera. Respondeo negando minorem, quia suficit proportio maioris inequalitatis (58) respectu partis. Sed tunc dico contra: etiam arguitur quod equale non possit agere in equale nec minus in maius, quia tante potentie in proportione equalitatis vel maioris in proportione maioris inequa(59)litatis est agens ad resistendum quam minus vel equale ad agendum. Igitur non agent. Dicendum negando consequentiam. Nec similiter sequitur: *a* maioris potentie quam *b*, igitur suficit resistere ne (60) *b* agat.

Ad tertiam instantiam dico quod quamvis *a* et *b* sint eque gravia, non tamen sunt eque gravia secundum situm et ideo non est ibi equalitas proportionis et ideo ad equalitatem revertuntur.

(61) Ad quartam nego consequentiam. Ad probationem, aut talis reductio sit ab equali aut ab excedenti. Dico ab equali, non tamen ex merito proportionis equalitatis sed merito dominii unius supra partem (62) alterius, non enim sit actio ex proportione totius ad totum, et hoc in casu isto, sed ex proportione totius ad partem.

Ad aliam dico, illa ratio est regula Aristotelis quam idem reprobat (63) 7° physicorum, non enim dicitur dupla potentia quia possit movere duplum mobile sed quia duplum illius tante est potentie in resistendo sicut istud in agendo.

Ad aliam nego consequentiam. (64) Ad probationem autem eius dico quod non est proportio bona quia una est comparatio virtutis movere, que in gravitate consistit, ut corporis gravioris ad corpus minus grave eius (1), ut iste est potentia motiva. In alio autem sit comparatio potentie motive ad resistentiam. Sed contra istam solutionem arguitur, ponendo casum quod illa duo sint duo dividentia, et tunc fiat ratio ut (2) prius. Et ideo dicendum aliter ut diceretur in solutione rationis penultime ante oppositum questionis facte.

Ad aliam nego antecedens. Ymo magnes non atraheret parvum ferrum equale, ut patet.

(3) Tertio vado ad instantias contra tertiam conclusionem factas.

Ad primam dico quod ratio non convenit quod quelibet proportio maioris inequalitatis sit suficiens ad causandum motum, cum in casu posito (4) sit excessus suficiens. Et hoc probo, nam mixtio fit per actionem fieri

equalium inequalium, cum dominans tamen elementum elementum domina-
tum non suficit in propriam (5) naturam convertere, tunc enim non foret
mixtio. Licet hec ratio probat quod non quilibet excessus suficit ad motum
qui est generatio vel coruptio, tamen hoc est verum etiam de aliis motibus.
(6) Et ideo est [verum cum] datur minimus excessus qui non suficit ad causandum motum. Hoc tamen non concedo nisi in casibus preassignatis, in aliis
autem non reputo fore verum (7) ut potest ex precedentibus elici manifeste. Ex quo immediate patet solutio secunde cum non sit verum in casu
quem sumit secunda ratio, ut generaliter loquendo quod conclusio posita
vera sit (8) solum in agentibus non gradualiter excedentibus passa et in
agentibus non distinctis loco et situ ut sunt equalia in mixto. Ista igitur
brevissime suficiant de (9) questione principali, ultima enim ratio nihil
convenit, ut patet.

[6ᵐ principale]

Nunc igitur ultimo accedo ad 6ᵐ principale ante propositum, ad solvendum videlicet rationes ad oppositum questionis factas.
(10) Ad primam nego antecedens. Ad probationem enim admito casum. Ad processum nego quod e continue se habeat in maiori proportione
ad suam resistentiam quam f ad suam quod de(11)claro, nam f se habet
uniformiter in toto motu et ideo uniformiter motum per totum motum sed
e difformiter se habet ad suam resistentiam et ideo difformiter movetur et
aliquando movetur velocius (12) et aliquando tardius, quia diminutio a
lucidi in fine vel versus finem non ita cito diminuit unbram suam sicut
diminutio opaci diminuit suam, quod patet ponendo quod ille (13) diminu-
tiones fiant per aliquod tempus, ut verbi gratia per horam precise, et quod
diminutio c opaci per medietatem hore fiat in duplo velocior quam diminutio d opaci. Tunc in (14) instanti medio hore date unbra d erit multo
longior unbra c et in equali tempore ponatur utraque corumpi quia in
medietate hore igitur unbra d velocius corumpetur (15) unbra c et f
insequens ipsum velocius minuetur ut patet per diffinitionem velocioris;
quare ratio non convenit.
Ad secundam que fortior est ceteris que movit quintam (16) opinionem dicendo quod a et b equalem et cetera, pro cuius solutione notandum
quod proportio est duplex, scilicet qualitativa, que atenditur secundum
virtutem agendi quando aliquod agens equaliter potest alia duo movere
(17) equaliter. Est autem alia proportio quando non solum potest illa duo
equali velocitate movere sed in equali etiam tempore. Tunc igitur dico
quod a equalem proportionem habet ad c sicut b ad d in (18) virtute agendi
sed non quantitative quia dum b pertransibit d, a pertransibit de c partem
equalem d quia licet c et d sint inequalia in quantitate possunt tamen esse
equalia in (19) virtute resistendi. Et sicut non differunt in qualitate
resistendi sed in quantitate, sic nec motus differunt per illam qualitatem
motus que est velocitas et tarditas, sed nisi (20) in quantitate motus que
est longitudo et brevitas temporis. Sed contra: tunc sequitur quod equalis
esset proportio inter c et d proportioni inter a et partem b, dicendum quod
non est inconveniens eandem (21) proportionem in virtute agendi, sed in
quantitate non esset verum. Que quidem solutio nihil vult ad dicere nisi
quod dicit. Quare a non dividit totum c est quia c est maioris (22)
quantitatis. Et ideo si foret quantitatis equalis cum d sumpta equali

resistentia virtuali, tunc pertransiret *a* totum *c* sicut *b* totum *d*. Et ideo quia *c* maioris est quantitatis *d*, ideo (23) tantam quantitatem pertransibit de *c* quanta est *d*.

Ad tertiam dico quod *a* non continue movetur in duplo velocius *b* quia quando *a* erit in fine spacii quiescit cum motum reflexum (24) faciat ut dicit Phylosophus 6° physicorum et 5°; *b* autem tunc movetur vel aliter sumpto quod motus duo reflexi faciant motum continuum ut aliqui ponunt. Dicendum quod non continue (25) in duplo velocius movetur quia non ita cito deveniet ad terminum reflectendo sicut non reflectendo. Et ideo dicendum quod ratio non convenit quia illud spacium in motu reducto equivalet (26) duobus spaciis equalibus.

Ad quartam concedo antedecens et nego consequentiam, ut in corelariis patuit, si enim dupla velocitas [fit] per duplationem sequeretur quod prima potentia sit precise (27) dupla ut ibidem superius clare ostensum est.

Ad quintam dico quod 9 et 8 non movebuntur equaliter, quia cum ternarius moveat per spacium pedale et dimidium, 9 propter tri(28)plicem proportionem movebit per spacium plus quam tripedale ut patet numeranti, 8 autem per tripedale precise. Quare ratio non convenit.

Ad sextam dico ut ad quartam concedendo (29) assumptum sed negando consequentiam ut similiter in corelariis dictum est. Sed regula Phylosophi intendit ubi primo potentia motoris dupla fuerit; modo hic non est dupla sed minus dupla, quia 10 (30) ad 6 est proportio multiplex superpartiens, proportio 5 autem ad 3 alia superpartiens, quare et cetera.

Ad septimam concedo antecedens et nego consequentiam, ut enim tactum fuit (31) supra in tertio principali, ut conclusio verificetur oportet cetera paria presuponi, ut videlicet quod mobilia sint figure et dispositionis consimilis et quod unum non impediat motum al(32)terius, quia tunc (?) est inopportuna (?), sicut patet posito pondere magno in uno brachio equilibre et uno parvo pondere in alio latere tunc magnum trahet minus, sic quod (33) minus eque velociter movebitur cum maiori, minus tamen in ascendendo, maius autem in descendendo; quare non valet.

Ad octavam nego antecedens. Ad probationem concesso casu (34) concedo quod punctus in circumferentia *a* movet velocius sed nego quod quadruplum spacium pertranseat, quia quamvis *a* excedeat *b* circulum in quadrupla proportione, tamen solum excedet circumferentiam in (35) duplo et punctus in circumferentia *a* non pertransibit totum *a* sed solum circumferentiam et similiter *b*. Nunc autem non sequitur: *a* est triplum et punctus in circumferentia *a* movetur ad motum *a*; igitur triplum spacium pertran(36)sit; quare et cetera.

Ad nonam nego assumptum. Ad probationem: admisso casu dico cum aliis sumptis quod aut pondera pendent perpendiculariter aut unum est elevatum (37) supra aliud. Si primo modo, dico quod sunt eque gravia secundum situm et equalem proportionem habent ad proprias resistentias et sic inteligit supositio de ponderibus allegata. Si vero (38) unum sit elevatum supra aliud, dico quod licet sint eque gravia, non tamen sunt eque gravia secundum situm et ideo non equalem proportionem habent ad suas resistentias et *c* maiorem pro(39)portionem habebit ad suam resistentiam quam *d* ad suam.

Ad decimam rationem dicendum negando antecedens. Ad probationem dico quod eque facile elevare magnetem cum ferro continuato (40) magneti sicut magnetem per se sine ferro, ellevans enim magnetem non ellevat ferrum, sed ferrum movetur a magnete, ipso ellevato, et ideo dico quod *a* et (41) *c* non movebunt duo mobilia coniuncta quia ipsi non movebunt nisi ferrum cum resistentia. Ad confirmationem ex cuius solutione ista clarius patebit, dicendum quod magnes non (42) atrahit ferrum modo quo alia trahentia trahunt. Sed ferrum certam dispositionem recipit a magnete; qua recepta movetur ad illam. Et ita movetur ferrum parvum sicut magnes. (43) Sed contra: magnes illa alteratione potest fortius alterare ferrum parvum quam magnum igitur [et cetera]. Dicendum quod non sequitur, quia in tali motu ferrum non movet secundum ultimum potentie sue. Nam (44) illa dispositio nihil facit nisi ferrum apetere coniungi magneti et ideo sive illa dispositio fuerit fortior sive debilior. Si foret ei coniunctum non movebitur ferrum, (45) igitur non resistit elevari magnetem ex causa dicta. Quare nulla istarum rationum procedit.

Ad alias rationes. Ad primam quod maior est proportio ignis ad corpus (46) minus calidum, licet velocius alteret corpus magis calidum, hoc enim non facit virtute maioris proportionis sed virtute maioris similitudinis; quia igitur cetera non sunt paria ut presuponi debet, (47) ratio non convenit. Vel arguitur quod maiorem proportionem habet ad corpus magis calidum quia minus eidem resistit, Nunc autem penes proportionem ad resistentiam atenditur motus, nec (48) sequitur est maius, ergo minorem proportionem habet cum non sit maius in resistentia.

Ad secundam dico quod in proportione qualitativa de qua loquimus maiorem proportionem habet ad (49) terram maiorem quam minorem cum minus resistat ut in precedentis solutione dicebatur et hoc dico in motu naturali quia in motu violento velocius movet minorem quam ma(50)iorem. Cum igitur cetera non sint paria, igitur, quia non equalis inclinatio, ratio non convenit.

Ad tertiam dico quod a nulla proportione agit, sed agit vel incipit agere ab aliqua proportione; que (51) autem sit illa proportio non significari debet cum ibi proportio confuse tantum suponat in qua supositione suposita significari non debet, ut in alia scientia patet.

Idem simile dicitur ad (52) quartam sequentem.

Ad quintam. Dicendum primo negando consequentiam. Vel arguitur pro nunc quod unum si infinitam proportionem haberet ad aliud quod eiusdem speciei specialissime foret, cuiusmodi non est homo respectu asini. (53) Similiter quod non est verum de corporis et superficiei quantitate. Vel pro isto secundo dicendum quod infinitum potest dupliciter ut presenti proposito pertinet destingui. Primo proprie pro corpore exten(54)so sine terminis et sic nullum corpus est infinitum. Secundo minus proprie pro corpore habente infinitas partes proportionales quarum quelibet est maior aliqua certa quantitate data quecunque illa sint (55) sive eiusdem sive speciei diverse, et sic ipsum corpus propter in infinitum secundo modo excedere superficiem ponitur <in> infinitum.

Ad sextam [dico quod nullum agens corporeum, et per consequens nec ignis, (56) est vigoris infiniti]. Similiter respondeo ut ad rationem primam harum.

Ad septimam nego consequentiam. Ad probationem dico negando quod ignis summus descriptus in secunda parte (57) proportionali in duplo velocius agat quem in prima; nec per consequens quod duplam caliditatem inducat. Ratio huius est primo quia agens debilitatur secundo concedendo reactionem ignis repatitur; (58) tertio quia tempus est in duplo minus. Si igitur agens remaneret virtutis equalis et passum resistentie in duplo minoris ut existit in casu posito et tempus equale equalem caliditatem induceret (59) et non maiorem nedum remisso agente et tempore in duplo minori existente.

Ad rationes igitur hanc solutionem probantes. Ad primam dico quod illud iuvamentum in virtute (60) sua non potest facere ignem equalem caliditatem inducere in secunda parte proportionali nedum duplam. Ex qua solutione concedo quod nullum agens fatigabile nec remissibile potest uniformiter (61) agere. Et si dicatur, igitur non erit dare quantitatem uniformem, nego hanc consequentiam. Nam licet ex parte agentis et eius actionis non sit causa ponere, tamen ex par(62)te quantitatis repugnaret. Ex quo concludo quod aliquod agens continue per totum tempus quo aget aget uniformiter difformiter inducendo quantitatem uniformiter difformen et tamen in fine erit illa quantitas uni(63)formis. Patet hoc corelarium de igne agente in aquam usque quo in ea inducat caliditatem summam ut superius in isto proposito contra solutionem arguebatur prefatam.

Ad secundam dico quod (64) Aristoteles inteligit eadem potentia in vigore eodem stans vel eadem, id est equalis.

Ad tertiam patet solutio concedendo consequens quod concluditur ut patet in corelario immediate dicto. Et secundo vado (65) ad rationes probantes reactionem esse non posse.

Ad primam nego consequentiam. Ad probationem eius dico primo quod pars non repassa non potest agere in partem repassam. (1) Ratio huius est quia, ut dictum, nullum corpus potest ab intrinseco alterari. Et ideo sicut in isto casu nego actionem principalem sic nego reactionem. Et concedo multos (2) casus in quibus nego reactionem. Primus est in corpore uniformiter difformi in quo nec pars intensior remissiorem valet remittere nec econtra, propter causam superius in ultimo (3) corelario assignatam. Secundus casus est quando agens difforme aproximatur secundum eius extremum remissius passo uniformi, quia quelibet pars remissior in extremo est continue intendi a parte (4) intensiori et hoc sit verum sumpto quod agens tale sit mixtum ex igne et aqua. Tertius casus qui coincidit cum secundo est exemplum positum et ratio huius quia passum in principio agit (5) in aliam partem agentis principalis et in illo instanti primo actionis agens principale est sic dispositum quod nulla pars eius est contraria alteri, ideo pro illo instanti nulla pars eius incipit agere (6) in aliam. Nunc autem illa non stant simul: quod pro illo instanti nulla pars eius non sit contraria alteri nec agat et quod postea agat. Et si dicas contra: aliqua pars non impassa *a* in processu actionis (7) est contraria parti repasse et supra eam dominans et maioris virtutis quam passum principale in eam igitur aget ipsam alterando, dico quod non sequitur. Ratio est quia passum (8) prius incipit agere in agens principale, nunc autem licet pars non repassa maioris virtutis existat passo principali non potest actio illa per tempus confirmata dependens ab agente (9) debiliori pro quolibet instanti impediri ab agente fortiori, sicut patet sensibiliter in exemplo si ponatur unus homo debilis ad trahendum cordam deinde ponitur homo fortior ad trahendum (10) eandem cordam ad aliquam differentiam positionis quia licet fortior per totum illud tempus nitatur trahere cordam ita fortiter sicut potest, non tamen propter impedit debilius agens ne (11) trahat.

Ad secundam nego consequentiam, non enim equale agit in equale ex

proportione equalitatis nec minus in maius ex proportione minoris inequalitatis quia non agunt nec equale (12) nec minus ex proportione totius ad
totum sed secundum proportionem totius supra partem supra quam habent
proportionem inequalitatis maioris.

Ad tertiam dico primo negando consequentiam. Ad probationem dico
dupliciter: (13) et primo quod illa regula inteligitur: si fuerit proportio
maioris inequalitatis primi ad secundum et cetera, tunc proportio primi ad
tertium erit dupla quod non est verum in proposito; secundo suposito
quod regula generaliter (14) loquatur. Dicendum quod non inteligitur
quod sit dupla proportio sed quod est duplicata; nunc autem dupla et
duplata non convertuntur sed se habent adinvicem ut superius et inferius
quia omnis (15) proportio dupla est duplata, sed non econverso, et quod
sit duplata patet per aliam regulam ponentem quod si fuerit proportio
maioris inequalitatis primi ad secundum et secundi ad tertium (16) proportio primi ad tertium componitur ex proportione primi ad secundum et
secundi ad tertium.

Ad quartam dico quod actio data inter *a* et *b* cessabit et cum queritur
quando dico quod in illo instanti in quo (17) utrumque induxerit reliquum
quemlibet gradum citra medium gradum sue totius latitudinis et cum
dicitur etiam: tunc *a* et *b* sunt equalia ut prius et suficienter ut prius
aproximata et equalem proportionem (18) habentia ut in principio, igitur
si in principio agebant et nunc agent ulterius, dicendum quod propter
defectum neutrus istorum cessant agere sed ob defectum contrarietatis,
tunc enim secundum extrema apro(19)ximata non suficienter sibi invicem
contrariantur.

Ad quintam diverse a variis assignantur solutiones. Primo quod possibile ex illo aere puro sic intercluso generari ignem et aquam (20) secundum se totum et si dicatur igitur contraria possunt simul intendi in
eandem, non sequitur quia ignis et aqua que sunt in aere non sunt ita
contraria sicut ignis et aqua existentia in mixto uniformi. (21) Secunda
solutio est quod non est inconveniens quod contraria simul intendantur,
dum tamen non omnia equalia. Unde dicunt quod in aere puro simul
generabuntur caliditas et frigiditas sed in tali in(22)tensione remitteretur
aliqua qualitas. Cogita.

Ad sextam dico quod in illo casu sicut principalis actio non conceditur
fit nec reactio, cuius ratio est superius assignata.

Istis solutis revertor (2) ad solvendum rationes principales alias.

Ad octavam igitur que bene est dificilis ex cuius solutionis declaratione manifestabitur magis solutio prime principalis rationis. Et sic nego (24)
consequentiam. Ad probationem respondent quidam concedendo quod quanto alique sunt magis similia tanto est potentie motive supra resistentiam
proportio minor et quanto minus similia tanto proportio maior (25) et
quod frigidum remissum datum non magis resistit igitur quam calidum
remissum, non sumpto quod ignis ceteris paribus haberet inducere caliditatem in isto frigido, nec haberet inducere caliditatem (26) illam corumpendo frigiditatem, tunc in eodem tempore induceret caliditatem in corpus
frigidum sicut in remisse calidum. Modo autem solutio per illam conclusionem: si due quantitates et cetera, illius quedam ad am(27)bas ad
maiorem quedam proportio minor, ad minorem vero proportio maior erit.
Sed illa solutio non videtur inteligere quid per proportionem inteligatur,
videtur enim inteligere proportionem (28) quantitativam respectu resisten-

tie quantitative sed hanc proportionem hoc modo sumptam non semper ponere debemus, sed aliquando respectu resistentie qualitative debemus proportionem prefatam describere. Possibile enim (29) esset quod resistentia quantitativa esset magna — ut in proposito resistentia calidi maioris — et qualitative parva et ideo quia minor est resistentia qualitative in maiori calido ideo maior (30) est proportio ignis ad calidum maius quam minus et sic solvitur et inteligitur prefata conclusio. Et ideo dicendum quod ignis datus quantitative maiorem proportionem habet (31) ad frigidum remissum quam ad calidum remissum sed qualitative non. Tunc igitur ad rationem nego consequentiam. Ad probationem dico quod ratio non convenit, non enim circa hic suponuntur paria, nam non (32) maior proportio est causa immediata velocitatis maioris sed similitudo. Prima tamen responsio improbata est satis bona fuga et secundum aliquorum mentem melior illa.

Ad novam dico (33) quod movens principale non est idem mobile, licet enim medium sit minus, proportio qualitativa agentis dati non est maior ut satis clare patet.

Ad decimam dicendum negando (34) consequentiam. Ad probationem vero eius admisso casu dico quod c et d sunt equalis resistentie extrinsece, intrinsece non. Nunc autem si duo forent equalis resistentie extrinsece (35) precise, intrinsece vero non, non omnia debet poni proportio agentis ad illa duo, nec quicquid, ut arguebatur, potest agere in unum potest agere in reliquum. Ubi notatur (36) quod resistentia est triplex scilicet quantitativa, qualitativa et composita ex utraque. Qualitativa aduc triplex: intrinseca, extrinseca et mixta ex utraque. Resistentia qualitativa intrinseca quia (37) anime satis innotesciunt reperitur in hiis per que quo ad dempsitatem et raritatem et alias conditiones intrinsecas sunt equalia quo ad motum. Extrinseca reperitur in (38) hiis < que > per iuvamenta extrinseca equaliter resistunt et ideo pars inexistens toti et eadem separata a toto non equaliter resistunt, licet sint virtutis equalis quia pars a toto separata per (39) iuvamentum extrinsecum non operatur, sed pars inexistens toti per iuvamentum et auxilium partium eidem continuatarum; sic est in proposito quia illa duo, scilicet aer et lignum, primo non sunt equalis (40) resistentie quantitative nec qualitative intrinsece, ideo non sequitur omnia proportio, nec requiritur omne potens uni dominari et reliquo, si enim aer dictus in tantum foret condempsatus sicut lignum, (41) equalis foret dificultas aerem descriptum dividere sicut lignum.

Ad ultimam concedo totam rationem et nego consequentiam quia in casu dato a alterabit a maiori (42) proportione et maiori comparando secundam ad primam et sic continue velocius alterabit secundum illam comparationem sed continue alterabit tardius comparando secundam ad se ipsam et primam ad se (43) ipsam quia secunda in comparatione ad primam est maior et tertia ad secundam sed secunda respectu sui est tardior et minor quam prima respectu sui.

Et in hoc terminetur (44) sententia huius questionis que mihi magno tempore intulit multa dubia et magnas dificultates. Nunc vero animus meus claram (45) in eisdem habet evidentiam que tamen si non forte veritatem complete ostenderit et nec invenerit, viam tamen dat utilio(46)rem veritatem inquirendi. Siqua tamen dixerimus non verisimilia nec multum consona veritati diligens studentium et al(47)terorum virorum perspicax intellectus sapientia corrigat. Amen.

Francischus.

VARIANT READINGS

58r
4 earundem que *MS*
5 sit que *MS*
7 sepessimas *supra scr. MS*
13 [Rationes Quod Non] *addidi*
16-28 *pro* Sed... f *est figura in mg*:

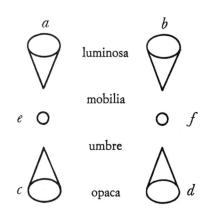

17 scilicet c et d *supra scr. MS*
18 *post* equales *add. et del. MS* scilicet
    c et d
24 scilicet c *supra scr. MS / post*
    quam[2] *scr. et del. MS* f
28 si... sequeretur *MS indicat deletio-*
    nem sed est necessarium; supra si
    scr. MS va- *et supra* sequeretur
    scr.-cat / [quod] *supplevi /* [habe-
    ret] *supplevi / post* velocius *scr.*
    et del MS c / *post* b[2] *scr. et del.*
    MS d
30 *post* sicut *del. MS* d
32 *post* veloces[1] *scr. et del. MS* tunc
    sic
35 *ante* Sed *del. MS* q
40 *ante* ut *del. MS* sed
42 *post* asumptum *del. MS* questio (?)
43 *post* idem[1] *del. MS* spacium
48 *post* bipedale *del. MS* ergo 8
52 *ante* movet *del. MS* et eius
56 *post* gravia *del. MS* mota
57 *ante* c *del. MS* idem
58 *post* et[2] *del. MS* c

58v
1-2 *pro* tunc... gravia *est figura*:

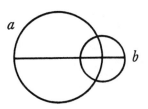

2 *ante* mota *del. MS* et tamen
5 concentrica: concentrici *MS /* dein-
    de: demum (?) *MS / post* est *del.*
    MS conclusio
7 *post* permutata *del. MS* que
12 *post* dupla *del. MS* 2
17 breviori *correxi ex* leviori / apen-
    datur *correxi ex* apendantur / *post*
    quare *del. MS* aliquid / *post* gra-
    via[2] *del. MS* vel (?)
20 *post* dupla *del. MS* tunc
21 c[2] *supra scr. MS, et del.* d
22 *ante* a *del. MS* d
23 *ante* ambo *del. MS* mo
26 aliis *correxi ex* alliis
29 inequales *correxi ex* equales
33 *post* ubi *del. MS* ca / *post* alicui
    del. MS magis quam aliud *et forte*
    etiam dicitur
34 *post* oppositam *del. MS* habet
35 *post* modo *del. MS* magis
42 plus[1,2] *correxi ex* prius
43 *post* a *del. MS* dupla / plus *correxi*
    ex prius
44 et similiter quantitas *supra scr. MS*
45 *post* nam *del. MS* in / *ante* habet
    del. MS quia
47 *post* proposito *del. MS* idem
51 ec et fd *in mg.*
52 *ante* a[1] *del. MS* et ab / scilicet a
    et b *supra scr. MS*
53 d[2] *corr. MS ex* f / c[2] *corr. MS ex*
    d / d[3] *corr. MS ex* f
56 *post* inducendo *del. MS* medietatem
57 *post* caliditatis *del. MS* et sic
59 secunda *corr. MS ex* prima

59r
1 *ante* caliditatem *del. MS* velocitate
4 *post* igitur *del. MS* p / *ante* seu
del. MS quod
7 *post* inducet *del. MS* in
10 *post* summam *del. MS* tunc
15 calidum *supra scr. MS* / Deinde *vel*
demum (?) *MS* / frigidum *supra*
scr. MS
16 c[1] *corr. MS ex* d / *ante* b *del.*
MS d
17 c *corr. MS ex* d
19 si *corr. MS ex* sic
20 deinde *vel* demum (?) *MS*
25 *post* erunt *del. MS* c
26 *ante* sic *del. MS* tunc
30 *ante* Sexto *del MS* 8° principaliter
probo quod motus non sequatur
proportionem
31 *post* prius *del. MS* 9 (?) / *ante*
corpus *del. MS* calidum
33 *post* corporis *add. MS* est *et delevi*
/ remissiorem *del. MS*
36 [Septimo] *addidi*
39 tale *corr. MS ex* talem
40 habet[1] *supra scr. MS*
47 Decimo: 10 *MS*
48 Deinde *vel* demum (?) *MS*
50 fuerit *correxi ex* fuerint
53 sit *supra scr. MS*
54 *post* Sed *del. MS* continue
57 potest *supra scr. MS*

59v
1 *hab. mg. MS* Ad oppositum / ponente *correxi ex* potē
2 Hec *correxi ex* Hoc
6 *hab. mg. MS* Ad questionem
11 Primum principale *mg. MS*
17 numero *supra scr. MS* / immediate
supra scr. MS / rationalem *supra*
scr. MS
19 *post* vocatur *del. MS* ratio
24 Ratio... cetera *mg. MS* / *post* proportio *del. MS* numeri / ideo[2] *forte*
del. MS
31 *post* est *del. MS* quando
32 superparticularis *corr. MS ex* superpartiens / *post* superparticularis
del. MS illam multotiens et eius
partem aliquotam continens que si

ipsum *addendo* va- *supra* illam et
-cat *supra* ipsum
33 continentis *correxi ex* continens
35 *post* non[2] *del. MS* ad
36 superpartiens *corr. ex* superparticularis
37 continentis *correxi ex* continens
39 *post* una *del. MS* est
41 continentis *correxi ex* continens
42 continentis *correxi ex* continens
44 arismetica *corr. MS ex* geometrica /
ante numerali *del. MS* est / *post*
ut *del. MS* 12, 6 et 6, 3
46 *post* idem *del. MS* ex
48 *post* proportio *add. MS* excessus
sed delevi / *post* gratia *del. MS*
6 / *post* scilicet *del. MS* 4
49 *ante* est[1] *del. MS* ex / 3[3, 4] *cor-*
rexi ex 2 / 6 ad 4 [excessum]:
6 ad 4 cum (?) *habet et del. MS*
50 *post* 3 *del. MS* est si
51 Ex *corr. MS ex* Expedit / *post*
causa *del. MS* velocitatis in
53 2[m] principale *mg. MS*
54 *post* penes *del. MS* proportionem
potentie / moventis *vel* motoris (?)
MS / *post* quod[2] *del. MS* causa
quare
56 Recipit *correxi ex* Recit
57 *post* omnis *del. MS* excessus
58 dicente *correxi ex* dicentes
59 restat *correxi ex* restatur

60r
3 2 excedunt 1 *correxi ex* 4 excedunt 2
4 *post* per[1] *del. MS* n / *post* spacium
del. MS eadem potentia
7 *post* excedit[2] *del. MS* 1'
9 separata *correxi ex* separati
16 *post* quia *del. MS* subtilitate medio
ad duplum / *post* pleno *del. MS*
non simul
17 *post* quod[1] *del. MS* quidem / *post*
quod[2] *del. MS* ad duplum
18 *post* medium *del. MS* per
21 debilis *correxi ex* debilius (?) /
post notando *del. MS* ad
24 *ante* Ad *scr. MS* Est igitur secunda
opinio que ponit quod velocitas *et*

del. Est igitur *sed totum delevi* /
super *supra scr.* MS
30 precedentes *correxi ex* precedens
37 *ante* Ex *del.* MS que regula
41 *post* levium *del.* MS Et 7° physicorum
42 gravitates *correxi ex* gravitatem /
media *corr.* MS *ex* medietas
43 velocitatis *correxi ex* velocitas
44 *post* maioris *del.* MS quam
51 *ante* mobile *del.* MS aliquod
60 *post* vel *del.* MS passam / *post*
parvam *add.* MS *et delevi* ad passum *quia in seq. pag. repetitur*

60v
5 *post* motoris *delevi* p'
6 sit dupla illius *interavit* MS
7 duplam *supra scr.* MS / primum
correxi ex primam
10 velocitatis *correxi ex* velocitas
14 *ante* passorum *del.* MS age
23 *ante* 4° *del.* MS 3
27 motivas et resistentias *correxi ex*
motivam et resistentiam / *post* nam
del. MS opo
28 diatessaron *correxi ex* diatertson
33 propositionem *correxi ex* proportionem
38 proportionabiliam MS
41 *ante* seu *del.* MS est
42 *post* autem *del.* MS exce
44 in duplo *correxi ex* vel duplus
47 *post* motores *del.* MS mo
50 motum: mo^m MS (*forte* modum
sive motivum?)
51 *post* Sed *del.* MS ma
53 suspensis *correxi ex* suspenso
54 *post* movebit *del.* MS plus
55 *ante* in^1 *del.* MS quam / *post* ve-
locius^1 *del.* MS Similiter expo (!)
56 eiusdem *correxi ex* eadem / *post*
in^2 *del.* MS equali vel / minori et
supra scr. eodem / [adendo... tem-
pore] *addidi*

61r
(*Tit.*) 3^m principale *mg.* MS
1 secundo *correxi ex* tertio
3 *post* numero *del.* MS relico (?)
c (?)

4 *post* equalitatis *del.* MS nec / nec^2...
inequalitatis *mg.* MS
5 *ante* Primam *del.* MS 4^a conclusio
quod
7 Nunc autem *corr* MS *ex* Nec *et*
verbo quod non legere possum
15 *ante* per *del.* MS q
19 motoris *correxi ex* motorum
31 *post* motus *del.* MS velocior quocunque motu
32 *post* quia *del.* MS aut ille
33 *post* procedens *del.* MS a procedens
34 equalitatis *corr.* MS *ex* minoris
inequalitatis
37 maioris *correxi ex* minoris
38 nominata a *del.* MS
41 *post* duplo *del.* MS ita quod
44 *ante* aliud *del.* MS ad
45 dupla *corr.* MS *ex* duplam
50 c^1 *et* a *et* et sit d *supra scr.* MS
51 *post* dupla^1 *del.* MS ad proportionem b ad d (*cum* c *supra*) sed probo
igitur ad proportionem et d ad c
52 *ante* per *del.* MS quia

61v
1 pro... volo *supra scr.* MS
3 autem *corr.* MS *ex* etiam / alterationis *corr. ex* alterationem
6 Hec *correxi ex* hoc (?) / locali *supra*
scr. MS
7 *ante* extremo^2 *del.* MS no (?)
14 *post* isto *del.* MS non
15 *post* casu *del.* MS ex excessu
18 4^m principale *mg.* MS
19 *post* hoc *del.* MS 'c (?)
22 omni *corr.* MS *ex* omnis
23 *post* motiva *del.* MS dupla precise
ad
26 a^2 *corr.* MS *ex* ad / *in mg. add.*
MS: vere ista —— non inte ——
(*quod non bene legere possum*)
28 minus *corr.* MS *ex* maius / *post*
ad^2 *del.* MS b et a / *post* dupla
del. MS igitur
29 aliquis diceret *corr. ex* diceret quis
33 *in mg. sunt pauca verba quae non*
legere possum
34 *ante* minus *del.* MS plus
37 *in mg. sunt pauca verba quae non*
legere possum

38 *post* casibus *scr. MS sed delevi* consimiliter ac ibidem ponente / promo *corr. ex* prima

39 secundo *corr. ex* secunda / aliis *corr. ex* alliis / *ante* regulis *del. MS* r^{iis} / *post* Ad *del. MS* eius autem

43 *In mg. add. MS* Ista duo corelaria sunt... (*quod non bene legere possum*)

46 a^1 *corr. MS ex* ab / b$^{1-4}$ *corr. MS ex* c / c^{1,2} *corr. MS ex* b / a^2 *corr. MS ex* b / precise^1 *corr. MS ex* plusquam / *post* a^3 *del. MS* moveret c

47 a^1 *corr. MS ex* ad b / *ante* magis *scr. MS et delevi* minor quam

48 *ante* a *del. MS* p

52 *post* sit *del. MS* mai

54 potentiarum *correxi ex* potentiam

55 quia *correxi ex* quare

56 autem *supra scr. MS*

61 extrema *supra scr. MS*

62r

(Tit.) 5^m principale *mg. MS*

6 primam *correxi ex* prima

9 *post* cetera *del. MS* 2° principaliter contra

10 non *supra scr. MS*

11 Tertio *corr. ex* secundo

12 *ante* canonum *scr. MS et delevi* eius / pendulum *corr. MS ex* simetrum

15-16 si... equilibra: Cum equilibris De pond. Prop. 2

16 positio *correxi ex* po^a Cf. De pond. Prop. 2 / separetur *correxi ex* non discedet Cf. De pond. Prop. 2

23 scilicet a *supra scr. MS*

35 *ante* pro *del. MS verbum quod non legere possum*

36 *ante* ad *del. MS* non

45 *post* aere *scr. MS et delevi* triplicem / *post* resistit *del MS* nam *et signum quod non legere possum*

49 Cuius *vel* eius? *in MS* / sit *correxi ex* sint

53 *ante* Similiter *del. MS* Ratio / Similiter... inequalitatis *delendum?* /

ante Similiter *del. MS* 7 (? a ?) Ratio

54 *post* equalia *del. MS* in

58 tunc dico: t d *MS*

62 *post* ex proportione *add. MS* ex ac^e (*del.*) proportione *quod totum delevi*

64 corpus *bis MS*

62v

2 *post* magnes *del. MS* citius *et supra scr.* non

4 *post* non *del. MS* fit

6 minimus *correxi ex* minimum

9 [6^m principale] *addidi* / rationes *correxi ex* rationis

16 *post* quod^2 *del. MS* resistentia est

17 in^2 *bis MS*

19 *post* est *del. MS* loquendo et brevitas

22 *post* tunc *del. MS* divideret (?)

23 de c *supra scr. MS*

26 si *correxi ex* oportet (?) / [fit] *addidi*

28 proportionem *correxi ex* proportiones / *post* quartam *del. MS* quod

29 minus *correxi MS ex* plusquam

30 superpartiens^2 *correxi ex* superparticularis (*et ante* superparticularis *del. MS* de eiusdem)

31 *ante* oportet *del. MS* tc (?) / paria *bis MS*

32 inopportuna: ino^a *MS*

37 *post* resistentias *del. MS* Si vero

39 *post* dicendum *del. MS* quod

40 non *supra scr. MS*

44 *post* nihil *delevi* ad

47 *ante* Vel *del. MS* ad

48 *ante* minorem *del. MS* minus

52 Ad quintam *mg. MS*

55-56 dico... infiniti *del. MS scribendo* va- *supra* dico *et* -cat *supra* infiniti

57 *post* Ratio *del. MS* est

61 *ante* consequentiam *del. MS* consimilem / *ante* eius *scr. MS et delevi* m

64 corelario *correxi ex* corelarium

63r

1 Et^2 *correxi ex* Ex

3 *ante* intendi *scr. MS et delevi* in

6 non[2] *supra scr. MS* / non impassa *supra scr. MS* / a *del. MS*?
10 *post* cordam *del. MS* secundum
11 *ante* Ad *del. MS* quartus
12 *post* dico[1] *del. MS* dupliciter
15 *post* tertium *del. MS* erit
21 *post* tali *del. MS* te *et signum verticale*
23 *ante* rationes *del. MS* 8^{am} / octavam *correxi ex* ultimam
25 non[2] *correxi ex* nam
26 *post* si *del. MS* fuerint
26-27 ambas *correxi ex* anbas

27 minor *corr. MS ex* minorem / inteligatur: inteligentius (?) *MS*
29 maior *correxi MS ex* minor
30 *ante* dicendum *del. MS* ad rationem / *post* quod *del. MS* illa velocitas
33 decimam: 10 *MS*
36 *post* triplex[2] *delevi* vel
38 separata[2] *correxi ex* separato
40 dictus *supra scr. MS*
41 *post* maiori *scr. MS et delevi* pro
44 tempore *correxi ex* tenpore
45 *post* tamen *del. MS* n
46 veritati *corr. MS ex* veritatem

INDEX VERBORUM

14, 31, 35, 40-41, 56; 60v: 24;
62r: 3.
casus: 58r: 16, 21, 39; 59r: 53; 61v:
15, 38, 52, 60; 62r: 36, 39, 41-42,
44, 62; 62v: 1, 3, 6-7, 10, 33, 36,
58; 63r: 1-4, 22, 34, 41.
causa: 58r: 4, 10; 59v: 51, 55; 60v:
56, 60; 61r: 17, 21, 29; 61v: 59;
62r: 33; 62v: 45, 61; 63r: 2, 32.
causaliter: 61r: 3.
causare: 61r: 5; 62r: 30, 48-49; 62v:
3, 6.
celo, De - see Aristoteles
centuplus: 58v: 46.
cessare: 62r: 55; 63r: 16, 18.
circularis: 59r: 23.
circulus: 58v: 6, 11, 18; 62v: 34.
circumferentia: 58v: 6-10,   12-13;
62v: 34-35.
cito: 62v: 12, 25.
clare: 58r: 7; 59r: 34; 59v: 53; 60r:
53; 60v: 7; 61r: 42; 61v: 9, 37,
58; 62v: 27, 41; 63r: 33.
clarus: 58r: 21, 28, 44; 58v: 50;
59v: 46; 60r: 21, 51; 61r: 38;
63r: 44.
cogitare: 63r: 22.
coincidere: 63r: 4.
collatio: 61v:2.
commensurabilis: 59v: 22-24.
Commentator (= Averroes); 59v: 2,
5; 60r: 24, 39; 60v: 10, 28, 34,
36, 38, 56; 61r: 6-7, 9, 16-17;
61v: 3; and see Averroes.
commentator de ponderibus: 60r: 42;
60v: 10.
commentum: 59v: 2-3, 24, 57-59;
60r: 26; 60v: 16, 28-29; 61r: 17,
20.
commotus: 58r: 7.
communis: 59r: 14; 61r: 27; 61v: 12.
communiter: 59v: 12, 14, 16, 27;
60v: 36, 38, 42.
comparabilis: 60v: 20.
comparare: 58v: 29-30; 59v: 13-14,
44-45 ,48; 60v: 21-22; 63r: 42.
comparatio: 59v: 12-13, 15; 60v: 19,
22, 39; 61r: 27; 62r: 35, 64; 62v:
1; 63r: 42-43.
completus and complete: 61v: 42;
63r: 45.

componere: 59v: 30, 38-40, 42; 60r:
10; 61r: 15; 63r: 16, 36.
concedere: 59r: 8; 60v: 40; 61v: 1-2;
62r: 39, 41, 43; 62v: 6, 26, 28, 30,
33-34, 57, 60, 64; 63r: 1, 22, 24, 41.
concentricus: 58v: 5.
concludere: 59v: 7, 10, 51; 60r: 15;
60v: 12; 61r: 24, 37; 61v: 36, 53,
57; 62r: 8; 62v: 62, 64.
conclusio: 58v: 6; 59r: 8, 12, 19;
59v: 10; 60r: 42, 53; 60v: 8, 13,
28-29; 61r: 2-6, 17, 22, 25, 30;
61v: 1-2, 5-6, 9, 16, 18, 24, 33, 53,
57-59; 62r: 1-2, 10, 23, 29, 33,
50-51; 62v: 3, 7, 31; 63r: 26, 30.
condempsatus: 63r: 40.
conditio: 63r: 37.
confirmare: 58v: 24; 59v: 8; 60r: 34,
44, 50; 60v: 10, 28, 52; 61r: 12,
16, 21; 63r: 8.
confirmatio: 60r: 40; 62v: 41.
confiteri: 62r: 3.
confuse: 62v: 51.
coniungere: 58v: 19, 22-23; 60r: 9,
11; 61r: 14; 62v: 41, 44.
consequens: 58v: 24, 27, 32, 40,
43-45, 55; 59r: 7-9, 14, 18, 38, 47;
60r: 2, 5, 10, 13, 17, 21; 60v: 22,
27, 38, 60; 61r: 13, 26, 31, 38;
61v: 31, 36, 47; 62r: 3, 5, 8, 22,
27, 31, 37, 48; 62v: 55, 57, 64.
consequentia: 58r: 15, 28, 36, 40, 44,
50; 58v: 4, 14, 25, 28, 32, 37, 40,
43, 50, 55; 59r: 8, 10, 14, 18, 23,
34, 38, 46-47, 49, 53; 60r: 3, 6,
11, 13, 16, 20, 48, 51; 60v: 21, 48,
60; 61r: 32, 38; 61v: 9; 62r: 6, 8,
11, 23, 25, 31, 48, 51, 59, 61, 63;
62v: 26, 29-30, 52, 56, 61, 65;
63r: 11-12, 24, 31, 34, 41.
consimilis: 58v: 37; 61r: 36; 62v:
31.
consimiliter: 61v: 49.
consistere: 60v: 28; 61v: 60; 62r:
64.
consonare: 60r: 42.
consonus: 59v: 9; 63r: 46.
constare: 58r: 35; 59r: 51; 60v: 32,
39.
continere: 59v: 31, 33, 35, 37, 41-42;
60v: 43; 61v: 3; 62r: 26.

contingere: 59v: 4; 60v: 5, 18; 61v: 14.

continuatus: 62v: 39; 63r: 39.

continue: 58r: 14, 19-27, 38; 58v: 4; 59r: 8-11, 52, 54-56, 57; 62r: 31; 62v: 10, 23-24, 62; 63r: 3, 42.

continuus: 62v: 24.

contra: 60r: 5, 10; 60v: 10; 61r: 9, 16; 61v: 13; 62r: 6, 9-10, 29, 38, 41, 50, 56, 58; 62v: 1, 3, 20, 43, 63; 63r: 6.

contrariare: 59r: 35; 63r: 19.

contrarietas: 59r: 26-27; 62r: 56; 63r; 18.

contrarius: 59r: 14, 27, 32; 61v: 2, 57; 62r: 52, 54; 63r: 5-7, 20-21.

conus: 58r: 18-19, 21, 58.

convenire: 62v: 3, 9, 15, 25, 28, 47, 50; 63r: 31.

conversim: 59v: 3.

convertere: 62v: 5; 63r: 14.

corda: 63r: 9-10.

corelarium: 59v: 10, 61v: 18-21, 23-24, 32-33, 36, 43, 45, 48, 50, 57, 58; 62r: 1; 62v: 26, 29, 63-64; 63r: 3.

corporalitas: 60r: 57.

corporeus: 60v: 25, 35; 62v: 55.

corpus: 58r: 19; 58v: 34, 47-49; 59r: 20, 31, 33, 36, 44, 46; 60v: 17, 19, 35; 61v: 7, 10, 14, 16; 62r: 8, 22-24, 27, 42, 64; 62v: 45-47, 53-55; 63r: 1-2, 26.

corrigere: 63r: 47.

corumpere: 58r: 20, 26; 58v: 56-57; 62v: 14; 63r: 26.

coruptio: 58r: 21; 62v: 5.

costa: 59v: 19-20, 23-24.

cotatio: 58v: 6.

creare: 58r: 18.

credere: 61v: 6.

crescere: 58r: 19.

dare: 58v: 48; 59r: 24, 48-49, 51; 60r: 15; 61r: 11, 31; 61v: 21-22; 62r: 5-6, 21, 38; 62v: 14, 54; 63r: 16, 25, 30, 33, 41, 45.

debere: 58r: 30; 59v: 13; 60v: 9, 15, 21, 34; 61r: 21, 23-25, 27; 62r: 50; 62v: 46, 51; 63r: 28, 35.

debilis: 60r: 20-21; 60v: 16; 62v:

44; 63r: 9-10.

debilitare: 59r: 3, 8; 62v: 57.

debilitas: 60v: 13.

debilitatio: 59r: 4-5.

debiliter: 58r: 37.

declarare: 58r: 16, 29, 37; 58v: 5, 15, 55; 59r: 14, 18, 34, 46, 53; 59v: 5; 60r: 11, 23, 25; 60v: 3-4, 8, 10, 53; 61r: 2, 6-7, 30, 32, 44; 61v: 5, 9, 19, 24, 33, 39, 49-50; 62r: 6, 56; 62v: 10.

declaratio: 58r: 56; 61v: 43; 62r: 1; 63r: 23.

decrescere: 58r: 20.

deducere: 58r: 41, 50; 58v: 26, 33, 50, 52; 59r: 15, 47; 61r: 38; 61v: 45; 62r: 6, 23.

deductio: 58v: 37.

defectus: 59r: 27; 60r: 1; 61v: 40; 63r: 18.

deficere: 59r: 27.

deinceps: 59r: 40, 42; 59v: 34; 62r: 32.

demittere: 62r: 40.

demonstrative: 61r: 9.

dempsitas: 60r: 34; 62r: 4, 7, 45; 63r: 37.

denominare: 58r: 56; 59v: 17-18, 20-21.

denominatio: 58r: 51, 54-55; 59v: 17-19; 61v: 8.

deorsum: 58v: 35.

dependere: 58v: 44; 63r: 8.

deperdere: 60v: 27.

descendere: 58r: 57-58, 60-61; 60r: 43; 60v: 11; 61r: 26; 62r: 8, 17, 22, 47; 62v: 33.

descensus: 58r: 60.

describere: 58v: 18; 59v: 12; 62v: 56; 63r: 28, 41.

destingui: 62v: 53.

determinatio: 59v: 6; 61r: 36.

determinative: 61v: 59.

determinatus: 58r: 1.

Deus: 58r: 2; 62r: 9, 50.

devenire: 60r: 14; 62v: 25.

diapente: 60v: 28.

diatessaron: 60v: 28.

dicere: 58v: 33; 59r: 1, 3; 59v: 2-4, 9-10, 13, 24, 34, 57-59; 60r: 24, 26, 35, 40, 42, 45, 54, 56, 59; 60v:

39-45.
magnitudo: 60v: 18.
magnus: 58r: 12, 14, 16, 22-23, 25, 31; 58v: 5, 18, 23, 25, 28-32, 36, 39 *et passim*.
manere: 58r: 19; 59r: 33-34, 36; 60r: 29-30, 34, 37-38, 55; 60v: 13-14.
manifestare: 62r: 28; 63r: 23.
manifestatio: 60r: 50.
manifeste: 62v: 7.
manifestus: 58r: 59; 58v: 14, 37; 61r: 7, 17.
materia: 58r: 3; 59v: 7; 60v: 39; 61r: 2.
maxime: 59r:32; 59v: 24.
medians: 62r: 35.
medietas: 58r: 49-50, 52-54, 55*; 58v: 19, 42, 58; 59r: 6-7; 59v: 21, 33; 60r: 4-8, 36, 42; 60v: 2-4, 7, 44, 49, 52, 56-57; 61r: 7-8, 10-12; 61v: 20-21; 62v: 13-14.
medium: 58r: 57; 58v: 2; 59r: 27, 29-30, 46; 60r: 12-14, 17-18, 50; 61r: 26, 44; 61v: 14, 50; 62r: 3-4, 7-9, 22-23, 37, 40, 44-45, 49; 63r: 33.
medius: 59r: 26; 59v: 24; 60r: 42; 60v: 3; 61v: 8; 62v: 14; 63r: 17.
mens: 59v: 1-2, 9; 60r: 39, 59; 60v: 10; 61r: 1, 17; 61v: 3; 62r: 4, 43; 63r: 32.
mensura: 59v: 22-23.
mensurare: 59v: 22-24.
meritus: 61v: 57; 62r: 53, 61.
minoratio: 62r: 32.
minuere: 62v: 15.
mirus: 61v: 15.
mixtio: 62v: 4-5.
mixtus: 60r: 17; 61v: 3, 16; 62v: 8; 63r: 4, 20, 36.
mobile: 58r: 18-19, 22, 35, 40, 43, 49; 58v: 1, 4, 19, 21; 59r: 6; 60r: 4-5, 9, 11, 18, 21-22, 29, 32, 34, 36, 39-40, 44, 47-49, 51-52; 60v: 2, 52, 54-56; 61r: 6, 13-14; 61v: 19, 37, 41-42, 44; 62r: 7, 9, 20-21, 25, 35, 37, 63; 62v: 31, 41; 63r: 33.
moderni doctores: 58r: 5.
modicus: 62r: 13.

modo (*adv.*): 58r: 31, 42; 58v: 34; 59r: 22; 59v: 49; 60v: 37, 46, 52; 61r: 49; 61v: 27, 54; 62r: 36; 62v: 29; 63r: 26.
modulatio: 60v: 27.
modus: 58v: 35; 59r: 15; 59v: 16; 60v: 6, 18-19, 24, 27; 62r: 5; 62v: 37, 42, 55; 63r: 28.
motivus: 58r: 15; 58v: 20; 59v: 8-9; 60r: 27, 45, 47-49, 54; 60v: 4, 7, 15, 18-19, 23-25, 27, 33, 36-37, 40, 44; 61r: 20, 27; 61v: 18-19, 23-25, 32-34, 38, 40, 44-45, 48; 62r: 15; 62v: 1; 63r: 24.
motor: 58r: 14, 23, 43-44; 58v: 19, 21; 59v: 4, 58; 60r: 4, 9-11, 19-21, 23, 26-28, 46; 60v: 5, 15, 47; 61r: 7-8, 10-13, 15-16, 18-19; 62v: 29.
motus: (*noun*) 58r: 3, 9, 15, 21, 32-33; 58v: 12, 22, 25-26, 33-34; 59r: 18-19, 22-23, 37-38, 44^2, 45; 59v: 1, 4^1, 5, 51, 54^1, 57, 58^2; 60r: 6^2, 8, 13, 15, 22, 25, 26^2, 29, 34, 38, 46, 54-55; 60v: 1, 5^1, 14, 29, 45-46, 50, 54; 61r: 3-4, 5, 17, 19^3, 20-21, 25, 28, 31-34; 61v: 1-6, 16-17, 21$^{3, 4}$, 54, 56; 62r: 3-4, 15-16, 19, 22, 24, 30, 33, 36, 38, 41, 44, 47, 52, 54; 62v: 3, 5-6, 11$^{1, 3}$, 19-20, 23-25, 31, 35, 43, 47, 49; 63r: 37.
motus: (*p.p.*) 58r: 57; 58v: 2; 59r: 44^1, 46; 59v: 3, 4$^{2, 3}$, 54^2, 58^1, 59; 60r: 5, 6^1, 10, 23, 26^1, 27-28; 60v: 4, 5$^{2, 3, 4}$, 6-7, 15, 25, 59; 61r: 7-8, 11-12, 15-16, 18, 19$^{1, 2}$; 61v: 18, 20, 21$^{1, 2}$; 62v: 11^2.
movens: 58r: 3, 9, 39; 58v: 19-21, 31, 36; 59v: 3, 54-55; 60r: 5-6, 13-14, 37, 52; 60v: 4, 6, 30-33, 57, 61; 61r: 25-26, 29; 61v: 20, 54; 62r: 35; 63r: 33.
movere: 58r: 10-11, 13-14, 16, 20, 22, 26-28, 31, 36-37, 39-49, 52-53, 57, 59; 58v: 2-4, 10, 12-14, 18-19, 22-25, 31-32, 36; 59r: 5-6, 14, 44, 46, *et passim*.
multiplex: 59v: 16, 29-31, 35, 39-40, 42; 62v: 30.
multipliciter: 59v: 37; 60v: 8; 62r:

10.
multotiens: 59v: 31, 36, 40, 42.
multum (adv.): 58r: 4-5, 58; 59v: 51; 60r: 27; 63r: 46.
multus: 60r: 23, 43; 61r: 21; 62r: 28, 40; 63r: 1, 44.
musca: 60r: 21.
musica: 60v: 27.

natura: 62r: 20, 30; 62v: 5.
naturalis: 58r: 4; 59r: 45, 57; 60v: 15; 61v: 54-55; 62r: 50; 62v: 49.
naturaliter: 58v: 34.
natus: 62r: 54.
necessario: 59v: 4; 60v: 5, 7.
necessarius: 60v: 13.
necesse: 61r: 39; 62r: 26.
negare: 62r: 51, 54, 57, 59, 61, 63; 62v: 2, 10, 26, 29-30, 33-34, 36, 39, 52, 56, 61, 65; 63r: 1-2, 11-12, 23, 31, 33, 41.
neuter: 63r: 18.
nimis: 59v: 38, 51.
niti: 63r: 10.
nomen: 58r: 51, 56; 59v: 53.
nominatus: 61r: 38.
notanter: 59v: 13, 34.
notare: 59r: 13; 60r: 21; 61r: 22, 24, 27; 61v: 1; 62r: 36, 45, 51; 62v: 16; 63r: 35.
notificatio: 59v: 11, 53; 62r: 34.
notus: 59r: 53; 62r: 25.
numeralis: 58r: 51; 59v: 20, 44.
numerans: 62v: 28.
numerus: 59v: 17-22, 24; 61r: 3, 42.

obicere: 60v: 25.
obscure: 58r: 6.
obstaculum: 58r: 20.
obstans: 62r: 33.
obtinere: 58v: 29.
octonarius: 60r: 7.
octuplus: 58v: 43; 60r: 49.
omnimodus: 61v: 4.
omnis: 58r: 51; 58v: 6, 11; 59v: 14, 24, 57; 60r: 54; 60v: 9, 12, 21, 24-25, 40, 43; 61r: 25, 31; 61v: 21-22, 53; 62r: 9, 48, 54, 56; 63r: 14, 21, 35, 40.
opacus: 58r: 17, 20; 62v: 12-13.
operare: 63r: 39.

opinio: 59r: 3; 59v: 5, 8-9, 53-54, 56; 60r: 1, 8, 25, 27, 29, 30, 34, 38, 44-47, 50-51; 60v: 9, 12, 14, 19, 26, 34-35, 45-46, 52, 57; 61r: 1-2, 16, 23; 62r: 5-6, 8; 62v: 16.
oportet: 59v: 57; 60v: 33; 62r: 5, 40, 48; 62v: 31.
oppositus: 58v: 34; 59r: 18, 45; 59v: 1, 11, 19; 60r: 45; 60v: 39; 62r: 10; 62v: 2, 9.
ordo: 60r: 43; 60v: 11; 61v: 43.
orologium: 60v: 53, and see horologium.
ostendere: 58r: 45, 56; 58v: 3; 59r: 23; 59v: 23, 53; 60r: 1, 53; 60v: 7; 61r: 32; 61v: 59; 62v: 27; 63r: 45.

Padua: 58r: 1.
par: 61r: 25, 27-28; 62v: 31, 46, 50; 63r: 25, 31.
pars: 58r: 23, 26; 58v: 30, 48, 57-59; 59r: 2, 4, 6-7, 12, 15-17, 32-36; 59v: 32-37, 41; 60r: 30, 34, 38; 60v: 3, 7, 30-32; 61v: 7, 59; 62r: 12, 14-15, 45, 48, 52, 55-58, 61-62; 62v: 18, 20, 54, 56, 60-61, 65; 63r: 2-3, 5-8, 12, 38-39.
parvus: 58r: 61; 58v: 24-25; 60r: 12, 21, 60; 60v: 1; 62v: 42-43; 63r: 29.
passivus: 60v: 20, 23.
passus: 58r: 3, 12, 16, 41; 59v: 1, 25; 60r: 29-30, 34, 38, 55; 60v: 1, 5, 13-14, 24, 30-32, 40, 46, 51, 57; 61r: 3; 62r: 12; 62v: 8, 58; 63r: 3-4, 7-8.
patere: 58r: 34, 39, 43, 55, 59-61; 58v: 1-2, et passim.
pati: 58v: 34; 60r: 35, 59; 61v: 3, 60.
pedalis: 58r: 46, 48; 59v: 22, 26, 29; 62v: 27.
pendere: 62v: 36.
pendulum: 62r: 12-15, 17.
penultimus: 62v: 2.
perfectio: 58v: 45, 47.
perfectus: 58v: 46.
permutatim: 58v: 8; 60v: 31; 61r: 10.
permutatus: 58v: 7; 61r: 11.

perpendiculariter: 62v: 36.
perpetuus: 59r: 22.
perspicax: 63: 47.
perspicuus: 58r: 6.
pertinere: 62v: 53.
pertractare: 58r: 7.
pertransire: 58r: 27, 30, 32-36, 39;
 58v: 4, 10, 13, 18; 60r: 33; 62v:
 18, 22-23, 34-35.
phylosophi naturales moderni: 58r: 4.
phylosophia: 58r: 4; 60r: 1.
Phylosophus or Philosophus: 58r: 34;
 59r: 23; 59v: 1, 9, 56; 60r: 31,
 35-36, 39-41, 53, 56, 59; 60v: 1-2,
 8, 10, 21, 33, 54, 60; 61r: 6, 8,
 16; 62r: 29; 62v: 24, 29, and see
 Aristoteles.
Physica: see Aristoteles
piramidalis or pyramidalis: 58r:
 58-59; 61r: 26.
planus: 58r: 50.
plenum: 60r: 15-16, 19; 62r: 6, 39,
 42.
plenus: 61r: 22.
plumbum: 58r: 61.
ponderosus: 60r: 52.
pondus: 58v: 15, 17, 21; 60r: 43,
 53; 60v: 11, 53; 61v: 3, 17; 62r:
 14, 16; 62v: 32, 36.
ponere: 58r: 18-19, 59; 58v: 16, 38,
 41, 46; 59r: 11, 18, 20, 24, 26, 45;
 59v: 1-2, 5, 8-9, 13, 24, 50-51;
 60r: 17, 25, 36, 47; 60v: 2, 34,
 45, 58; 61r: 2, 6, 22-23, 49; 61v:
 3, 59; 62r: 2, 4, 17-18, 24, 33;
 62v: 1, 12, 14, 24, 55, 61; 63r: 9,
 15, 28, 35.
portare: 60r: 21.
portio: 58v: 18.
positio: 60r: 49; 61r: 39; 62r: 16;
 63r: 10.
positus: 58r: 16, 23, 27, 41, 50, 61;
 58v: 20, 55; 59r: 23, 28, 53, 56;
 59v: 7; 60r: 52; 60v: 9; 61r: 1,
 21-22, 47, 52; 62r: 22, 25, 29, 33,
 36, 44, 50; 62v: 3, 7, 32, 58; 63r:
 4.
posse: 58r: 8, 46; 58v: 1, et passim.
possibilis: 59r: 14, 30; 59v: 34; 60v:
 3; 61r: 39; 61v: 60; 62r: 10; 63r:
 19, 28.

potens: 63r: 40.
potentia: 58r: 9, 15, 29, 39-40, 49,
 52; 58v: 20, 44; 59r: 3, 5-6, 25;
 59v: 3-4, 54, 57-59; 60r: 2, 4, 23,
 26-27, 36, 39, 44-45, 47-49, 51-52;
 60v: 2, 4-5, 14, 16-20, 22-25, 27,
 33, 35, 37, 40, 44, 52, 54, 56, 59;
 61r: 6, 20, 27, 29; 61v: 18-25,
 32-34, 37-38, 40-42, 44-45, 48,
 54-55; 62r: 15, 20-21, 58-59, 63;
 62v: 1, 26, 29, 43, 64; 63r: 24.
preassignatus: 62v: 6.
prebere: 58v: 3.
precedens: 58v: 37, 50; 59v: 41;
 60r: 27-28, 30; 61v: 33, 45, 49;
 62v: 7, 49.
precise: 58v: 9; 59v: 14; 60r: 14,
 18, 45-46; 61r: 43, 45-47, 51-52;
 61v: 26-28, 30, 34-35, 40, 46; 62r:
 21, 38; 62v: 13, 26, 28; 63r: 35.
precisus: 62r: 36.
predictus: 61v: 24.
predominans: 62r: 30.
prefatus: 61v: 18; 62r: 33; 62v: 63;
 63r: 28, 30.
prenotatus: 62r: 5.
presens: 62v: 53.
presumere: 61v: 4.
presuponere: 61r: 25; 62v: 31, 46.
prex: 58r: 1, 7.
principale (as principal sections of
 treatise): 59v: 11, 53; 60r: 61;
 61v: 17, 61; 62v: 9*.
principalis: 59r: 15; 59v: 11; 61r:
 1-2, 30; 61v: 15; 62r: 1, 23, 34;
 62v: 9, 31; 63r: 1, 5, 7-8, 22-23,
 33.
principaliter: 58r: 27; 59r: 37; 62r:
 1, 10, 29.
principium: 58r: 38; 59r: 28, 45-46;
 63r: 4, 18.
prior: 58r: 40; 58v: 4; 59r: 25, 31;
 60r: 18; 61v: 42.
probare: 58r: 23, 26, 42, 44, 54, 57;
 58v: 10, 20, 28, 37, 41-42, 44;
 59r: 2, 8, 10, 12, 23, 31, 36, 38
 41, 43, 52; 60r: 8, 13, 38, 45;
 60v: 7, 19, 22, 28-29, 46, 49, 54,
 56; 61r: 12, 16, 35-36, 41, 48;
 61v: 2-3, 32, 36, 39, 59; 62v: 4-5,
 59, 65.

probatio: 62r: 51, 54, 61, 64; 62v: 10, 33, 36, 39, 56, 65; 63r: 12, 24, 31, 34.

procedere: 58r: 37; 59v: 7; 61r: 33; 62v: 45.

processus: 62v: 10; 63r: 6.

producere: 58v: 34; 59r: 9, 29, 38.

prolixus: 59v: 38.

proponere: 58r: 7, 8.

proportio: 58r: 3, 12, 47-49, 51, 53-56; 58v: 7, 11, 12, *et passim*; arismetica: 59v: 43-44; 61r: 13-14, 16; armonica: 59v: 44, 47; dupla: 58r: 31; 58v: 12, 42; 59v: 4, 18, 21, 46, 50; 60r: 3; 60v: 7, 49; 61r: 29, 49; 61v: 18, 20-21, 23, 27, 44, 46, 48; 62r: 25, 38; 63r: 15; duplicata: 58v: 11-12; equalis *or* equalitas proportionis: 58r: 28, 30, 34, 37, 43-45, 49; 58v: 17, 49, 52; 59r: 45, 49; 59v: 48-49; 60v: 47, 49, 56; 61r: 25, 28; 61v: 56; 62r: 18, 25, 60; 62v: 17, 37-38; 63r: 17; equalitatis: 59r: 17; 59v: 26, 52; 61r: 30, 32-35, 37, 39, 41-42, 47, 49, 52; 61v: 15; 62r: 12, 22, 27, 51, 58, 61; 63r: 11; finita: 58v: 41; geometrica: 59v: 43, 45; 61r: 23-24; inequalitatis *and* maioris inequalitatis *and* minoris inequalitatis: 59v: 27-31, 36, 38, 40, 42, 52; 60r: 24, 28; 60v: 40; 61r: 4-5, 30, 33-38, 40-45, 47-50, 52-53; 61v: 5, 8, 22, 58; 62r: 11, 30, 35, 53, 57-58; 62v: 3; 63r: 11-13, 15; infinita: 58v: 40, 45, 47; 59r: 38, 40; 59v: 38; 62v: 52; irrationalis: 59v: 18, 22; maior: 58r: 12, 14, 16, 22-23, 25; 58v: 25, 28, 30-31, 39; 59r: 2, 39, 42, 46, 55-56; 60r: 2; 60v: 23; 61r: 28; 62r: 7, 49; 62v: 10, 38, 46-48; 63r: 24, 27, 32, 41; minor: 58v: 30-31, 36; 59r: 54; 61r: 28; 62v: 48; 63r: 24; mobilis: 62r: 7-8, 37; potentie: 58r: 9, 15; 60v: 5, 14; 61r: 27, 29; 61v: 54; rationalis: 58r: 16; 59v: 17, 20-21, 25; tripla: 58r: 43; 59r: 42*.

proportionabilis: 58v: 6, 8, 16; 60v: 18, 38.

proportionalis: 58v: 48, 57-59; 59r: 2, 4, 6-7; 60v: 17; 61r: 10; 62v: 54, 57, 60.

proportionaliter: 59v: 56; 60r: 35, 59-60; 62r: 40.

propositio: 60v: 33; 62r: 5, 15.

propositus: 58v: 17, 40, 47; 59r: 6, 24; 59v: 9, 15, 51; 60r: 45; 61r: 2, 45; 62r: 19; 62v: 9, 53, 63; 63r: 13, 29, 39.

proprie: 59v: 14; 60v: 37, 42; 61v: 55; 62v: 53-54.

proprius: 59v: 58; 62v: 4, 37.

prosillio: 58r: 33.

provenire: 59v: 17, 22; 61r: 31-32, 34; 61v: 5, 15; 62r: 11, 22, 51, 53, 56.

punctus: 58v: 9, 10, 12-13, 62v: 34-35.

purus: 63r: 19, 21.

puta: 60r: 42.

pyramidalis: 61r: 26, *and* also *see* piramidalis.

quadrangularis: 58r: 58.

quadratum: 59v: 24.

quadruplus: 58r: 33, 35; 58v: 4, 10, 12-13, 42; 59v: 18; 60r: 49; 61r: 43; 61v: 27; 62v: 34.

qualis: 58v: 6, 8; 59r: 21; 59v: 19; 61r: 10-11.

qualitas: 58v: 34; 59r: 9, 12; 59v: 26; 62v: 19; 63r: 22.

qualitative: 62r: 29; 63r: 29, 31, 40.

qualitativus: 61v: 16; 62v: 16, 48; 63r: 28, 33, 36.

quamplures: 60r: 1; 60v: 34; 62r: 29.

quantitas: 58v: 8, 29, 44, 48; 59r: 50; 59v: 2, 14-15, 21-24, 26, 28, 31-32, 37, 40, 42, 59; 60v: 15,17, 25-26, 34, 37; 61r: 9, 39; 62r: 26, 40, 45; 62v: 18-23, 53-54, 61-62; 63r: 26.

quantitative: 62v: 18; 63r: 30.

quantitativus: 60v: 3, 27; 61v: 16; 63r: 28-29, 36, 40.

quaternarius: 60r: 7.

querere: 58r: 10; 59r: 9; 60v: 61; 62r: 23, 34; 63r: 16.

quesitus: 59v: 8, 53; 61r: 30; 62r:

45; 63r: 32-33, 37.
scientia: 58r: 5; 62v: 51.
scindere: 59r: 47.
scolar: 58r: 1.
sectio: 59v: 43.
sensibiliter: 63r: 9.
sensus: 60v: 43; 62r: 11.
sententia: 60r: 35; 61r: 1; 63r: 44.
separare: 58v: 24; 60r: 9, 12; 60v: 17, 19; 62r: 16; 63r: 38.
sequi: 58r: 21, 28; 58v: 26, 31, 37, 40, 43, 46, 54; 59r: 3, 7-8, 18, 22, 28, 31, 37, 47; 59v: 2, 21, 53, 58; 60r: 1-2, 4, 8, 12, 19, 38, 47, 58; 60v: 7, 12, 14-15, 24, 47, 54, 58; 61r: 31, 37; 61v: 25, 28, 31, 57; 62r: 3, 6, 8, 10, 21, 38, 41, 43, 59; 62v: 20, 26, 35, 43, 48, 52; 63r: 7, 20, 40.
sermo: 59v: 16, 24.
sexquialterus: 59v: 24, 33, 47; 60r: 3; 60v: 28.
sexquioctavus: 60v: 28.
sexquiquartus: 59v: 33-34.
sexquitertius: 59v: 24, 33-34; 60v: 28.
signare: 59v: 24, 52; 61r: 42; 61v: 27.
significare: 62v: 51.
signum: 58r: 6.
similis: 58v: 27, 53-54; 59r: 44; 59v: 13; 62r: 31; 62v: 51; 63r: 24.
similiter: 58r: 17, 24, 47; 58v: 35, 42, 44, 47-48; 59r: 6, 33, 36, 51, 57; 59v: 7, 38, 41, 43, 50; 60r: 5, 10, 15, 18, 28, 34, 36, 50; 60v: 55; 61r: 13, 36, 53; 61v: 1, 16, 52, 58, 60; 62r: 53, 55, 59; 62v: 29, 35, 53, 56.
similitudo: 62v: 46; 63r: 32.
simplex: 58v: 50; 59v: 30, 39, 41; 60r: 10-11, 13; 61r: 15-16.
simpliciter: 59r: 50; 60r: 26; 61r: 39; 62r: 26, 46.
simul: 59r: 14, 17, 28-31; 61r: 11; 63r: 6, 20-21.
singularis: 59v: 38.
sinonimus: 59v: 15.
situs: 58v: 17; 62r: 16, 60; 62v: 8, 37-38.

solum: 60r: 24; 60v: 36; 62r: 47; 62v: 35.
solutio: 59v: 6; 60r: 54; 62r: 5, 28, 34-35, 38, 51; 62v: 1-2, 7, 16, 21, 41, 49, 59-60, 63-64; 63r: 19, 21, 23, 26-27.
solutus: 62r: 50; 63r: 22.
solvere: 59v: 11; 60v: 39; 61r: 30; 62v: 9; 63r: 23, 30.
sonare: 60r: 43, 61r: 8.
spacium: 58r: 27, 30, 32-33, 36, 38-41, 44, 46-48, 52-53; 58v: 4, 10, 13, 18-21; 59r: 6; 60r: 4-5, 9, 32, 36-40, 51; 60v: 2-3, 53-54, 56; 61r: 6-7, 14; 62v: 23, 25-28, 34-35.
specialis: 59v: 16, 43.
specialiter: 60r: 35, 59; 60v: 21-22; 62v: 52.
species: 59v: 30, 42; 60v: 21-22, 39; 62v: 52, 55.
stare: 59r: 5; 62r: 22; 62v: 64; 63r: 6.
statim: 59v: 6, 46; 60r: 53; 62r: 17.
stilus: 59v: 7.
strictim: 59v: 14.
studiens: 58r: 7; 63r: 46.
subdividere: 59v: 27, 38.
subduplatio: 59v: 24.
subduplus: 61r: 43, 46-47; 62r: 7.
subiectum: 60v: 21.
subintelegere: 61v: 41.
substantia: 60v: 21-22, 39.
substantialis: 58v: 53-54.
subtiliare: 60r: 17.
subtiliatio: 60r: 14, 50; 62r: 7.
subtiliatus: 62r: 3, 6.
subtilitas: 60r: 56; 60v: 26; 62r: 4.
subtilis: 60r: 31, 33; 61v: 2; 62r: 41.
subtiliter: 60r: 45; 61r: 53.
suficere: 59r: 25; 60r: 23; 60v: 45, 60; 61r: 5; 62r: 28, 30, 36, 56-57, 59; 62v: 3-6, 8.
suficienter: 59r: 12, 26; 60r: 25; 62r: 28, 32, 52; 63r: 17, 19.
sumere: 61r: 9, 38, 44; 62v: 7.
summus: 58v: 30, 52, 56; 59r: 10-11, 13, 23-25, 28, 37-38; 61v: 6; 62r: 31; 62v: 56, 63.
sumptus: 58r: 29, 37, 39, 45, 50;

58v: 5, 50; 59r: 10, 18, 31, 47; 59v: 14, 16; 60r: 43; 60v: 11, 42, 49; 61r: 35; 61v: 6, 40, 60; 62r: 21; 62v: 22, 24, 36; 63r: 4, 25, 28.

superbipartiens: 58r: 56.

superbitertius: 58r: 55.

superficies: 58v: 5, 47-49; 62v: 53, 55.

superfluus: 58r: 8.

superparticularis: 59v: 30, 32, 35*, 39-40, 42.

superpartiens: 59v: 30, 36, 39-40; 62v: 30.

suponere or subponere: 58r: 37, 39, 52, 55; 58v: 5; 59r: 10; 60r: 35, 38, 59; 61r: 27; 62r: 40; 62v: 51; 63r: 13, 31.

supositio: 58r: 50; 58v: 5, 7; 61r: 9, 11, 38, 41, 44, 51; 62v: 37, 51.

supra: 59v: 54, 58-59; 60r: 23, 26-27; 60v: 36; 61r: 16; 61v: 9, 60; 62r: 5, 12, 50, 55-56, 61; 62v: 31, 37-38; 63r: 7, 12, 24.

sursum: 58v: 35; 62r: 18, 22.

suscipere: 60r: 30-31.

suspensus: 60v: 53.

tactus: 60v: 36; 62v: 30.

tarde: 58r: 11; 58v: 32, 36; 59r: 53, 56-57; 60r: 41; 60v: 9, 12, 61; 62v: 12; 63r: 42.

tardare: 59r: 45.

tarditas: 58r: 3, 9; 58v: 26, 44; 59v: 1-2, 5, 25, 52, 54; 60r: 22, 25-26, 29, 34, 48-49, 54; 60v: 13-14, 45, 47-48, 58; 61r: 3-4, 20-24, 27; 61v: 53, 62r: 2, 4; 62v: 19.

tardus: 58v: 33; 60v: 1; 61r: 28, 31, 34; 61v: 21, 56; 63r: 43.

temperamentum: 62r: 19-20, 55.

tempus: 58r: 26-27, 32-34, 36-37, 40, 42, 49, 53, 57; 58v: 2, 4, 18-19, 22, 57; 59r: 6-8, 24, 52; 59v: 2-3, 59; 60r: 4-5, 9-10, 33, 35-42, 51, 59-60; 60v: 1-3, 30, 32, 52, 54, 56; 61r: 6-7, 13-14; 61v: 37, 41, 56; 62v: 13-14, 17, 20, 58-59, 62; 63r: 8, 10, 26, 44.

tenuis: 58r: 58.

tenuitas: 58r: 60.

terminare: 61v: 6-7; 63r: 43.

terminus: 59v: 7, 11, 15, 48; 61v: 5, 12, 33; 62r: 34; 62v: 25, 54.

ternarius: 58r: 46-47; 62v: 27.

terra: 58v: 50; 60r: 12; 62v: 49.

testari: 61r: 21.

titulus: 58r: 8.

tonus: 60v: 27.

totus: 58r: 50, 53; 58v: 42, 59r: 9, 12-13, 26, 29, 32-33; 59v: 36; 60r: 5-6, 8, 13; 60v: 4, 6-7, 27, 31-32, 51-52, 57; 61r: 8, 10-11; 61v: 21; 62r: 12, 39, 47, 53-56, 62; 62v: 11, 21-22, 35, 62; 63r: 10, 12, 20, 38, 41.

tractatus: 58r: 6.

Tractatus de proportionibus motuum (*i.e. references to Bradwardine's tract*): 58r: 13, 50; 59r: 19; 60r: 55; 60v: 4. (*N.B. There are copious other implied references since this tract is based on Bradwardine's.*)

trahere: 58v: 22, 24; 62v: 32, 42; 63r: 9-11.

transformare: 62r: 20.

transitus: 61v: 4, 14.

translatio: 61v: 40.

transmutare: 58v: 59; 62r: 54-55.

tripedalis: 58r: 48; 59v: 22; 62v: 28.

triplex: 62v: 27; 63r: 36.

triplicare: 58r: 48.

triplus: 58r: 47-48, 51; 59r: 40, 42; 59v: 18, 20, 31; 62v: 35; in triplo: 58r: 40, 42; 58v: 47, 59; 59r: 42; 61r: 32; 61v: 22; 62r: 31.

ullatenus: 61r: 5.

ultimus: 58v: 55; 59v: 2; 62v: 9, 43; 63r: 2, 41.

unbra (= umbra): 58r: 18-19, 21; 62v: 12, 14-15.

uniformis: 58r: 58; 59r: 9, 11; 61v: 61; 62v: 61-62; 63r: 3, 20.

uniformitas: 59r: 9.

uniformiter: 58r: 20; 58v: 56; 59r: 8, 31, 36; 61v: 6, 10, 58, 61; 62v: 11, 60, 62; 63r: 2.

unitas: 58r: 45-46, 48; 59v: 45, 50.

universaliter: 61r: 17, 22.

LEONARDO DA VINCI: MECHANICS

Although Leonardo was interested in mechanics for most of his mature life, he would appear to have turned more and more of his attention to it from 1508 on. It is difficult to construct a unified and consistent picture of his mechanics in detail, but the major trends, concepts, and influences can be delineated with some firmness.[1] Statics may be considered first, since this area of theoretical mechanics greatly attracted him and his earliest influences in this field probably came from the medieval science of weights. To this he later added Archimedes' *On the Equilibrium of Planes*, with a consequent interest in the development of a procedure for determining the centers of gravity of sundry geometrical magnitudes.

In his usual fashion Leonardo absorbed the ideas of his predecessors, turned them in practical and experiential directions, and developed his own system of nomenclature—for example, he called the arms of balances "braccia" while in levers "lieva" is the arm of the lever to which the power is applied and "contralieva" the arm in which the resistance lies. Influenced by the Scholastics, he called the position of horizontal equilibrium the "sito dell'equalità" (or sometimes the position "equale allo orizzonte"). Pendent weights are simply "pesi," or "pesi attacati," or "pesi appiccati"; the cords supporting them—or, more generally, the lines of force in which the weights or forces act or in which they are applied—are called "appendicoli." Leonardo further distinguished between "braccia reali o linee corporee," the actual lever arms, and "braccia potenziali o spirituali o semireali," the potential or effective arms. The potential arm is the horizontal distance to the vertical (that is, the "linea central") through the center of motion in the case of bent levers, or, more generally, the perpendicular distance to the center of motion from the line of force about the center of motion. He often called the center of motion the "centro del circunvolubile," or simply the "polo."

The classical law of the lever appears again and again in Leonardo's notebooks. For example, *Codex Atlanticus*, folio 176v-d, states, "The ratio of the weights which hold the arms of the balance parallel to the horizon is the same as that of the arms, but is an inverse one." In *Codex Arundel*, folio 1v, the law appears formulaically as $W_2 = (W_1 \cdot s_1)/s_2$: "Multiply the longer arm of the balance by the weight it supports and divide the product by the shorter arm, and the result will be the weight which, when placed on the shorter arm, resists the descent of the longer arm, the arms of the balance being above all basically balanced."

A few considerations that prove the influence of the medieval science of weights upon Leonardo are in order. The medieval science of weights consisted essentially of the following corpus of works:[2] Pseudo-Euclid, *Liber de ponderoso et levi*, a geometrical treatment of basic Aristotelian ideas relating forces, volumes, weights, and velocities; Pseudo-Archimedes, *De ponderibus Archimenidis* (also entitled *De incidentibus in humidum*), essentially a work of hydrostatics; an anonymous tract from the Greek, *De canonio*, treating of the Roman balance or steelyard by reduction to a theoretical balance; Thâbit ibn Qurra, *Liber karastonis*, another treatment of the Roman balance; Jordanus de Nemore, *Elementa de ponderibus* (also entitled *Elementa super demonstrationem ponderum*), a work existing in many manuscripts and complemented by several reworked versions of the late thirteenth and the fourteenth centuries, which was marked by the first use of the concept of positional gravity *(gravitas secundum situm)*, by a false demonstration of the law of the bent lever, and by an elegant proof of the law of the lever on the basis of the principle of virtual displacements; an anonymous *Liber de ponderibus* (called, by its modern editor, Version P), which contains a kind of short Peripatetic commentary to the enunciations of Jordanus; a *Liber de ratione ponderis*, also attributed to Jordanus, which is a greatly expanded and corrected version of the *Elementa* in four parts and is distinguished by a more correct use of the concept of positional gravity, by a superb proof of the law of the bent lever based on the principle of virtual displacements, by the same proof of the law of the straight lever given in the *Elementa*, and by a remarkable and sound proof of the law of the equilibrium of connected weights on adjacent inclined planes that is also based on the principle of virtual displacements—thereby constituting the first correct statement and proof of the inclined plane law—and including a number of practical problems in statics and dynamics, of which the most noteworthy are those connected with the bent lever in part III; and Blasius of Parma, *Tractatus de ponderibus*, a rather inept treatment of the problems of statics and hydrostatics based on the preceding works.

The whole corpus is marked methodologically by its geometric form and conceptually by its use of dynamics (particularly the principle of virtual velocities or the principle of virtual displacements in one form or another) in application to the basic statical problems and conclusions inherited from Greek antiquity. Leonardo was much influenced by the corpus' general dynamical approach as well as by particular conclusions of specific works. Reasonably conclusive evidence exists to show that Leonardo had read Pseudo-Archimedes, *De canonio*, and Thâbit ibn Qurra, while completely conclusive evidence reveals his knowledge of the *Elementa, De ratione ponderis*, and Blasius. It is reasonable to suppose that he also saw the other works of the corpus, since they were so often included in the same manuscripts as the works he did read.

The two works upon which Leonardo drew most heavily were the *Elementa de ponderibus* and the *Liber de ratione ponderis*. The key passage that shows decisively that Leonardo read both of them occurs in *Codex Atlanticus*, folios 154v-a and r-a. In this passage, whose significance has not been properly recognized before, Leonardo presents a close Italian translation of the postulates from the *Elementa* (E.01–E.07) and the enunciations of the first two propositions (without proofs), together with a definition that precedes the proof of proposition E.2.[3] He then shifts to the *Liber de ratione ponderis* and includes the enunciations of all the propositions of the first two parts (except R2.05) and the first two propositions of the third part.[4] It seems likely that Leonardo made his translation from a single manuscript that contained both works and, after starting his translation of the *Elementa*, suddenly realized the superiority of the *De ratione ponderis*. He may also have translated the rest of part III, for he seems to have been influenced by propositions R3.05 and R3.06, the last two propositions in part III, and perhaps also by R3.04 in passages not considered here (see MS *A*, fols. 1r, 33v). Another page in *Codex Atlanticus*, folio 165v-a and v-c, contains all of the enunciations of the propositions of part IV (except for R4.07, R4.11–12, and R4.16). The two passages from the *Codex Atlanticus* together establish that Leonardo had complete knowledge of the best and the most original work in the corpus of the medieval science of weights.

The passages cited do not, of course, show what Leonardo did with that knowledge, although others do. For example in MS *G*, folio 79r, Leonardo refutes the incorrect proof of the second part of proposition E.2 (or of its equivalent, R1.02). The proposition states: "When the beam of a balance of equal arm lengths is in the horizontal position, if equal weights are suspended [from its extremities], it will not leave the horizontal position, and if it should be moved from the horizontal position, it will revert to it."[5] The second part is true for a material beam supported from above, since the elevation of one of the arms removes the center of gravity from the vertical line through the fulcrum and, accordingly, the balance beam returns to the horizontal position as the center of gravity seeks the vertical. The medieval proof in E.2 and

R1.02, however, treats the balance as if it were a theoretical balance (with weightless beam) and attempts to show that it would return to the horizontal position. Based on the false use of the concept of positional gravity, it asserts that the weight above the line of horizontal equilibrium has a greater gravity according to its position than the weight depressed below the line, for if both weights tended to move downward, the arc on which the upper weight would move more intercepts more of the vertical than the arc on which the lower weight would tend to move. Leonardo's refutation is based on showing that because the weights are connected, the actual arcs to be compared are oppositely directed and so have equal obliquities—and thus the superior weight enjoys no positional advantage. In MS *E*, folio 59r, Leonardo seems also to give the correct explanation for the return to horizontal equilibrium of the material beam.

A further response to the *Liber de ratione ponderis* is in *Codex Atlanticus*, folio 354v-c, where Leonardo again translates proposition R1.09 and paraphrases proposition R1.10 of the medieval work: "The equality of the declination conserves the equality of the weights. If the ratios of the weights and the obliquities on which they are placed are the same but inverse, the weights will remain equal in gravity and motion" ("La equalità della declinazione osserva la equalità de' pesi. Se le proporzioni de' pesi e dell' obbliqua dove si posano, saranno equali, ma converse, essi pesi resteranno equali in gravità e in moto"). The equivalent propositions from the *Liber de ratione ponderis* were "R1.09. Equalitas declinationis identitatem conservat ponderis" and "R1.10. Si per diversarum obliquitatum vias duo pondera descendant, fueritque declinationum et ponderum una proportio eodum ordine sumpta, una erit utriusque virtus in descendendo."[6] While Leonardo preserved R1.09 exactly in holding that positional weight on the incline is everywhere the same as long as the incline's declination is the same, his rephrasing of R1.10 indicates that he adopted a different measure of "declination."

R. Marcolongo believed that Leonardo measured obliquity by the ratio of the common altitude of the inclines to the length of the incline (that is, by the sine of the angle of inclination), while Jordanus had measured obliquity by the length of the incline that intercepts the common altitude (that is, by the cosecant of the angle).[7] Thus, if p_1 and p_2 are connected weights placed on the inclined planes that are of lengths l_1 and l_2, respectively, with common altitude h, then, according to Marcolongo's view of Leonardo's method, $p_1/p_2 = (h/l_2)/(h/l_1)$, and thus $p_1/p_2 = l_1/l_2$, as Jordanus held. Hence, if Marcolongo is correct, Leonardo had absorbed the correct exposition of the inclined plane problem from the *Liber de ratione ponderis*. Perhaps he did, but it is not exhibited in this passage, for the figure accompanying the passage (Figure 1), with its numerical designations of 2 and 1

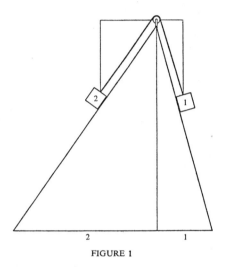

FIGURE 1

on the bases cut off by the vertical, shows that Leonardo believed the weights on the inclines to be inversely proportional to the tangents of the angles of inclination rather than to the sines—and such a solution is

clearly incorrect. The same incorrect solution is apparent in MS *G*, folio 77v, where again equilibrium of the two weights is preserved when the weights are in the same ratio as the bases cut off by the common altitude (thus implying that the weights are inversely proportional to the tangents).

Other passages give evidence of Leonardo's vacillating methods and confusion. One (*Codex on Flight*, fol. 4r) consists of two paragraphs that apply to the same figure (Figure 2). The first is evidently an explanation of a proposition expressed elsewhere (MS *E*, fol. 75r) to the effect that although equal weights balance each other on the equal arms of a balance, they do not do so if they are put on inclines of different obliquity. It states:

> The weight *q*, because of the right angle *n* [perpendicularly] above point *e* in line *df*, weighs 2/3 of its natural weight, which was 3 pounds, and so has a residual force ["che resta in potenzia"] [along *nq*] of 2 pounds; and the weight *p*, whose natural weight was also 3 pounds, has a residual force of 1 pound [along *mp*] because of right angle *m* [perpendicularly] above point *g* in line *hd*. [Therefore, *p* and *q* are not in equilibrium on these inclines.]

The bracketed material has been added as clearly implicit, and so far this analysis seems to be entirely correct. It is evident from the figure that Leonardo has applied the concept of potential lever arm (implying static moment) to the determination of the component of weight along the incline, so that $F_1 \cdot dn = W_1 \cdot de$, where F_1 is the component of the natural weight W_1 along the incline, *dn* is the potential lever arm through which F_1 acts around fulcrum *d*, and *de* is the lever

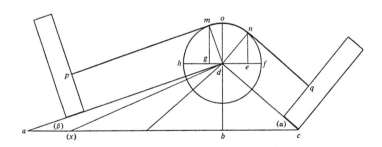

FIGURE 2

217

arm through which W_1 would act when hanging from n. Hence, $F_1 = W_1 \cdot (de/dn) = W_1 \cdot \sin \angle dne$. But $\angle dne = \angle \alpha$, and so we have the correct formulation $F_1 = W_1 \cdot \sin \alpha$. In the same way for weight p, it can be shown that $F_2 = W_2 \cdot \sin \beta$. And since Leonardo apparently constructed $de = 2df/3$, $dg = hd/3$, and $W_1 = W_2 = 3$, obviously $F_1 = 2$ and $F_2 = 1$ and the weights are not in equilibrium.

In the second paragraph, which also pertains to the figure, he changes p and q, each of which was initially equal to three pounds, to two pounds, and one pound, respectively, with the object of determining whether the adjusted weights would be in equilibrium:

> So now we have one pound against two pounds. And because the obliquities da and dc on which these weights are placed are not in the same ratio as the weights, that is, 2 to 1 [the weights are not in equilibrium]; like the said weights [in the first paragraph?], they alter natural gravities [but they are not in equilibrium] because the obliquity da exceeds the obliquity dc, or contains the obliquity dc, $2\frac{1}{2}$ times, as is demonstrated by their bases ab and bc, whose ratio is a double sesquialterate ratio [that is, 5:2], while the ratio of the weights will be a double ratio [that is, 2:1].

It is abundantly clear, at least in the second paragraph, that Leonardo was assuming that for equilibrium the weights ought to be directly proportional to the bases (and thus inversely proportional to the tangents); and since the weights are not as the base lines, they are not in equilibrium. In fact, in adding line $d[x]$ Leonardo was indicating the declination he thought would establish the equilibrium of p and q, weights of two pounds and one pound, respectively, for it is obvious that the horizontal distances $b[x]$ and bc are also related as 2 : 1. This is further confirmation that Leonardo measured declination by the tangent.

Assuming that both paragraphs were written at the same time, it is apparent that Leonardo then thought that both methods of determining the effective weight on an incline—the method using the concept of the lever and the technique of using obliquities measured

by tangents—were correct. A similar confusion is apparent in his treatment of the tensions in strings. But there is still another figure (Figure 3), which has accompanying it in MS H, folio 81(33)v, the following brief statement: "On the balance, weight ab will be as weight cd." If Leonardo was assuming that the weights are of the same material with equal thickness and, as it seems in the figure, that they are of the same width, with their lengths equal to the lengths of the vertical and the incline, respectively, thus producing weights that are proportional to these lengths, then he was indeed giving a correct example of the inclined-plane principle that may well reflect proposition R1.10 of the *Liber de ratione ponderis*. There is one further solution of the inclined-plane problem, on MS A, folio 21v, that is totally erroneous, and so far as is known, unique. It is not discussed here; the reader is referred to Duhem's treatment, with the caution that Duhem's conclusion that it was derived from Pappus' erroneous solution is questionable.[8]

As important as Leonardo's responses to proposition R1.10 are his responses to the bent-lever proposition, R1.08, of the *Liber de ratione ponderis*. In *Codex Arundel*, folio 32v, he presents another Italian translation of it (in addition to the translation already noted in his omnibus collection of translations of the enunciations of the medieval work). In this new translation he writes of the bent-lever law as "tested": "Tested. If the arms of the balance are unequal and their juncture in the fulcrum is angular, and if their termini are equally distant from the central [that is, vertical] line through the fulcrum, with equal weights applied there [at the termini], they will weigh equally [that is, be in equilibrium]"—or "Sperimentata. Se le braccia della bilancia fieno inequali, e la lor congiunzione nel polo sia angulare, se i termini lor fieno equalmente distanti alla linea central del polo, li pesi appiccativi, essendo equali, equalmente peseranno," a translation of the Latin text, "R1.08. Si inequalia fuerint brachia libre, et in centro motus angulum fecerint, si termini eorum ad directionem hinc inde equaliter accesserint, equalia appensa, in hac dispositione equaliter ponderabunt."[9] Other, somewhat confused passages indicate that Leonardo had indeed absorbed the significance of the passage he had twice translated.[10]

More important, Leonardo extended the bent-lever law beyond the special case of equal weights at equal horizontal distances, given in R1.08, to cases in which the more general law of the bent lever—"weights are inversely proportional to the horizontal distances"— is applied. He did this in a large number of problems to which he applied his concept of "potential lever arm." The first passage notable for this concept is in

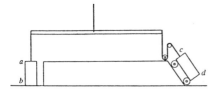

FIGURE 3. (Note: The horizontal base line and the continuation of the line from point c beyond the pulley wheel to the balance beam are in Leonardo's drawing but should be deleted.)

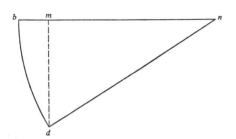

FIGURE 4. (Note: This is the essential part of a more detailed drawing.)

MS *E*, folio 72v, and probably derives from proposition R3.05 of the *Liber de ratione ponderis*.[11] Here Leonardo wrote (see Figure 4): "The ratio that space *mn* has to space *nb* is the same as the ratio that the weight which has descended to *d* has to the weight it had in position *b*. It follows that, *mn* being 9/11 of *nb*, the weight in *d* is 9/10 (! 9/11) of the weight it had in height *b*." In this passage *n* is the center of motion and *mn* is the potential lever arm of the weight in position *d*. His use of the potential lever arm is also illustrated in

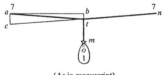

(As in manuscript)

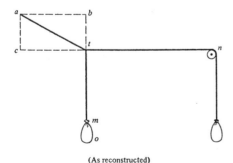

(As reconstructed)

FIGURE 5

a passage that can be reconstructed from MS *E*, folio 65r, as follows (Figure 5). A bar *at* is pivoted at *a*; a weight *o* is suspended from *t* at *m* and a second weight acts on *t* in a direction *tn* perpendicular to that of the first weight. The weights at *m* and *n* necessary to keep the bar in equilibrium are to be determined. In this determination Leonardo took *ab* and *ac* as the potential lever arms (and so labeled them), so that the weights *m* and *n* are inversely proportional to the potential lever arms *ab* and *ac*. The same kind of problem is illustrated in a passage in *Codex Atlanticus*, folio 268v-b, which can be summarized by reference to Figure 6. Cord *ab* supports a weight *n*, which is

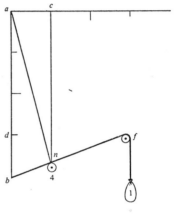

FIGURE 6

pulled to the position indicated by a tangential force along *nf* that is given a value of 1 and there keeps *an* in equilibrium. The passage indicates that weight 1, acting through a distance of four units (the potential lever arm *an*), keeps in equilibrium a weight *n* of 4, acting through a distance of one unit (the potential counterlever *ac*). A similar use of "potential lever arm" is found in problems like that illustrated in Figure 7, taken from MS *M*, folio 40r. Here the potential lever arm is *an*, the perpendicular drawn from the line of force *fp* to the center of motion *a*. Hence the weights suspended from *p* and *m* are related inversely as the distances *an* and *am*. These and similar problems reveal Leonardo's acute awareness of the proper factors of horizontal distance and force determining the static moment about a point.

The concept of potential lever arm also played a crucial role in Leonardo's effort to analyze the tension

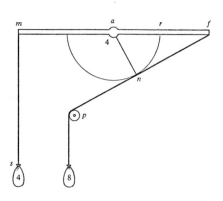

FIGURE 7

and it will be in the direct continuation of cord *ag*, and the ratio which space *df* has to space *db*, the weight [that is, tension] in cord *ba* will have to the weight [that is, tension] in cord *fa*.

Thus, with *da* the common altitude and *df* and *db* used as the measures of the angles at *a*, Leonardo is actually measuring the tensions by the inverse ratio of the tangents. This same incorrect procedure appears many times in the notebooks (for instance, MS *E*, fols. 67v, 68r, 68v, 69r, 69v, 71r; *Codex Arundel*, fol. 117v; MS *G*, fol. 39v).

In addition to this faulty method Leonardo in some instances employed a correct procedure based on the concept of the potential lever arm. In *Codex Arundel*, folio 1v, Leonardo wrote that "the weight 3 is not distributed to the real arms of the balance in the same [although inverse] ratio of these arms but in the [inverse] ratio of the potential arms" (see Figure 9).

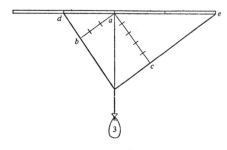

FIGURE 9

in strings. Before examining that role, it must be noted that Leonardo often used an incorrect rule based on tangents rather than sines, a rule similar to that which he mistakenly applied to the problem of the inclined plane. An example of the incorrect procedure appears in MS *E*, folio 66v (see Figure 8):

The heavy body suspended in the angle of the cord divides the weight to these cords in the ratio of the angles included between the said cords and the central [that is, vertical] line of the weight. Proof. Let the angle of the said cord be *bac*, in which is suspended heavy body *g* by cord *ag*. Then let this angle be cut in the position of equality [that is, in the horizontal direction] by line *fb*. Then draw the perpendicular *da* to angle *a*

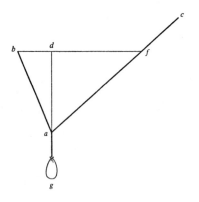

FIGURE 8

The cord is attached to a horizontal beam and a weight is hung from the cord. In the figure *ab* and *ac* are the potential arms. This is equivalent to a theorem that could be expressed in modern terms as "the moments of two concurrent forces around a point on the resultant are mutually equal." Leonardo's theorem, however, does not allow the calculation of the actual tensions in the strings. By locating the center of the moments first on one and then on the other of the concurrent forces, however, Leonardo discovered how to find the tensions in the segments of a string supporting a weight. In *Codex Arundel*, folio 6r (see Figure 10), he stated:

Here the potential lever *db* is six times the potential counterlever *bc*. Whence it follows that one pound placed in the force line "appendiculo" *dn* is equal in power to six pounds placed in the semireal force line

220

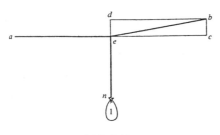

FIGURE 10

ca, and to another six pounds of power conjoined at *b*. Therefore, the cord *aeb* by means of one pound placed in line *dn* has the [total] effect of twelve pounds.

All of which is to say that Leonardo has determined that there are six pounds of tension in each segment of the cord. The general procedure is exhibited in MS *E*, folio 65r (see Figure 11). Under the figure is the caption "*a* is the pole [that is, fulcrum] of the angular balance [with arms] *ad* and *af*, and their force lines ['appendiculi'] are *dn* and *fc*." This applies to the figure on the left and indicates that *af* and *da* are the lever arms on which the tension in *cb* and the weight hanging at *b* act. Leonardo then went on to say: "The greater the angle of the cord which supports weight *n* in the middle of the cord, the smaller becomes the potential arm [that is, *ac* in the figure on the right] and the greater becomes the potential counter-arm [that is, *ba*] supporting the weight." Since Leonardo has drawn the figure so that *ab* is four times *ac* and has marked the weight of *n* as 1, the obvious implication is that the tension in cord *df* will be 4.

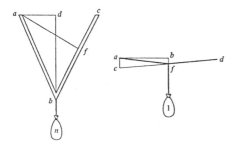

FIGURE 11

In these last examples the weight hangs from the center of the cord. But in *Codex Arundel*, folio 6v, the same analysis is applied to a weight suspended at a point other than the middle of the cord (Figure 12).

Here the weight *n* is supported by two different forces, *mf* and *mb*. Now it is necessary to find the potential levers and counterlevers of these two forces *bm* and *fm*. For the force [at] *b* [with *f* the fulcrum of the potential lever] the [potential] lever arm is *fe* and the [potential] counterlever is *fa*. Thus for the lever arm *fe* the force line ["appendiculo"] is *eb* along which the motor *b* is applied; and for the counterlever *fa* the force line is *an*, which supports weight *n*. Having arranged the balance of the power and resistance of motor and weight, it is necessary to see what ratio lever *fe* has to counterlever *fa*; which lever *fe* is 21/22 of counterlever *fa*. Therefore *b* suffers 22 [pounds of

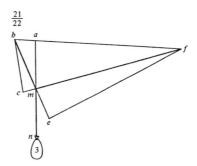

FIGURE 12. (Note: There is some discrepancy between Leonardo's text and the figure as concerns the relative length of the lines.)

tension] when *n* is 21. In the second disposition [with *b* instead of *f* as the fulcrum of a potential lever], *bc* is the [potential] lever arm and *ba* the [potential] counterlever. For *bc* the force line is *cf* along which the motor *f* is applied and weight *n* is applied along force line *an*. Now it is necessary to see what ratio lever *bc* has to counterlever *ba*, which counterlever is 1/3 of the lever. Therefore, one pound of force in *f* resists three pounds of weight in *ba*; and 21/22 of the three pounds in *n*, when placed at *b*, resist twenty-two placed in *ba*. . . . Thus is completed the rule for calculating the unequal arms of the angular cord.

In a similar problem in *Codex Arundel*, folio 4v, Leonardo determined the tensions of the string segments; in this case the strings are no longer attached or fixed to a beam but are suspended from two pulley wheels from which also hang two equal weights (Figure 13). Here *abc* is apparently an equilateral

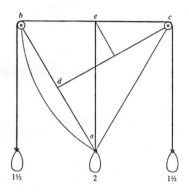

FIGURE 13

triangle, *cd* is the potential arm for the tension in string *ba* acting about fulcrum *c* and *ec* is the potential counterlever through which the weight 2 at *a* acts. If an equilateral triangle was intended, then, by Leonardo's procedure, the tension in each segment of the string ought to be $2/\sqrt{3} = 1.15$, rather than 1.5, as Leonardo miscalculated it. It is clear from all of these examples in which the tension is calculated that Leonardo was using a theorem that he understood as: "The ratio of the tension in a cord segment to the weight supported by the cord is equal to the inverse ratio of the potential lever arms through which the tension and weight act, where the fulcrum of the potential bent lever is in the point of support of the other segment of the cord." This is equivalent to a theorem in the composition of moments: "If one considers two concurrent forces and their resultant, the moment of the resultant about a point taken on one of the two concurrent forces is equal to the moment of the other concurrent force about the same point." The discovery and use of this basic concept in analyzing string tensions was Leonardo's most original development in statics beyond the medieval science of weights that he had inherited. Unfortunately, like most of Leonardo's investigations, it exerted no influence on those of his successors.

Another area of statics in which the medieval science of weights may have influenced Leonardo was that in which a determination is made of the partial forces in the supports of a beam where the beam itself supports a weight. Proposition R3.06 in the *Liber de ratione ponderis* states: "A weight not suspended in the middle [of a beam] makes the shorter part heavier according to the ratio of the longer part to the

shorter part."[12] The proof indicates that the partial forces in the supports are inversely related to the distances from the principal weight to the supports. In *Codex Arundel*, folio 8v (see Figure 14), Leonardo arrived at a similar conclusion:

> The beam which is suspended from its extremities by two cords of equal height divides its weight equally in each cord. If the beam is suspended by its extremities at an equal height, and in its midpoint a weight is hung, then the gravity of such a weight is equally distributed to the supports of the beam. But the weight which is moved from the middle of the beam toward one of its extremities becomes lighter at the extremity away from which it was moved, or heavier at the other extremity, by a weight which has the same ratio to the total weight as the motion completed by the weight [that is, its distance moved from the center] has to the whole beam.

Leonardo also absorbed the concept, so prevalent in the medieval science of weights (particularly in the *De canonio*, in Thābit ibn Qurra's *Liber karastonis*, and in part II of the *Liber de ratione ponderis*), that a

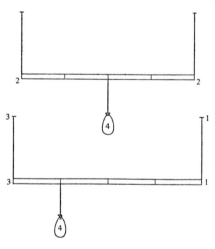

FIGURE 14

segment of a solid beam may be replaced by a weight hung from the midpoint of a weightless arm of the same length and position. For example, see Leonardo's exposition in MS *A*, folio 5r (see Figure 15):

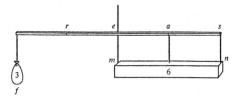

FIGURE 15

If a balance has a weight which is similar [that is, equal] in length to one of its arms, the weight being *mn* of six pounds, how many pounds are to be placed in *f* to resist it [that is, which will be in equilibrium with it]? I say that three pounds will suffice, for if weight *mn* is as long as one of its arms, you could judge that it may be replaced in the middle of the balance arm at point *a;* therefore, if six pounds are in *a*, another six pounds placed at *r* would produce resistance to them [that is, be in equilibrium with them], and, if you proceed as before in point *r* [but now] in the extremity of the balance, three pounds will produce the [necessary] resistance to them.

This replacement doctrine was the key step in solving the problem of the Roman balance in all of the above-noted tracts.[13]

The medieval science of weights was not the only influence upon Leonardo's statics, however, since it may be documented that he also, perhaps at a later time, read book I of Archimedes' *On the Equilibrium of Planes*. As a result he seems to have begun a work on centers of gravity in about 1508, as may be seen in a series of passages in *Codex Arundel*. Since these passages have already been translated and analyzed in rather complete detail elsewhere,[14] only their content and objectives will be given here. Preliminary to Leonardo's propositions on centers of gravity are a number of passages distinguishing three centers of a figure that has weight (see *Codex Arundel*, fol. 72v): "The first is the center of its natural gravity, the second [the center] of its accidental gravity, and the third is [the center] of the magnitude of this body." On folio 123v of this manuscript the centers are defined:

The center of the magnitude of bodies is placed in the middle with respect to the length, breadth and thickness of these bodies. The center of the accidental gravity of these bodies is placed in the middle with respect to the parts which resist one another by standing in equilibrium. The center of natural gravity is that which divides a body into two parts equal in weight and quantity.

It is clear from many other passages that the center of natural gravity is the symmetrical center with respect to weight. Hence, the center of natural gravity of a beam lies in its center, and that center would not be disturbed by hanging equal weights on its extremities. If unequal weights are applied, however, there is a shift of the center of gravity, which is now called the center of accidental gravity, the weights having assumed accidental gravities by their positions on the unequal arms. The doctrine of the three centers can be traced to Scholastic writings of Nicole Oresme, Albert of Saxony, Marsilius of Inghen (and, no doubt, others).[15] In all of these preliminary considerations, Leonardo assumed that a body or a system of bodies is in equilibrium when supported from its center of gravity (be it natural or accidental). It should also be observed that Leonardo assumed the law of the lever as being proved before setting out to prove his Archimedean-like propositions.

The first Archimedean passage to note is in *Codex Arundel*, folio 16v, where Leonardo includes a series of statements on equilibrium that is drawn in significant part from the postulates and early propositions of book I of *On the Equilibrium of Planes*. His terminology suggests that when Leonardo wrote this passage, he was using the translation of Jacobus Cremonensis (*ca*. 1450).[16]

More important than this passage are the propositions and proofs on centers of gravity, framed under the influence of Archimedes' work, which Leonardo specifically cites in a number of instances. The order for these propositions that Leonardo's own numeration seems to suggest is the following:

1. *Codex Arundel*, folio 16v, "Every triangle has the center of its gravity in the intersection of the lines which start from the angles and terminate in the centers of the sides opposite them." This is proposition 14 of book I of *On the Equilibrium of Planes*, and Leonardo included in his "proof" an additional proof of sorts for Archimedes' proposition 13 (since that proposition is fundamental for the proof of proposition 14), although the proof for proposition 13 ignores Archimedes' superb geometrical demonstration. Depending, as it does, on balance considerations, Leonardo's proof is more like Archimedes' second proof of proposition 13. Leonardo's proof of proposition 14 is for an equilateral triangle, but at its end he notes that it applies as well to scalene triangles.

2. *Codex Arundel*, folio 16r: "The center of gravity of any two equal triangles lies in the middle of the

line beginning at the center of one triangle and terminating in the center of gravity of the other triangle." This is equivalent to proposition 4 of *On the Equilibrium of Planes*, but Leonardo's proof differs from Archimedes' in that it merely shows that the center of gravity is in the middle of the line because the weights would be in equilibrium about that point. Archimedes' work is here cited (under the inaccurate title of *De ponderibus*, since the Pseudo-Archimedean work of that title was concerned with hydrostatics rather than statics).

3. *Codex Arundel*, folio 16r: "If two unequal triangles are in equilibrium at unequal distances, the greater will be placed at the lesser distance and the lesser at the greater distance." This is similar to proposition 3 of Archimedes' work. Leonardo, however, simply employed the law of the lever in his proof, which Archimedes did not do, since he did not offer a proof of the law until propositions 6 and 7.

4. *Codex Arundel*, folio 17v: "The center of gravity of every square of parallel sides and equal angles is equally distant from its angles." This is a special case of proposition 10 of *On the Equilibrium of Planes*; but Leonardo's proof, based once more on a balancing procedure, is not directly related to either of the proofs provided by Archimedes.

5. *Codex Arundel*, folio 17v: "The center of gravity of every corbel-like figure [that is, isosceles trapezium] lies in the line which divides it into two equal parts when two of its sides are parallel." This is similar to Archimedes' proposition 15, which treated more generally of any trapezium. Leonardo made a numerical determination of where the center of gravity lies on the bisector. Although his proof is not close to Archimedes', like Archimedes he used the law of the lever in his proof.

6. *Codex Arundel*, folio 17r: "The center of gravity of every equilateral pentagon is in the center of the circle which circumscribes it." This has no equivalent in Archimedes' work. The proof proceeds by dividing the pentagon into triangles that are shown to balance about the center of the circle. Again Leonardo used the law of the lever in his proof, much as he had in his previous propositions. The proof is immediately followed by the determination of the center of gravity of a pentagon that is not equilateral, in which the same balancing techniques are again employed. Leonardo here cited Archimedes as the authority for the law of the lever, and the designation of Archimedes' proposition as the "fifth" is perhaps an indication that he was using William of Moerbeke's medieval translation of *On the Equilibrium of Planes* instead of that of Jacobus Cremonensis, in which the equivalent proposition is number 7. Although these exhaust those propositions of Leonardo's that are directly related to *On the Equilibrium of Planes*, it should be noted that Leonardo used the same balancing techniques in his effort to determine the center of gravity of a semicircle (see *Codex Arundel*, fols. 215r-v).

It should also be noted that Leonardo went beyond Archimedes' treatise in one major respect—the determination of centers of gravity of solids, a subject taken up in more detail later in the century by Francesco Maurolico and Federico Commandino. Two of the propositions investigated by Leonardo may be presented here to illustrate his procedures. Both propositions concern the center of gravity of a pyramid and appear to be discoveries of Leonardo's. The first is that the center of gravity of a pyramid (actually, a regular tetrahedron) is, at the intersection of the axes, a distance on each axis of 1/4 of its length, starting from the center of one of the faces. (By "axis" Leonardo understood a line drawn from a vertex to the center of the opposite face.) In one place (*Codex Arundel*, fol. 218v) he wrote of the intersection of the pyramidal axes as follows: "The inferior [interior?] axes of pyramids which arise from [a point lying at] 1/3 of the axis of their bases [that is, faces] will intersect in [a point lying at] 1/4 of their length [starting] at the base."

Despite the confusion of singulars and plurals as well as of the expression "inferior axes," the prop-

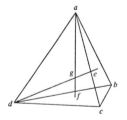 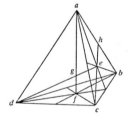 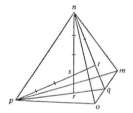

FIGURE 16

osition is clear enough, particularly since Leonardo provided both the drawing shown in Figure 16 and its explanation and, in addition, the intersection of the pyramidal axes is definitely specified as the center of gravity in another passage (fol. 193v): "The center of gravity of the [pyramidal] body of four triangular bases [that is, faces] is located at the intersection of its axes and it will be in the 1/4 part of their length." The proof of this is actually given on the page of the original quotation about the intersection of the axes (fol. 218v), but only as a proof following a more general statement about pyramids and cones:

The center of gravity of any pyramid—round, triangular, square, or [whose base is] of any number of sides—is in the fourth part of its axis near the base. Let the pyramid be *abcd* with base *bcd* and apex *a*. Find the center of the base *bcd*, which you let be *f*, then find the center of face *abc*, which will be *e*, as was proved by the first [proposition]. Now draw line *af*, in which the center of gravity of the pyramid lies because *f* is the center of base *bcd* and the apex *a* is perpendicularly above *f* and the angles *b*, *c* and *d* are equally distant from *f* and [thus] weigh equally so that the center of gravity lies in line *af*. Now draw a line from angle *d* to the center *e* of face *abc*, cutting *af* in point *g*. I say for the aforesaid reason that the center of gravity is in line *de*. So, since the center is in each [line] and there can be only one center, it necessarily lies in the intersection of these lines, namely in point *g*, because the angles *a*, *b*, *c* and *d* are equally distant from this *g*.

As noted above, the proof is given only for a regular tetrahedron, none being given for the cone or for other pyramids designated in the general enunciation. It represents a rather intuitive mechanical approach, for Leonardo abruptly stated that because the angles *b*, *c*, and *d* weigh equally about *f*, the center of gravity of the base triangle, and *a* is perpendicularly above *f*, the center of gravity of the whole pyramid must lie in line *af*. This is reminiscent of Hero's demonstration of the equilibrium of a triangle supported at its center with equal weights at the angles. This kind of reasoning, then, seems to be extended to the whole pyramid by Leonardo at the end of the proof in which he declared that each of the four angles is equidistant from *g* and, presumably, equal weights at the angles would therefore be in equilibrium if the pyramid were supported in *g*. It is worth noting that a generation later Maurolico gave a very neat demonstration of just such a determination of the center of gravity of a tetrahedron by the hanging of equal weights at the angles.[17] Finally, one additional theorem (without proof), concerning the center of gravity of a tetrahedron, appears to have been Leonardo's own discovery (*Codex Arundel*, fol. 123v):

The pyramid with triangular base has the center of its natural gravity in the [line] segment which extends from the middle of the base [that is, the midpoint of one edge] to the middle of the side [that is, edge] opposite the base; and it [the center of gravity] is located on the segment equally distant [from the termini] of the [said] line joining the base with the aforesaid side.

Despite Leonardo's unusual and imprecise language (an attempt has been made to rectify it by bracketed additions), it is clear that he has here expressed a neat theorem to the effect that the center of gravity of the tetrahedron lies at the intersection of the segments joining the midpoint of each edge with the midpoint of the opposite edge and that each of these segments is bisected by the center of gravity. Again, it is possible that Leonardo arrived at this proposition by considering four equal weights hung at the angles. At any rate the balance procedure, whose refinements he learned from Archimedes, no doubt played some part in his discovery, however it was made.

Leonardo gave considerable attention to one other area of statics, pulley problems. Since this work perhaps belongs more to his study of machines, except for a brief discussion in the section on dynamics, the reader is referred to Marcolongo's brief but excellent account.[18]

Turning to Leonardo's knowledge of hydrostatics, it should first be noted that certain fragments from William of Moerbeke's translation of Archimedes' *On Floating Bodies* appear in the *Codex Atlanticus*, folios 153v-e, 153r-b, and 153r-c.[19] These fragments (which occupy a single sheet bound into the codex) are not in Leonardo's customary mirror script but appear in normal writing, from left to right. Although sometimes considered by earlier authors to have been written by Leonardo, they are now generally believed to be by some other hand.[20] Whether the sheet was once the property of Leonardo or whether it was added to Leonardo's material after his death cannot be determined with certainty—at any rate, the fragments can be identified as being from proposition 10 of book II of the Archimedean work. Whatever Leonardo's relationship to these fragments, his notebooks reveal that he had only a sketchy and indirect knowledge of Archimedean hydrostatics, which he seems to have drawn from the medieval tradition of *De ponderibus Archimenidis*. Numerous passages in the notebooks reveal a general knowledge of density and specific weight (for instance, MS *C*, fol. 26v, MS *F*, fol. 70r; MS *E*, fol. 74v). Similarly, Leonardo certainly knew that bodies weigh less in water than in air (see MS *F*, fol. 69r), and in one passage (*Codex Atlanticus*, fol. 284v) he proposed to measure the relative resistance of water as compared with air by plunging the weight on one arm of a balance held in aerial equilibrium into water and then determining how much extra weight must be added to the weight in the water to maintain the balance in equilibrium. See also MS *A*, folio 30v: "The weight in air exhibits the truth of its weight, the weight in water will appear to be less weight by the amount the water is heavier than the air."

So far as is known, however, the principle of Archimedes as embraced by proposition 7 of book I of the genuine *On Floating Bodies* was not precisely stated by Leonardo. Even if he did know the principle, as some have suggested, he probably would have learned it from proposition 1 of the medieval *De ponderibus Archimenidis*. As a matter of fact, Leonardo many times repeated the first postulate of the medieval work, that bodies or elements do not have weight amid their own kind—or, as Leonardo put it (*Codex Atlanticus*, fol. 365r-a; *Codex Arundel*, fol. 189r), "No part of an element weighs in its element" (cf. *Codex Arundel*, fol. 160r). Still, he could have gotten the postulate from Blasius of Parma's *De ponderibus*, a work that Leonardo knew and criticized (see MS *Ashburnham* 2038, fol. 2v). It is possible that since Leonardo knew the basic principle of floating bodies—that a floating body displaces its weight of liquid (see *Codex Forster* II₂, fol. 65v)—he may have gotten it directly from proposition 5 of book I of *On Floating Bodies*. But even this principle appeared in one manuscript of the medieval *De ponderibus Archimenidis* and was incorporated into John of Murs' version of that work, which appeared as part of his widely read *Quadripartitum numerorum* of 1343.[21] So, then, all of the meager reflections of Archimedean hydrostatics found in Leonardo's notebooks could easily have been drawn from medieval sources, and (with the possible exception of the disputed fragments noted above) nothing from the brilliant treatments in book II of the genuine *On Floating Bodies* is to be found in the great artist's notebooks. It is worth remarking, however, that although Leonardo showed little knowledge of Archimedean hydrostatics he had considerable success in the practical hydrostatic questions that arose from his study of pumps and other hydrostatic devices.[22]

The problems Leonardo considered concerning the hydrostatic equilibrium of liquids in communicating vessels were of two kinds: those in which the liquid is under the influence of gravity alone and those in which the liquid is under the external pressure of a piston in one of the communicating vessels. In connection with problems of the first kind, he expressly and correctly stated the law of communicating vessels in *Codex Atlanticus*, folio 219v-a: "The surfaces of all liquids at rest, which are joined together below, are always equal height." Leonardo further noted in various ways that a quantity of water will never lift another quantity of water, even if the second quantity is in a narrower vessel, to a level that is higher than its own, whatever the ratio between the surfaces of the two communicating vessels (*Codex Atlanticus*, fol. 165v)—although he never gave, as far as can be seen, the correct explanation: the equality of the air pressure on both surfaces of the water in the communicating vessels. In some but not all passages (see *Codex Leicester*, fol. 25r; *Codex Atlanticus*, fol. 321v-a) Leonardo did free himself from the misapplication to hydrostatic equilibrium of the principle of the equilibrium of a balance of equal arms bearing equal weights (*Codex Atlanticus*, fol. 206r-a; *Codex Arundel*, fol. 264r). His own explanations are not happy ones, however. He also correctly analyzed the varying levels that would result if, to a liquid in a U-tube, were added a specifically lighter liquid which does not mix with the initial liquid (see Figure 17). In MS *E*, folio

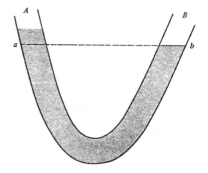

FIGURE 17

74v, he indicated that if the specific weight of the initial liquid were double that of the liquid added, the free surface of the heavier liquid in limb *B* would be at a level halfway between the level of the free surface of the lighter liquid and the surface of contact of the two liquids (the last two surfaces being in limb *A*).

In problems of the second kind, in which the force of a piston is applied to the surface of the liquid in one of the communicating vessels (Figure 18), despite some passages in which he gave an incorrect or only partially true account (see MS *A*, fol. 45r; *Codex Atlanticus*, fol. 384v-a), Leonardo did compose an entirely correct and generally expressed applicable statement or rule (*Codex Leicester*, fol. 11r). Assuming the tube on the right to be vertical and cylindrical, Leonardo observed that the ratio between the pressing weight of the piston and the weight of water in the tube on the right, above the upper level of the water in the vessel on the left, is equal to the ratio between the area under pressure from the piston and the area

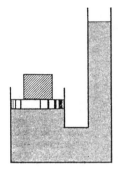

FIGURE 18

of the tube on the right (see *Codex Atlanticus*, fols. 20r, 206r-a, 306v-c; *Codex Leicester*, fol. 26r). And indeed, in a long explanation accompanying the rule in *Codex Leicester*, folio 11r, Leonardo approached, although still in a confused manner, the concept of pressure itself (a concept that appears in no other of his hydrostatic passages, so far as is known) and that of its uniform transmission through a liquid. Leonardo did not, however, generalize his observations to produce Pascal's law, that any additional pressure applied to a confined liquid at its boundary will be transmitted equally to every point in the liquid. Incidentally, the above-noted passage in *Codex Leicester*, folio 11r, also seems to imply a significant consequence for problems of the first kind: that the pressure in an enclosed liquid increases with its depth below its highest point.

Any discussion of Leonardo's knowledge of hydrostatics should be complemented with a few remarks regarding his observations on fluid motion. From the various quotations judiciously evaluated by Truesdell, one can single out some in which Leonardo appears as the first to express special cases of two basic laws of fluid mechanics. The first is the principle of continuity, which declares that the speed of steady flow varies inversely as the cross-sectional area of the channel. Leonardo expressed this in a number of passages, for example, in MS *A*, folio 57v, where he

stated: "Every movement of water of equal breadth and surface will run that much faster in one place than in another, as [the water] may be less deep in the former place than in the latter" (compare MS *H*, fol. 54(6)v; *Codex Atlanticus*, fols. 80r-b, 81v-a; and *Codex Leicester*, fols. 24r and, particularly, 6v). He clearly recognized the principle as implying steady discharge in *Codex Atlanticus*, folio 287r-b: "If the water is not added to or taken away from the river, it will pass with equal quantities in every degree of its breadth [length?], with diverse speeds and slownesses, through the various straitnesses and breadths of its length."

The second principle of fluid motion enunciated by Leonardo was that of equal circulation. In its modern form, when applied to vortex motion, it holds that the product of speed and length is the same on each circle of flow. Leonardo expressed it, in *Codex Atlanticus*, folio 296v-b, as:

> The helical or rather rotary motion of every liquid is so much the swifter as it is nearer to the center of its revolution. This that we set forth is a case worthy of admiration; for the motion of the circular wheel is so much the slower as it is nearer to the center of the rotating thing. But in this case [of water] we have the same motion, through speed and length, in each whole revolution of the water, just the same in the circumference of the greatest circle as in the least. . . .

It could well be that Leonardo's reference to the motion of the circular wheel was suggested by either a statement in Pseudo-Aristotle's *Mechanica* (848A)—that on a rotating radius "the point which is farther from the fixed center is the quicker"—or by the first postulate of the thirteenth-century *Liber de motu* of Gerard of Brussels, which held that "those which are farther from the center or immobile axis are moved more [quickly]. Those which are less far are moved less [quickly]."[23] At any rate, it is worthy of note that Leonardo's statement of the principle of equal circulation is embroidered by an unsound theoretical explanation, but even so, one must agree with Truesdell's conclusion (p. 79): "If Leonardo discovered these two principles from observation, he stands among the founders of western mechanics."

The analysis can now be completed by turning to Leonardo's more general efforts in the dynamics and kinematics of moving bodies, including those that descend under the influence of gravity. In dynamics Leonardo often expressed views that were Aristotelian or Aristotelian as modified by Scholastic writers. His notes contain a virtual flood of definitions of gravity, weight, force, motion, impetus, and percussion:

1. MS *B*, folio 63r: "Gravity, force, material motion and percussion are four accidental powers

with which all the evident works of mortal men have their causes and their deaths."

2. *Codex Arundel*, folio 37r: "Gravity is an invisible power which is created by accidental motion and infused into bodies which are removed from their natural place."

3. *Codex Atlanticus*, folio 246r-a: "The power of every gravity is extended toward the center of the world."

4. *Codex Arundel*, folio 37v: "Gravity, force and percussion are of such nature that each by itself alone can arise from each of the others and also each can give birth to them. And all together, and each by itself, can create motion and arise from it" and "Weight desires [to act in] a single line [that is, toward the center of the world] and force an infinitude [of lines]. Weight is of equal power throughout its life and force always weakens [as it acts]. Weight passes by nature into all its supports and exists throughout the length of these supports and completely through all their parts."

5. *Codex Atlanticus*, folio 253r-c: "Force is a spiritual essence which by accidental violence is conjoined in heavy bodies deprived of their natural desires; in such bodies, it [that is, force], although of short duration, often appears [to be] of marvelous power. Force is a power that is spiritual, incorporeal, impalpable, which force is effected for a short life in bodies which by accidental violence stand outside of their natural repose. 'Spiritual,' I say, because in it there is invisible life; 'incorporeal' and 'impalpable,' because the body in which it arises does not increase in form or in weight."

6. MS *A*, folio 34v: "Force, I say to be a spiritual virtue, an invisible power, which through accidental, external violence is caused by motion and is placed and infused into bodies which are withdrawn and turned from their natural use...."

On the same page as the last Leonardo indicated that force has three "offices" ("ofizi") embracing an "infinitude of examples" of each. These are "drawing" ("tirare"), "pushing" ("spignere"), and "stopping" ("fermare"). Force arises in two ways: by the rapid expansion of a rare body in the presence of a dense one, as in the explosion of a gun, or by the return to their natural dispositions of bodies that have been distorted or bent, as manifested by the action of a bow.

Turning from the passages on "force" to those on "impetus," it is immediately apparent that Leonardo has absorbed the medieval theory that explains the motion of projectiles by the impression of an impetus into the projectile by the projector, a theory outlined in its most mature form by Jean Buridan and repeated by many other authors, including Albert of Saxony, whose works Leonardo had read.[24] Leonardo's

dependence on the medieval impetus theory is readily shown by noting a few of his statements concerning it:

1. MS *E*, folio 22r: "Impetus is a virtue ["virtù"] created by motion and transmitted by the motor to the mobile that has as much motion as the impetus has life."

2. *Codex Atlanticus*, folio 161v-a: "Impetus is a power ["potenzia"] of the motor applied to its mobile, which [power] causes the mobile to move after it has separated from its motor."

3. MS *G*, folio 73r: "Impetus is the impression of motion transmitted by the motor to the mobile. Impetus is a power impressed by the motor in the mobile.... Every impression tends toward permanence or desires to be permanent."

In the last passage Leonardo's words are particularly reminiscent of Buridan's "inertia-like" impetus. On the other hand, Buridan's quantitative description of impetus as directly proportional to both the quantity of prime matter in and the velocity of the mobile is nowhere evident in Leonardo's notebooks. In some passages Leonardo noted the view held by some of his contemporaries (such as Agostino Nifo) that the air plays a supplementary role in keeping the projectile in motion (see *Codex Atlanticus*, fol. 168v-b): "Impetus is [the] impression of local motion transmitted by the motor to the mobile and maintained by the air or the water as they move in order to prevent a vacuum" (cf. *ibid.*, fol. 219v-a). In *Codex Atlanticus*, folio 108r-a, however, he stressed the role of the air in resisting the motion of the projectile and concluded that the air gives little or no help to the motion. Furthermore, in *Codex Leicester*, folio 29v, he gives a long and detailed refutation of the possible role of the air as motor, as Buridan had before him. And not only is it the air as resistance that weakens the impetus in a projectile; the impetus is also weakened and destroyed by the tendency to natural motion. For example, in MS *E*, folio 29r, Leonardo says: "But the natural motion conjoined with the motion of a motor [that is, arising from the impetus derived from the motor] consumes the impetus of the motor."

Leonardo also applied the impetus theory to many of the same inertial phenomena as did his medieval predecessors—for instance, to the stability of the spinning top (MS *E*, fol. 50v), to pendular and other kinds of oscillating motion (*Codex Arundel*, fol. 2r), to impact and rebound (*Codex Leicester*, fols. 8r, 29r), and to the common medieval speculation regarding a ball falling through a hole in the earth to the center and rising on the other side before falling back and oscillating about the center of the earth (*Codex Atlanticus*, fol. 153v-b). In a sense, all of these last are embraced by the general statement in MS

E, folio 40v: "The impetus created in whatever line has the power of finishing in any other line."

In the overwhelming majority of passages, Leonardo applied the impetus theory to violent motion of projection. In some places, however, he also applied it to cases of natural motion, as did his medieval predecessors when they explained the acceleration of falling bodies through the continuous impression of impetus by the undiminished natural weight of the body. For example, in *Codex Atlanticus*, folio 176r-a, Leonardo wrote: "Impetus arises from weight just as it arises from force." And in the same manuscript (fol. 202v-b) he noted the continuous acquisition of impetus "up to the center [of the world]." Leonardo was convinced of the acceleration of falling bodies (although his kinematic description of that fall was confused); hence, he no doubt believed that the principal cause of such acceleration was the continual acquisition of impetus.

One last aspect of Leonardo's impetus doctrine remains to be discussed—his concept of compound impetus, defined in MS *E*, folio 35r, as that which occurs when the motion "partakes of the impetus of the motor and the impetus of the mobile." The example he gave is that of a spinning body moved by an external force along a straight line. When the impetus of the primary force dominates, the body moves along a simple straight line. As that impetus dies, the rotary motion of the spinning body acts with it to produce a composite-curved motion. Finally, all of the impetus of the original motion is dissipated and a simple circular motion remains that arises only from the spinning body. A series of passages (*Codex Arundel*, fols. 143r-144v) are further concerned with the relationship of transversal and natural motions, which, if the passages and diagrams have been understood correctly, concur to produce resultant composite motions. It appears later in the same manuscript (fol. 147v) that Leonardo thought that the first part of a projectile path was straight until the primary impetus diminished enough for the natural motion to have an effect:

> The mobile is [first] moved in that direction ["aspetto"] in which the motion of its motor is moved. The straightness of the transversal motion in the mobile lasts so long as the internal power given it by the motor lasts. Straightness is wanting to the transversal motion because [that is, when] the power which the mobile acquired from its motor diminishes.

A beautiful instance of compound motion upon which Leonardo reported more than once is that of an arrow or stone shot into the air from a rotating earth, which arrow or stone would fall to the ground with rectilinear motion with respect to the rotating earth because it receives a circular impetus from the earth. But with respect to a stationary frame, the descent is said to be spiral, that is, compounded of

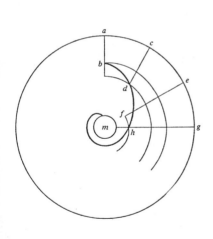

(As in manuscript)

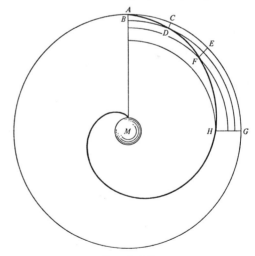

(Reconstructed)

FIGURE 19

229

rectilinear and circular motions. The longer of the passages in which Leonardo described this kind of compound motion is worth quoting (MS *G*, fol. 55r; see Figure 19):

> On the heavy body descending in air, with the elements rotating in a complete rotation in twenty-four hours. The mobile descending from the uppermost part of the sphere of fire will produce a straight motion down to the earth even if the elements are in a continuous motion of rotation about the center of the world. Proof: let *b* be the heavy body which descends through the elements from [point] *a* to the center of the world, *m*. I say that such a heavy body, even if it makes a curved descent in the manner of a spiral line, will never deviate in its rectilinear descent along which it continually proceeds from the place whence it began to the center of the world, because when it departs from point *a* and descends to *b*, in the time in which it has descended to *b*, it has been carried on to [point] *d*, the position of *a* having rotated to *c*, and so the mobile finds itself in the straight line extending from *c* to the center of the world, *m*. If the mobile descends from *d* to *f*, the beginning of motion, *c*, is, in the same time, moved from *c* to *f* [!*e*]. And if *f* descends to *h*, *e* is rotated to *g;* and so in twenty-four hours the mobile descends to the earth [directly] under the place whence it first began. And such a motion is a compounded one. [In margin:] If the mobile descends from the uppermost part of the elements to the lowest point in twenty-four hours, its motion is compounded of straight and curved [motions]. I say "straight" because it will never deviate from the shortest line extending from the place whence it began to the center of the elements, and it will stop at the lowest extremity of such a rectitude, which stands, as if to the zenith, under the place from which the mobile began [to descend]. And such a motion is inherently curved along with the parts of the line, and consequently in the end is curved along with the whole line. Thus it happened that the stone thrown from the tower does not hit the side of the tower before hitting the ground.

This is not unlike a passage found in Nicole Oresme's *Livre du ciel et du monde*.[25]

In view of the Aristotelian and Scholastic doctrines already noted, it is not surprising to find that Leonardo again and again adopted some form of the Peripatetic law of motion relating velocity directly to force and inversely to resistance. For example, MS *F*, folios 51r-v, lists a series of Aristotelian rules. It begins "1º if one power moves a body through a certain space in a certain time, the same power will move half the body twice the space in the same time. . . ." (compare MS *F*, fol. 26r). This law, when applied to machines like the pulley and the lever, became a kind of primitive conservation-of-work principle, as it had been in antiquity and the Middle Ages. Its sense was

that "there is in effect a definite limit to the results of a given effort, and this effort is not alone a question of the magnitude of the force but also of the distance, in any given time, through which it acts. If the one be increased, it can only be at the expense of the other."[26] In regard to the lever, Leonardo wrote (MS *A*, fol. 45r): "The ratio which the length of the lever has to the counterlever you will find to be the same as that in the quality of their weights and similarly in the slowness of movement and in the quality of the paths traversed by their extremities, when they have arrived at the permanent height of their pole." He stated again (*E*, fol. 58v):

> By the amount that accidental weight is added to the motor placed at the extremity of the lever so does the mobile placed at the extremity of the counterlever exceed its natural weight. And the movement of the motor is greater than that of the mobile by as much as the accidental weight of the motor exceeds its natural weight.

Leonardo also applied the principle to more complex machines (MS *A*, fol. 33v): "The more a force is extended from wheel to wheel, from lever to lever, or from screw to screw, the greater is its power and its slowness." Concerning multiple pulleys, he added (MS *E*, fol. 20v):

> The powers that the cords interposed between the pulleys receive from their motor are in the same ratio as the speeds of their motions. Of the motions made by the cords on their pulleys, the motion of the last cord is in the same ratio to the first as that of the number of cords; that is, if there are five, the first is moved one braccio, while the last is moved 1/5 of a braccio; and if there are six, the last cord will be moved 1/6 of a braccio, and so on to infinity. The ratio which the motion of the motor of the pulleys has to the motion of the weight lifted by the pulleys is the same as that of the weight lifted by such pulleys to the weight of its motor. . . .

It is not difficult to see why Leonardo, so concerned with this view of compensating gain and loss, attacked the speculators on perpetual motion (*Codex Forster* II₂, fol. 92v): "O speculators on continuous motion, how many vain designs of a similar nature have you created. Go and accompany the seekers after gold."

One area of dynamics that Leonardo treated is particularly worthy of note, that which he often called "percussion." In this area he went beyond his predecessors and, one might say, virtually created it as a branch of mechanics. For him the subject included not only effects of the impacts of hammers on nails and other surfaces (as in MS *C*, fol. 6v; MS *A*, fol.

53v) but also rectilinear impact of two balls, either both in motion or one in motion and the other at rest (see the various examples illustrated on MS *A*, fol. 8r), and rebound phenomena off a firm surface. In describing impacts in *Codex Arundel*, folio 83v, Leonardo wrote: "There are two kinds ["nature"] of percussion: the one when the object [struck] flees from the mobile that strikes it; the other when the mobile rebounds rectilinearly from the object struck." In one passage (MS *I*, fol. 41v), a problem of impacting balls is posed: "Ball *a* is moved with three degrees of velocity and ball *b* with four degrees of velocity. It is asked what is the difference ["varietà"] in such percussion [of *a*] with *b* when the latter ball would be at rest and when it [*a*] would meet the latter ball [moving] with the said four degrees of velocity."

In some passages Leonardo attempted to distinguish and measure the relative effects of the impetus of an object striking a surface and of the percussion executed by a resisting surface. For example, in *Codex Arundel*, folio 81v, he showed the rebound path as an arc (later called "l'arco del moto refresso") and indicated that the altitude of rebound is acquired only from the simple percussion, while the horizontal distance traversed in rebound is acquired only from the impetus that the mobile had on striking the surface, so that "by the amount that the rebound is higher than it is long, the power of the percussion exceeds the power of the impetus, and by the amount that the rebound's length exceeds its height, the percussion is exceeded by the impetus."

What is perhaps Leonardo's most interesting conclusion about rebound is that the angle of incidence is equal to that of rebound. For example, in *Codex Arundel*, folio 82v, he stated: "The angle made by the reflected motion of heavy bodies becomes equal to the angle made by the incident motion." Again, in MS *A*, folio 19r (Figure 20), he wrote:

> Every blow struck on an object rebounds rectilinearly at an angle equal ["simile"] to that of percussion. This proposition is clearly evident, inasmuch as, if you would strike a wall with a ball, it would rise rectilinearly at an angle equal to that of the percussion. That is, if

the ball *b* is thrown at *c*, it will return rectilinearly through the line *cb* because it is constrained to produce equal angles on the wall *fg*. And if you throw it along line *bd*, it will return rectilinearly along line *de*, and so the line of percussion and the line of rebound will make one angle on wall *fg* situated in the middle between two equal angles, as *d* appears between *m* and *n*. [See also *Codex Atlanticus*, fol. 125r-a.]

The transfer, and in a sense the conservation, of power and impetus in percussion is described in *Codex Leicester*, folio 8r:

> If the percussor will be equal and similar to the percussed, the percussor leaves its power completely in the percussed, which flees with fury from the site of the percussion, leaving its percussor there. But if the percussor—similar but not equal to the percussed—is greater, it will not lose its impetus completely after the percussion but there will remain the amount by which it exceeds the quantity of the percussed. And if the percussor will be less than the percussed, it will rebound rectilinearly through more distance than the percussed by the amount that the percussed exceeds the percussor.

Leonardo is here obviously groping for adequate laws of impact.

The last area to be considered is Leonardo's treatment of the kinematics of moving bodies, especially the kinematics of falling bodies. In *Codex Arundel*, folio 176v, he gave definitions of "slower" and "quicker" that rest ultimately on Aristotle:[27] "That motion is slower which, in the same time, acquires less space. And that is quicker which, in the same time, acquires more space." The description of falling bodies in respect to uniform acceleration is, of course, more complex. It should be said at the outset that Leonardo never succeeded in freeing his descriptions from essential confusions of the relationships of the variables involved. Most of his passages imply that the speed of fall is not only directly proportional to the time of fall, which is correct, but that it is also directly proportional to the distance of fall, which is not. In MS *M*, folio 45r, he declared that "the gravity [that is, heavy body] which descends freely, in every degree of time, acquires a degree of motion, and, in every degree of time, a degree of speed." If, like Duhem, one interprets "degree of motion," that is, quantity of motion, to be equivalent not to distance but to the medieval impetus, then Leonardo's statement is entirely correct and implies only that speed of fall is proportional to time of fall. One might also interpret the passage in MS *M*, folio 44r, in the same way (see Figure 21):

> Prove the ratio of the time and the motion together with the speed produced in the descent of heavy bodies

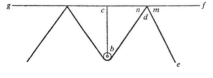

FIGURE 20

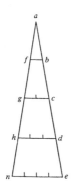

FIGURE 21

by means of the pyramidal figure, for the aforesaid powers ["potenzie"] are all pyramidal since they commence in nothing and go on increasing by degrees in arithmetic proportion. If you cut the pyramid in any degree of its height by a line parallel to the base, you will find that the space which extends from the section to the base has the same ratio as the breadth of the section has to the breadth of the whole base. You see that [just as] ab is 1/4 of ae so section fb is 1/4 of ne.

As in mathematical passages, Leonardo here used "pyramidal" where "triangular" is intended.[28] Thus he seems to require the representation of the whole motion by a triangle with point a the beginning of the motion and ne the final speed, with each of all the parallels representing the speed at some and every instant of time. In other passages Leonardo clearly coordinated instants in time with points and the whole time with a line (see Codex Arundel, fols. 176r-v). His triangular representation of quantity of motion is reminiscent of similar representations of uniformly difform motion (that is, uniform acceleration) in the medieval doctrine of configurations developed by Nicole Oresme and Giovanni Casali; Leonardo's use of an isosceles triangle seems to indicate that it was Casali's account rather than Oresme's that influenced him.[29] It should be emphasized, however, that in applying the triangle specifically to the motion of fall rather than to an abstract example of uniform acceleration, Leonardo was one step closer to the fruitful use to which Galileo and his successors put the triangle. Two similar passages illustrate this—the first (MS M, fol. 59v) again designates the triangle as a pyramidal figure, while in the second, in Codex Madrid I, folio 88v (Figure 22)[30] the triangle is divided into sixteen

equally spaced sections. Leonardo explained the units on the left of the latter figure by saying that "these unities are designated to demonstrate that the excesses of degrees are equal." Lower on the same page he noted that "the thing which descends acquires a degree of speed in every degree of motion and loses a degree of time." By "every degree of motion" he may have meant equal vertical spaces between the parallels into which the motion is divided. Hence, this phrase would be equivalent to saying "in every degree of time." The comment about the loss of time merely emphasizes the whole time spent during the completion of the motion.

All of the foregoing comments suggest a possible, even plausible, interpretation of Leonardo's concept of "degree of motion" in these passages. Still, one should examine other passages in which Leonardo seems also to hold that velocity is directly proportional to distance of fall. Consider, for example, MS M, folio 44v: "The heavy body ["gravità"] which descends, in every degree of time, acquires a degree of motion more than in the degree of time preceding, and similarly a degree of speed ["velocità"] more than

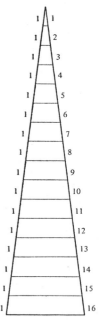

FIGURE 22

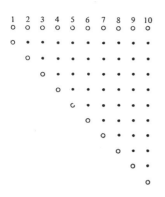

```
1   2   3   4   5   6   7   8   9   10
o   o   o   o   o   o   o   o   o   o
o   •   •   •   •   •   •   •   •   •
    o   •   •   •   •   •   •   •   •
        o   •   •   •   •   •   •   •
            o   •   •   •   •   •   •
                c   •   •   •   •   •
                    o   •   •   •   •
                        o   •   •   •
                            o   •   •
                                o   •
                                    o
```

FIGURE 23

in the preceding degree of motion. Therefore, in each doubled quantity of time, the length of descent ["lunghezza del discenso"] is doubled and [also] the speed of motion." The figure accompanying this passage (Figure 23) has the following legend: "It is here shown that whatever the ratio that one quantity of time has to another, so one quantity of motion will have to the other and [similarly] one quantity of speed to the other." There seems little doubt from this passage that Leonardo believed that in equal periods of time, equal increments of space are being acquired. One last passage deserves mention because, although also ambiguous, it reveals that Leonardo believed that the same kinds of relationships hold for motion on an incline as in vertical fall (MS *M*, fol. 42v): "Although the motion be oblique, it observes, in its every degree, the increase of the motion and the speed in arithmetic proportion." The figure (Figure 24) indicates that the motion on the incline is represented by the triangular section *ebc*, while the vertical fall is represented by *abc*. Hence, with this figure Leonardo clearly intended that the velocities at the end of both vertical and oblique descents are

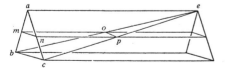

FIGURE 24

equal (that is, both are represented by *bc*) and also that the velocities midway in these descents are equal (that is, *mn* = *op*). The figure also shows that the times involved in acquiring the velocities differ, since the altitude of $\triangle ebc$ is obviously greater than the altitude of $\triangle abc$.

So much, then, for the most important aspects of Leonardo's theoretical mechanics. His considerable dependence on earlier currents has been noted, as has his quite significant original extension and development of those currents. It cannot be denied, however, that his notebooks, virtually closed as they were to his successors, exerted little or no influence on the development of mechanics.

NOTES

1. The many passages from Leonardo's notebooks quoted here can be found in the standard eds. of the various MSS. Most of them have also been collected in Uccelli's ed. of *I libri di meccanica*. The English trans. (with only two exceptions) are my own.
2. The full corpus has been published in E. A. Moody and M. Clagett, *The Medieval Science of Weights* (Madison, Wis., 1952; repr., 1960). Variant versions of the texts have been studied and partially published in J. E. Brown, "The 'Scientia de ponderibus' in the Later Middle Ages" (dissertation, Univ. of Wis., 1967).
3. Moody and Clagett, *op. cit.*, pp. 128–131. Incidentally, the definition that precedes the proof is the sure sign that Leonardo translated the postulates and first two enunciations from the *Elementa* rather than from version P (where the enunciations are the same), since the definition was not included in Version P.
4. *Ibid.*, pp. 174–207.
5. *Ibid.*, pp. 130–131.
6. *Ibid.*, pp. 188–191. Leonardo's translation of these same propositions in *Codex Atlanticus*, fols. 154v-a-r-a, is very literal: "La equalità della declinazione conserva la equalità del peso. Se per due vie di diverse obliquità due pesi discendano, e sieno medesimo proporzione, se della d[ecli-nazione] de' pesi col medesimo ordine presa sarà ancora una medesima virtù dell'una e d[ell'altra in discendendo]." The bracketed material has been added from the Latin text.
7. Marcolongo, *Studi vinciani*, p. 173.
8. Duhem, *Les origines de la statique*, I, 189–190.
9. Moody and Clagett, *op. cit.*, pp. 184–187. Leonardo's more literal translation of RI.08 in *Codex Atlanticus*, fol. 154v-a, runs: "Se le braccia della libra sono inequali e nel centro del moto faranno un angolo, s'e' termini loro s'accosteranno parte equalmente alla direzione, e' pesi equali in questa disposizione equalmente peseranno."
10. Marcolongo, *op. cit.*, p. 149, discusses one such passage (*Codex Arundel*, fol. 67v); see also pp. 147–148, discussing the figures on MS *Ashburnham* 2038, fol. 3r; and *Codex Arundel*, fol. 32v (in the passage earlier than the one noted above in the text).
11. Moody and Clagett, *op. cit.*, pp. 208–211.
12. *Ibid.*, pp. 210–211.
13. *Ibid.*, pp. 64–65, 102–109, 192–193.
14. Clagett, "Leonardo da Vinci . . .," pp. 119–140.
15. *Ibid.*, pp. 121–126.
16. *Ibid.*, p. 126.
17. Archimedes, *Monumenta omnia mathematica, quae extant . . . ex traditione Francisci Maurolici* (Palermo, 1685), De

momentis aequalibus, bk. IV, prop. 16, pp. 169–170. Mauro-lico completed the *De momentis aequalibus* in 1548.

18. Marcolongo, *op. cit.*, pp. 203–216.
19. See Clagett, "Leonardo da Vinci . . .," pp. 140–141. That account is here revised, taking into account the probability that the fragments were not copied by Leonardo.
20. *Ibid.*, p. 140, n. 65, notes the opinion of Favaro and Schmidt that the fragments are truly in Leonardo's hand. But Carlo Pedretti, whose knowledge of Leonardo's hand is sure and experienced, is convinced they are not. Arredi, *Le origini dell'idrostatica*, pp. 11–12, had already recognized that the notes were not in Leonardo's hand.
21. M. Clagett, *The Science of Mechanics in the Middle Ages* (Madison, Wis., 1959; repr., 1961), pp. 124–125.
22. Here Arredi's account is followed closely.
23. Clagett, *The Science of Mechanics*, p. 187.
24. *Ibid.*, chs. 8–9.
25. *Ibid.*, pp. 601–603.
26. Hart, *The Mechanical Investigations*, pp. 93–94.
27. Clagett, *The Science of Mechanics*, pp. 176–179.
28. Clagett, "Leonardo da Vinci . . .," p. 106, quoting MS *K*, fol. 79v.
29. M. Clagett, *Nicole Oresme and the Medieval Geometry of Qualities* (Madison, Wis., 1968), pp. 66–70.
30. I must thank L. Reti, editor of the forthcoming ed. of the Madrid codices, for providing me with this passage.

MEDIEVAL MATHEMATICS

VIII

The Medieval Latin Translations from the Arabic of the *Elements* of Euclid, with Special Emphasis on the Versions of Adelard of Bath*

THE question of the translation into Latin of Euclid's *Elements* in the middle ages can be resolved into two principal sub-questions. The first inquires into how much there remained of any translations from the Greek up to and including the twelfth century. The second seeks to find out the number and completeness of the translations of the *Elements* made from the Arabic. Friedlein, Heiberg, Bubnov,

* Thanks are due to Mr Thomas E. Brittingham, Jr and the Thomas E. Brittingham Trust Fund for financial assistance in the preparation of this paper.
§ University of Wisconsin.

Tannery, and others have addressed themselves to the first question with some success, although they have been unable to show that a *complete* translation was made from the Greek, either by Boethius or anybody else.[1] Research on the second question has been most incomplete and inconclusive, mainly because there has been no detailed study of the numerous and various manuscripts that lie in the libraries of Western Europe. In the hope of solving some of the problems connected with the translations from the Arabic, I have been collecting films of Euclid manuscripts.

In making this collection I was particularly attentive to the complex problem of how many versions of the *Elements* ought to be ascribed to Adelard of Bath. Charles Homer Haskins had already suggested that "it is not clear pending a comparison of the manuscripts, whether in its original form, his [i.e., Adelard's] own work was an abridgement, a close translation, or a commentary."[2] We are now, I believe, in a position to form at least preliminary conclusions on the Adelard question, as well as on the number and nature of the other translations of the *Elements* made from the Arabic.

The first and foremost conclusion is that there are *three* principal versions of the *Elements* which were attributed to, and with some probability were made by, Adelard. All of these versions can be traced back to the twelfth century. Version I is a literal translation. Version II is an abridgement, and Version III is a reworking or paraphrase (an "editio") using the enunciations of Version II, but whose proofs are full and complete in the manner of Version I. It has been my purpose in the first and main part of this article to describe each of these versions. In the second part I have treated briefly other translations made from the Arabic, principally those of Hermann of Carinthia and Gerard of Cremona. Finally, I have included two Appendixes in which I have given for each version of Adelard and the other translations enough of several propositions to permit the reader to distinguish one version from the other as to form and content.

---

[1] Evidence has been uncovered of the fragments of at least four antique and early medieval Latin translations of the *Elements* from the Greek. (1) There is a fragment of Censorinus, presumably dating from the third century A.D. which includes definitions, postulates, and axioms from Book I. It begins "Nota est cuius pars nulla est . . ." (See edition of Censorinus' works, editor F. Hultsch, Teubner Text, Leipzig, 1867, pp. 60–63.) (2) A palimpsest at Verona (Chapter Library No. 40?) contains fragments of another translation from the Greek dating from the fourth century. It contains parts of books XII and XIII (numbered in the codex Books XIV and XV). It has not been published, but see G. Cantor, *Vorlesungen u. Gesch. d. Math.*, *1* (3rd edit. 1907), p. 565. (3) Boethius made some kind of translation, of which apparently only excerpts from the first five books remain. Various parts of these excerpts are preserved by Cassiodorus, by the gromatici veteres, and by the anonymous editors of the five book and two book versions of the Pseudo-Boethian Geometry. Only three proofs (those of Book I, Propositions 1–3) have been preserved. J. L. Heiberg has prepared the text of these excerpts (*Zeit. f. Math. u. Phys.*, *35* (1890), Hist.-lit. Abth., pp. 48–58, 81–100; the text appears on pp. 86–98). N. Bubnov, *Gerberti . . . Opera mathematica*, Berlin, 1899, pp. 161–179, has done an excellent job tracing the history of the Boethian excerpts, complementing Heiberg and going beyond him to include definitions from Book V as given by Cassiodorus. Bubnov guesses, but without evidence, that Boethius had made a complete translation of the *Elements*. M. Curtze and others have specu-

lated on whether or not at least the propositions for all fifteen books without proofs circulated in the twelfth century to be consulted by Adelard of Bath and others (See Heiberg's article for the literature on this question; also consult the introduction of T. L. Heath in *The Thirteen Books of Euclid's Elements, 1,* 2nd edit., Cambridge, 1926). But none of these authors have examined and analysed the interesting twelfth century Chartres MS at Paris, BN latin 10257 which not only has all the Boethian excerpts given by Heiberg and Bubnov (admittedly contaminated by Version II of Adelard) but it has propositions through the 15th book many of which were clearly translated from the Greek rather than the Arabic. It thus constitutes the only evidence for the translation from the Greek, at least in the Boethian tradition, of propositions from books beyond Book IV. I hope to publish an article on this manuscript soon. (4) There is finally a fragment of an entirely different translation contained in the Munich Univ. Library MS 20752 of the tenth century. It has been published by M. Curtze in the Introduction to his *Supplementum* to Heiberg's Teubner edition of Euclid's *Opera omnia*, Leipzig, 1899. It contains I.37 to I.38 and II.8 to II.9. The translator, an Italian, is very literal in his translation. His knowledge of mathematics is poor, since he translates as numbers the letters used to mark the geometric figures. The result is completely unintelligible. For corrections to Curtze, see *Bibliotheca Mathematica*, Dritte Folge, *2* (1901), pp. 365–6 and *Hermes, 3* (1903), pp. 354–356.

[2] C. H. Haskins, *Studies in the History of Medieval Science*, Cambridge, 1927, p. 25.

A. THE VERSIONS OF ADELARD OF BATH

(See Appendix I)

Version I — *A Translation*

The version which we have selected as the first one made by Adelard of Bath is obviously a close translation from the Arabic of the *Elements* of Euclid, including the non-Euclidian Books XIV and XV. Version I was in fact the *first full* translation of the *Elements* into Latin. So far as I know it has never been singled out for notice or study. The reason for this is that there is no one manuscript containing all of this version. I have had to piece it together from several manuscripts on the basis of distinctive translating techniques and characteristics. Two twelfth century manuscripts contain almost all of the first eight books. These are Oxford, Trinity College 47, 139r–180v,[3] and Paris, Bibliothèque Nationale, Fonds latin 16201, 35r–82r.[4] The second is in all likelihood a copy of the first. Books VII and VIII of this version are found tucked away between two other versions in British Museum, Burney 275, 302r–308r [5] a manuscript catalogued as of the fourteenth century. Finally, Books X.36–XV.2 are contained in the thirteenth century Bodleian MS, D'Orville 70, 39r–71v.[6]

There are certain characteristics of translation and presentation which mark this version. It is the only version, either of Adelard or anybody else, in which the translator starts out most of his propositions with the phrase: *Nunc demonstrandum est* (or occasionally: *Nunc signandum est*). By this phrase the translator is presumably rendering the Arabic *nubaiyinu*.[7] Furthermore, the great majority of the demonstrations in all books are concluded by the phrase, also peculiar to this version: *Hoc est quod demonstrare intendimus* (Wa dhalika mā aradnā an nubaiyinu). The peculiarity lies particularly in using the verb: *intendimus* (instead of *"voluimus"* or *"proposuimus"*). A host of Arabicisms peek through the Latin of this version. And if there is any doubt that the various sections I have singled out all belong to the same version, it will be dispelled by the fact that the same Arabicisms peculiar only to this version appear throughout the fifteen books. Some of these transliterated Arabic words are: *alhamud* (for *perpendicularis*), *alkaida* (for *basis*), *mutekefia* (for *proportionalia*), and so on.[8] In this connection the reader is particularly invited to

---

[3] The leaves in Trinity 47 containing Version I have been bound in the wrong order. The following arrangement represents the correct order: Books I–II.5, ff. 171r–180v; II.5–III.24, 163r–170v; III.24–V.14, 155r–162v; V.14–VI.6, 147r–154v; VII.6–VIII.22, 139r–146v.

[4] Paris, BN Latin 16201 breaks off in VIII.22 at exactly the same place as Trinity 47. I suspect that it was copied from the Trinity MS which is more carefully done and contains many more drawings than the Paris MS. The scribe of BN 16201 concludes as follows: "Hic finitur liber octavius euclidis; deficuit residuum octavi."

[5] Folios 293r–302r contain I–VII.1 in Version II (except for VII.1 which is unlike either Version I or Version II). Folios 302r–309r contain VII.2–VIII in Version I. Folios 308r–335r contain IX–XV.2 in Version III.

[6] MS D'Orville 70 contains on folios 1r–39r, I–X.36 of a version different from any of the three versions here treated, and presumably is a thirteenth century commentary. The fragment of Version I begins on 39r with another proof of X.36. Also notice that folios 61v–62v contain a short work on the hyperbola translated from the Arabic by John of Palermo, which I have published in *Osiris*, *11* (1953).

[7] This is the word that commonly introduces the propositions in the al-Hajjāj text which appears to be the one used by Adelard. We do not have any early copy of this text, but Codex Leidensis 399.1 contains a copy of the first six books of this version with much additional commentary added by al-Narizi. It has been published by R. O. Besthorn, J. L. Heiberg, William Thomson, G. Junge, and J. Raeder, *Codex Leidensis 399.1. Euclidis elementa ex interpretatione al-Hadschdschadschii cum commentariis al Narizii etc.*, Copenhagen (i.e., Hauniae), 1893–1932. Besthorn and Heiberg are responsible for the text and Latin translation of the first four books. Then over a generation later Thomson did the text of Books V and VI (and the beginning of Book VII where the codex breaks off). Thomson also did an English translation of these books from which Raeder made a Latin translation. Junge reviewed the text for its mathematical soundness. For the use of the word "nubaiyinu," see practically any proposition, but particularly Prop. I.1, (Besthorn, edit. Fasc. I, p. 46).

[8] Cf. MS Trinity College 47, II.13 (f. 165v), IV.5 (159r), VI.13 (150r).

examine the *explicit* of this version given at the end of this section. It reveals almost as much transliterated Arabic as it does Latin.

I have accepted this version as anterior to the other versions partly because it seems likely that a translation preceded the abridgement and paraphrase of Versions II and III, and partly because the leaves of Trinity College 47 containing the first part of Version I appear to be the oldest of all the manuscripts of any Adelard version. I date these leaves before 1150. In fact, it would not surprise me to discover that these leaves came from an Adelard autograph, for in a number of places the Arabic word has been put in the text, and then later crossed out, I believe, by the same hand, and replaced by the Latin word. But not having any specimen of Adelard's hand before me, I hesitate to say any more. I would judge the handwriting of these leaves to be older than that used on leaves 104v–138r of the same manuscript, containing the abridgement I have called Version II.

The attribution of Version I to Adelard in Trinity College 47 runs as follows: "Institutio artis geometrice ab Euclide descripta XV libros continens, per Adelardum Batoniensam ex arabico in latinum sermonem translata, incipit." In Paris, BN Latin 16201, 35r, however, the title is given without reference to Adelard: "Geometrice Demonstracionis Euclidis Commentum. Incipit liber primus." The use of the term "commentum," I believe, stems from the fact that the Boethian excerpts from Euclid generally circulated without demonstrations, and quite probably Euclid was thought to be the author only of the definitions, postulates, axioms, and enunciations; and so the demonstrations were conceived of as commentary. This curious idea persisted down to the Renaissance when the commentary or "expositio" of Campanus was contrasted with the "expositio" of Theon, as if each was an original commentary, although in actuality Campanus' demonstrations were a reworking of the Arabic tradition and Theon's supposed exposition was pretty close to the Euclidian proofs themselves.[9] As a matter of fact, the enunciations of the propositions in Adelard's Version II often circulated without the demonstrations or with considerably different sets of demonstrations and so no doubt did much to foster this idea of a skeleton framework of enunciations by Euclid to which were added various commentaries.[10] This undoubtedly gave rise to the common statement on later manuscripts: "Euclidis liber Elementorum . . . translatus ab arabico in Latinum per Adelardum Bathoniensem sub commento Magistri Campani Novarriensis."[11] We shall discuss the popularity of the enunciations of Version II later.

In preparing this translation of Version I, Adelard possibly used some form of the so-called al-Hajjāj Arabic text, the first of the translations of the *Elements* from Greek into Arabic. Unfortunately the only extant copy of that text is fairly late and no doubt has been emended.[12] The whole question of the relationship of the known Arabic text to the Latin translation needs careful study. However, we can say there are similarities in Adelard's Version I and the extant copy of the al-Hajjāj text.[13]

[9] In the first edition of the Zamberti translation we have the attribution of the exposition to Theon: Euclidis megaresis . . . habent in hoc volumine . . . elementorum libros xiii cum expositione Theonis, Venetiis, in aedibus Joannis Tacuini, 1506. The first edition of the Campanus version had already been published by E. Ratdoldt in Venice in 1482. The two translations are presented together in editions of Paris, Henricus Stephanus, 1516; Basel, J. Herwagius, 1537, 1546, and 1557. For a discussion of this whole point and for the titles of the early editions, see H. Weissenborn, *Die Uebersetzungen des Euklid durch Campano und Zamberti*, Halle a/S, 1882, and also C. Thomas-Stanford, *Early Editions of Euclid's Elements*, London, 1926.

[10] For example Oxford, MS C.C.C. 224, Brit.

Museum Royal 15.A.27; Oxford, Trinity College 47 (Books VII–XV without proofs). See the detailed description of these MSS in the list of MSS at the end of the section on Version II.

[11] For example, Oxford, Bodl. Arch. Selden B. 13 (3r); Paris, BN latin 7213 (1r); and Paris, BN latin 16197.

[12] See the edition cited in footnote 7. Codex Leid. 399.1 from which this edition was made, obviously represents a somewhat later elaboration of the al-Hajjāj text and presumably it started with the so-called "al-Ma'mūni" version rather than the earlier "al-Hārūni" version.

[13] A detailed comparison of the Besthorn edition and the Adelard text still remains to be done. I did not have the Besthorn work at hand after I had found Version I of Adelard.

We can not date Adelard's translation with any surety. We know that it was made prior to his work on the astrolabe, for he tells us in that treatise that he has made such a translation: [14] "Et omnium quidem supradictorum simpliciter expositorum siquis rationum postulaverit, intelligat eam apud Euclidem a quindecim libris artis geometrice quos ex arabico in latinum convertimus sermonem esse conniciendam." Haskins is inclined to think that the work on the astrolabe was composed sometime between 1142 and 1146.[15] This then would place Adelard's translation of the *Elements* sometime before that period. We cannot help but notice the almost exactly similar reference here to his translation with that appearing in the title of Trinity College 47, particularly in using the same expression *latinum sermonem*.

INCIPIT: The incipit of Version I from both twelfth century MSS is "Punctus est illud cui pars non est. . . ." (See Appendix I (1)). For the question of whether Version I contained an introduction, see the discussion of Version III below.

EXPLICIT: It is not known whether Version I was completed through XV.5, for the only manuscript we have of the last books of Version I goes only through XV.2. But XV.2 in that MS (D'Orville 70, f. 71v) runs as follows: "Nunc signandum est basium corpus habens VIII alkaidas equalium laterum intra corpus IIII^{or} alkaidarum triangularum equalium laterum. Sit itaque corpus cui. . . . Itaque factum est in almugecem *abgd* IIII alkaidarum almugecem habens alkaidas triangulorum equalium laterum et hoc est quod demonstrare indendimus." (Note: actually D'Orville 70 has *alraida* for *basis*, while the transliteration in Trinity 47 in other places is *alkaida* for the Arabic word *al-qa'idat* which, of course, does mean "base.")

EXTANT MSS: See the first paragraph of this section for mention of all of the extant MSS.

Version II — *An Abridgement*

This version, almost completely different from Version I, is clearly an abridgement. That it was made by Adelard seems to follow from the fact that most of the many manuscripts of it clearly ascribe it to Adelard. This includes the twelfth century copy of Trinity College 47, ff. 104v–138r. I do not think that Adelard translated it from the Arabic, but rather prepared it after his translation of Version I and under the influence of the more common pseudo-Boethian propositions which still circulated. It is possible, however, that Version II was prepared before Version I, Adelard constructing his abridgements directly from the Arabic text. The fact that Version II includes at least one Arabicism not present in Version I (e.g., "*elmunharife*" for "irregularis" in the definition of Book I) seems at least to show that an Arabic text was at hand whether or not Version II was prepared before or after Version I.

Hence, I had to make a rather superficial comparison of Version I with a few notes I had taken on the Besthorn text. The Propositions I.1, I.2, II.1, and II.2 in the al-Hajjāj text agree quite well with the Adelard translation. Furthermore in Book I the two texts both have 47 propositions rather than the 48 of the Thabit text and the best Greek tradition. The Greek tradition and the Thabit text include Prop. I.45, missing in the Adelard and the Besthorn text of al-Hajjāj. Also both texts give the axiom to the effect that two straight lines cannot include a surface as a postulate, although the Besthorn text also includes it as an axiom as well as a postulate. On the other hand, the introductory material included in the Besthorn text defining the parts of a proof is not in the Adelard translation, although it is clearly in the Arabic basis of the later translation of Gerard of Cremona.

From which we should either conclude that (1) Gerard had both Arabic texts in front of him, i.e., the al-Hajjāj text and the improved version of Thabit-Ishāq, or (2) the Thabit-Ishāq text as well as the al-Hajjāj text included this same material. It should also be noticed that Adelard's text agrees with the Besthorn wording of proposition II.14 which differs markedly from the Greek and Thabit traditions (See the differences by consulting Appendix I (1) and II (2)). In one case the Besthorn version of al-Hajjāj seems markedly different from Adelard and closer to the Thabit and Greek traditions, namely in the wording and order of the definitions of Book III; but on the whole it seems fairly certain that Adelard used an al-Hajjāj text.

[14] See Haskins, *op. cit.* in note 2, p. 25.
[15] *Ibid.*, p. 29.

Version II was by far the most popular of the three versions we have assigned to Adelard. It is this version which Weissenborn incorrectly settled on as the Adelard translation,[16] and which most students of the Euclid problem have continued since his time to call the Adelard Euclid. Its popularity is not only shown by the considerable number of manuscripts of it which are still extant (See list below), but also by the fact that the enunciations of Version II were used by numerous scholars of the thirteenth and fourteenth centuries (including Campanus) who wished to make commentaries on Euclid or rework the proofs in their own style and language. (I have referred to some of these commentaries in note 31.)

It is a tempting theory to suggest that perhaps Adelard's student John (or Nicolas?) Ocreat had something to do with preparing Version II. Some copies of Version II add to the regular proof of proposition IV.16 a statement, which no doubt refers, however, to the definitions of Book V. It runs:

Ex Ocrea Johannis, quoniam proportionem habet minor numerus ad maiorem eandem habet, pars denominata a maiore ad partem denominatam a minore. Item ex eadem, partium cuiuslibet totius minor maioris cuiusque si non fuerit pars erit (?) partes in sua propria denominatione secundum numerum denominationis eiusdem maioris partis. (MS Brit. Museum, Add. 34018, f. 17r; Cf. Oxford, C.C.C. 251, f.28v).

Now this statement is not in the twelfth century copy of Version II in Trinity College 47. Nor is there any such statement in Versions I or III. Thus it appears that Ocreat did some version other than the Adelard versions. Further evidence of this sort is contained in a reference in the old *Catalogi librorum manuscriptorum Angliae* etc., Oxford, 1697, p. 247. For MS no. 8639 it has "Euclidis Elementa, ex Arab. in Lat. versa per Joan. Ocreatum." The cataloguers of the Royal MSS at the British Museum identify this MS with Royal 15.A.27, a twelfth century manuscript. But this latter manuscript now has the first folio missing, so we are no longer able to confirm the statement in the old catalogue attributing it to Ocreat. However, on examining the MS we find that the definitions, axioms, and enunciations are in the form of Adelard's Version II. Only in the first book do we find any indications of proof, and, as a matter of fact, the brief proofs or indications of proof are given in the margin, while additional space under each proposition has been left blank to fill in the proofs, no doubt of Version II. To indicate how brief the marginal notes are, we can cite that for *I.1* (f.2r): "Incipit a triangulo, quia triangulus est principium omnium scientie, et omnia scientie resolvuntur in eo, scilicet, regularia. Deinde ex circuli descriptione argumentum elice." It is clear that the author of the marginal notes of Royal 15.A.27 knew of the Boethian excerpts. For he has notes on the Arabic words used in the definitions to Book I, where he gives the Greek equivalents as contained in the Boethian version. Thus above *elmuain* he has written, "i.e., rumbus," (f.1v). Similarly he has the following note next to *elmuharifa*: "Elmuharifa, vel trapezia, idem quod irregularis. Elmuharifa enim arabice, trapezia grece, irregularis latine" (f.1v). Also he introduces from the Greek (possibly from Gerbert's *De disciplina geometrie*) the common names for the three kinds of triangles, i.e., *ysopleurus, ysocheles,* and *scalendus* (!) (f.1r).

If Royal 15 A. 27 really is the Ocreat version, or at least started out to be the Ocreat version, we must adopt one of three conclusions: (1) It was Ocreat who prepared Version II instead of Adelard (and this seems highly unlikely in view of the numerous other MSS including the twelfth Century Trinity College 47 which attribute this version to Adelard). Or (2) Ocreat did nothing more than contribute a few marginal notes to Adelard's Version II, the notes for the first book being still extant in the margins of Royal 15.A.27. Or (3) Ocreat made a substantially different trans-

---

[16] H. Weissenborn, "Die Ubersetzungen des Euklid aus dem Arabischen in das Lateinische durch Adelhard von Bath etc.," *Zeit. f. Math.* *u. Physik, 25* (1880), hist.-lit. Abth., pp. 143–166.

lation of the proofs which he grafted on to the propositions or enunciations of Version II, and, furthermore, the scribe of Royal 15.A.27 meant to fill in these Ocreat translated proofs but never got around to it, leaving the present blank spaces under the enunciations. The paragraph from Ocreat's Book V cited above would fit into either of the last two conclusions. Unfortunately the paragraph is missing in Royal 15.A.27 no doubt because of the skeleton form of that MS. However, it seems likely that this paragraph can be used to identify Ocreat's version in the future. It is supposed that our John Ocreat is identical with the N. Ocreat who addressed an arithmetical treatise which he had translated from the Arabic to his master Adelard of Bath.[17]

In conformity with the fact that we have only the general abridged directions for proofs rather than the detailed proofs (at least in most of the books), the drawings usually found in manuscripts of Version II are without letters. Trinity College 47 is an exception, but it will be seen immediately that the letters from Version I as given in that same manuscript are used. Another peculiarity of many manuscripts of Version II is that the proofs *precede* rather than *follow* the enunciations of the propositions. However, the early Trinity College 47 has the proofs in the margins next to the propositions, and I would doubt that the original copy of Version II contained such a peculiar arrangement.

INCIPIT: The incipit of Version II generally runs: "Punctus est cui [*or*, cuius] pars non est. . . ." [See Appendix I(2).]

EXPLICIT: The explicit of Version II varies, depending on whether the proof of XV.5 precedes or follows the enunciation. If it precedes the enunciation, then the text ends with the enunciation as follows: "Intra datum corpus 20 basium triangularum equalium laterum corpus 12 basium pentagonalium equalium laterum figuraliter componere." [For example, see Brit. Mus. Add. 34,018, f. 78v.] On the other hand, if the text ends with the commentary of proof, it ends as follows: ". . . intra datum corpus viginti basium figurarum, quod oportet ostendere." [MS Corp. Christi 251, f.83] The most common colophon is the following: Explicit liber Euclidis philosophi de arte geometrica continens cccclxv proposita et propositiones, et xi porismata preter anxiomata singulis libris premissa, proposita quidem infinitivis, propositiones (vero) indicativis explicans." (See MS Auct. F.5.28, f. 15r, or Brit. Museum Add. 34,018, 78v, etc.]

EXTANT MSS: Some of the many manuscripts of Version II follow:

1. Oxford, Trinity College 47, 12th cent.— ff. 104v–138r. Includes propositions for Books I–XV, but proofs only for I–VI.

2. Oxford, Bodl. Auct. F. 5. 28, 1r–55r (ii–15r), 13c, Books I–XV.

3. Oxford, Bodl. Auct. F. 3. 13, 1r–48v, 13c, Books I–XV.

4. Oxford, Corp. Christ. Coll. 251, 22r–83r, 13c, Books I–XV.

5. Oxford, CCC 224, 114r–138r, 13c, Books I–XV (but no proofs).

6. Oxford, CCC 283, 50r–65v, 13c, Books X–XI.4.

7. London, Brit. Museum, Add. 34,018, 13c, Books I–XV.

8. London, BM, Burney 275, 13 or 14c, 293r–302r, Book I–VI (then follows parts of Versions I and III, discussed elsewhere).

9. London, BM, Royal 15.A.XXVII, 12 or early 13c, Books I–XV (proofs only in Book I; see discussion above).

10. London, BM, Royal 15.B.IV, 13c, 154r–157v, Books X.95–XII.3 (proofs without propositions).

11. London, BM, Add. 33,381, 13c, 186r–v, Books III.36–IV.4, 187r–188v, Books V.17–VI.3 (The numbers used by Adelard are noted here).

12. London, BM, Sloane 1044, 13c, 80r, Books X1.30–XI.32.

13. London, BM, Royal 10.A.VII, 13c, 213r, Book I, Def., post., axioms and prop. 1.

14. Paris, Bibl. Nat. lat. 7374A, 13 or 14c, 1r–47v, Books I–VII.

15. Paris, BN lat. 7374, 13 or 14c, (Is Version II only through I.46; then follows a paraphrase resembling a little Version I).

---

[17] Paris, BN Latin 6626, ff. 84–87. See C. Henry, "Prologus N. Ocreati in Helceph ad Adelardum Batensem magistrum suum etc.," *Zeit. f. Math. u. Physik*, 25 (1880), hist.-lit. Abth., pp. 129–139.

16. Paris, BN lat. 11245, 13c, 1r–57v, Books I–X.

17. Erfurt, Stadtbibl., Amplon. Q. 23, 13c, 1r–70v, Books I–XV.

18. Erfurt, Stadtbibl., Amplon. Q. 352, 13c.

19. Florence, Bibl. Naz. J.I. 32, 49r–103r, 13c.

20. Florence, Bibl. Naz. J.I. 18, 137r–160v.

21, Bresden, Db 86, 14c, 1–48, Books I–XV. (Note: MMS 17–21 I have not examined personally.) Sample propositions of Version II are contained in Appendix (2).

Version III — An "*editio*"

The third version of the *Elements* attributed to Adelard of Bath constitutes a paraphrase or an edition in which the enunciations of the propositions were borrowed from Version II, but in which the proofs are given in a complete, specific, and formal way not unlike the translation of Version I. However the specific letters used in the proofs differ from those used in Version I, and each part of the proof is labeled with its appropriate name, such as *exemplum, dispositio, probatio*, etc. Furthermore, Version III contains an Introduction, partly derived from Arabic sources and partly derived from Latin sources. It also contains commentary for the definitions of the books including and succeeding Book II. We have called this version an *editio* because it is so designated in the only manuscript containing a colophon. (See the explicit and colophon given at the end of this section.) Furthermore, the colophon specifically distinguishes this *editio* from the translation made by Adelard.

I see no overriding reason to doubt that Version III was prepared by Adelard. Not only do the only manuscripts which include a title or colophon so attribute it, but it is assigned to Adelard (or rather Alardus) by Roger Bacon when he quotes from this version giving it the title of *editio specialis Alardi Bathoniensis*, and distinguishing it from what he calls the *commentum* (Version I or II). [18] It should be

---

[18] R. Steele in his edition of the *Communia mathematica Fratris Rogeri*, Oxford, 1940, pp. x–xi has noticed that Bacon used the same text as in Digby 174, f. 99 et seq., and particularly that Bacon quotes from the introduction verbatim. Steele did not realize the existence of Adelard III, however, and he is clearly in error in thinking that the term *editio specialis* applied by Bacon referred only to the introduction in Digby 174, f. 99. In our explicit for this version at the end of this section we note that the colophon of one of the manuscripts calls the whole version an *editio*, and as a matter of fact Bacon himself quotes from the fifth book of the *editio specialis*, as we shall see below. Certain of the citations of this edition by Bacon are of interest: ". . . . et Alardus Batoniensis in sua *edicione speciali super Elementa Euclidis* ait: 'Concepciones sunt que ultimo (aliter primo) occurrunt humane intelligencie, in quibus non est exigendum propter quid.'" (Steele edition, pp. 65–66.) Cf. our Appendix I (3) where this quotation is found in the introduction to Version III. In another place (Steele edition, pp. 78–79) Bacon quotes at length from this introduction: "De hiis vero sic scribit Alardus Batoniensis in *edicione speciali super librum Elementorum Euclidis* dicens: 'Excercitacionis geometrici geometricalia instrumenta sunt mensure geometrici, scilicet pertica cum palma, digitus, pes, passus et ulna. Istis sex adde duo cum stadio, miliare . . .' (See introduction to Version II in Appendix I (3); the quote goes on through the phrase:) . . . Leuca miliarium unum et dimidium." A third citation of Adelard by Bacon appears to have been drawn from the introduction to Version III. It runs (Steele, pp. 69–70) "Et istorum complexorum omnium

divisio una reperitur, quam Alardus Batoniensis ponit. Et est quod quedam dicuntur proposita, quedam dicuntur proposiciones: proposita per infinitum proponuntur . . . proposiciones proponuntur per induccionem (indicativum?) . . . Item ut ait 'proposiciones proponuntur quod quidem sic vel non sic, proposita proponuntur ad faciendum vel non faciendum.'" Cf. Appendix I(3). In another reference to Adelard, Bacon distinguishes between the *editio specialis* and a *commentum*, saying that *proportionalitates* are called *medietates* in the *editio specialis* but not in the *commentum*: "Et his considerandum quod tres proporcionalitates sunt medietates et secundum Boetium et Jordanum in *Arismetica* et secundum Alardum Batoniensem in exposicione *quinti Elementorum* Euclidis quam tradidit, non in *Commento*, sed in *edicione speciali* quamvis vulgus hoc ignoret." It is quite true that in the comment to the fourth definition of Book V in Version III the identification of *proportionalitates* and *medietates* is made (See MS Balliol College 257, f. 22r) and it is not made in either Version II or Version I. One cannot tell whether Bacon had Version I or Version II in mind when he distinguished the *Commentum* from the special edition, but my guess is that he is referring to Version II as a *Commentum*. By the way, this and the next references to the parts of the *editio specialis* beyond the introduction constitute an immediate refutation of Steele's idea that Bacon only meant to refer to the Introduction as an *editio specialis*. Two other references to the fifth book of the *editio specialis* by Bacon can be found in the Steele edition, pp. 86 and 91. I think it possible that Bacon actually used the copy of Paris BN 16648 (which, although it now has

24

pointed out, however, that in the case of the attribution to Adelard in both the twelfth century MSS (Balliol College 257 and Digby 174) Adelard is actually said to be the translator, without any further information as to who prepared the *editio*. It is only in the colophon (see below) that the specific reference to Adelard as the editor as well as the translator is given. And Bacon in all probability is merely repeating the information of the colophon when he attributes this *editio* to Adelard.

As an *editio*, Version III does not limit itself to the text of Euclid alone. This is particularly born out by the Introduction included in MS Digby 174. (See Appendix I (3).) This Introduction is not a translation from the Arabic. For it includes material on the liberal arts and the classification of the sciences that was particularly common in twelfth century Latin works, e.g., in tracts such as the *Didascalicon de studio legendi* of Hugh of St Victor and the *De divisione philosophie* of Domingo Gundisalvi.[19] The Introduction to Version III also draws a few sentences on units of geometric measures primarily from Gerbert's *De disciplina geometrie*, Chap. II. On the other hand, it has some material in it which is unquestionably of Arabic origin. For example, the definitions of the various parts of a proof which ultimately go back to Proclus' commentary on the first book of Euclid's *Elements*, have certainly been drawn by Adelard from an Arabic source, and probably from the material which accompanied the al-Hajjāj text.[20] These definitions of *proposito, exemplum, dispositio*, etc. as given here in Appendix I (3) should be compared with similar material translated from the Arabic by Hermann of Carinthia and Gerard of Cremona which we have given in Appendix II (1) and II (2). It is quite evident then that the Proclus definitions accompanied the Arabic texts of Euclid. In fact, one wonders whether or not Adelard's Version I, quite probably made from the al-Hajjāj text, did not also include some of the introductory definitions, although the two copies of it from the twelfth century contain no such material. We can also note the use in the Introduction to Version III of the Arabic word for trapezia, namely, *helmuarife* (actually, *al-munharifa*). This word was also used in the 34th definition of Book I in Adelard's Version II.

The definitions contained in the various books after Book I in Version III ordinarily have some additional commentary not in the text. We have already noted the commentary to Book V, definition 4, defining *proportionalitas*. Finally, we should notice that the Introduction is followed immediately by Proposition I.1, and thus the definitions, postulates, and axioms have been omitted, at least in Digby 174.

---

only the last five books, was no doubt originally intact). For that MS not only calls Version III an editio as does Bacon but it also uses the abbreviated form "Alardus" as does Roger (see colophon given at end of this section on Version III).

[19] Cf. M. Clagett, "Some General Aspects of Medieval Physics," *Isis*, *39* (1948), pp. 29–36 where the classification literature is cited.

[20] For the original form of the Proclus definitions, see the edition of G. Friedlein, *Procli Diadochi in primum Euclidis Elementorum librum commentarii*, Leipzig, 1873, p. 203, lines 1–15, and also p. 255, lines 12–24. These definitions are contained as an introduction (attributed to Euclid) in the text of al-Hajjāj (See Besthorn text cited in Note 7, Fasc. I, pp. 6–9. Cf. *Ibid.* pp. 34–39 for statements drawn from Simplicius). In Gerard of Cremona's translation of the commentaries of al-Narizi on Euclid the definitions of the parts of proof as assigned to Euclid in the Besthorn text are missing, but the somewhat different statements of Simplicius to which we have also referred are present. (See Curtze edition in *Supplementum* of the *Opera omnia* of Euclid, Teubner, Leipzig, 1899, pp. 40–42).

Now we ought to say a further word about the two twelfth century MSS of Version III. Balliol 257 is much the better and more carefully written manuscript. It contains 98 folios. It commences with two propositions from Book X (106 and 107 in Adelard's numbering), which, by the way, are later repeated in this manuscript in somewhat different form. After these two propositions the text proper begins in the same way as Version II. In fact it looks as if our scribe copied Version II right down through proposition I, for our manuscript contains Version II's abridged proof of proposition I *preceding* the enunciation, and then suddenly following the enunciation it gives an entirely different, detailed proof of proposition 1. Thus I suspect that Version III actually begins in this manuscript with the second proof of proposition 1 rather than with the definitions, postulates, and axioms presumably copied directly from some manuscript of Version II. Perhaps the scribe of Balliol 257 decided he would rather have the definitions, axioms, and postulates of Version II than the introduction without definitions which appears in the Digby 174 copy of the *editio* of Version III. That the introduction does truly belong to Version III we judge not alone from the fact that it is contained in Digby 174, but also from the fact that it was also in the copy of Version III used by Roger Bacon in his *Communia mathematica* (See footnote 18). While Balliol College 257 contains more of Version III than any other manuscript, it stops abruptly with proposition XV.3, a final leaf apparently being lost. Fortunately the rest of this version is contained in Paris, BN latin 16648 (See the list of MSS below).

The other twelfth century MS, Digby 174, is written in a difficult, crabbed hand without much care. It is the only MS containing the introduction to Version III, but it is incomplete, containing only eleven books. It will be noticed from examining Appendix I (3) that proposition I.1 follows immediately after the introduction, thus omitting the definitions, postulates, and axioms. Whether their omission was the fault of the scribe of Digby 174 or the original intention of Adelard, we do not know. But it should be noticed that in Digby 174 the enunciations are not given in full. Ordinarily they include just enough words so that the enunciation as given in Version II can be identified. The proofs, however, are complete, and are, for the most part, like those of Balliol 257. The rather confused folio pagination of the various parts of Version III, and the accompanying section from Gerard of Cremona's translation, also contained in Digby 174, is given below in the list of extant MSS.

Version III B given in part in Appendix I (5) is probably a commentary of the thirteenth century, although it agrees with the twelfth century Digby 174 in proposition I.1. It needs further study before its position is clarified.

INCIPIT: (of introduction) "Geometrie sicut et reliquarum. . . ." See Appendix I (3). (Of text in Balliol 257) "Punctus est illud cui pars non est. . . ." (Of text in Digby 174) "*Triangulum equilaterum etc.* Data dicuntur quibus equalia. . . ."

EXPLICIT: The *explicit* and colophon of Version III from BN latin 16648, f. 58r runs as follows: ". . . ergo pentagone bases sunt equilaterum, quod exigit duodecedros unum propositum tam propositi theorematis quam totius geometrice confirmatur. Explicit edit[i]o alardi bathoniensis in geometriam euclidis per eundem a. bathoniensem translatam."

EXTANT MSS:

1. Oxford, Balliol College 257, 12c, 2r–98v, Books I–XV.3

2. Oxford, Digby 174, 12c, 99r–132v, Books I–XI

    139r–145r repetition of 125r–132v

    146r–153r Books I–V.2 (the same translation with the propositions completed)

    154r–159v VI.11–X (VII.–X without proofs)

    160r–173v Gerard translation, which see.

3. Oxford, Bodl. D'orville 70, 13c, VII.7–X.36

4. Paris, BN lat. 16648, 13c, 2r–58r, X–XV

5. London, BM Burney 275, 14c, IX–XV.3 (See discussion of Versions I and II for other parts of Burney 275)

6. London, BM Royal 15, B.IV, 13c, X.95–XI.37

B. OTHER TRANSLATIONS OF THE *ELEMENTS* FROM THE ARABIC

(See Appendix II)

I. The Translation of Hermann of Carinthia

In addition to Adelard of Bath's close translation of the *Elements* as contained in Version I there were at least two others and perhaps more translations made from the Arabic. The first of these translations is that appearing in Paris BN Latin 16646. Birkenmajer has suggested that this MS which was left by Gerard d'Abbeville, archdeacon of the church at Amiens, to the Sorbonne in his legacy of 1271 is identical with item 37 of the *Biblionomia* composed by Richard de Fournival at Amiens some time close to 1246.[21] This item runs as follows: "37. Euclidis geometria, arismetrica, et stereometria ex commentario Hermanni secundi, in uno volumine cuius signum est littera D."[22] "Hermannus secundus" was the famous translator, Hermann of Carinthia.[23] Undoubtedly the word *commentarius* is used in this citation in the same fashion as the word *commentum* applied to Adelard's translation of Version I, that is, to single out Hermann's translation of the proofs. I think we can be reasonably sure that BN Latin 16646 is a translation or translation-paraphrase made with the Arabic text in the editorial hand, for the translator uses a number of Arabic transliterations not present in the versions of Adelard, such as *aelmam geme* (for *scientia communis*, i.e., an axiom), *almukadimas* (for *theoremata*), *chateti* (for *linee*), etc. The reader will notice all of these Arabicisms in the selection from Hermann's translation given in Appendix II (1). The last of these expressions is particularly indicative of the fact that we are dealing with a translation, for it occurs in the heart of the proof of *I.1* where the translator is obviously careless, and so instead of the Latin word he ends up with a transcription of the Arabic which he gives as *chateti* (from Arabic *khatt*, linea).

The same situation occurs in VI.19 (BN 16646, 39v) where we see in the middle of the proof "mutekefia i.e., mutue." The usual translation for *mutekefia* is *proportionalia*. At the end of the same proof we find the Arabic phrase in talking about similar figures: "Ale chelkatu wa tahtit, i.e., in creatione et lineatione." This same phrase is also found without proof in the enunciation of VI.31 (f.43v) except that we have the transliteration "chelka*t*a." This same Arabic phrase had also been translated by Adelard in Version II by the phrase "in creatione et lineatione," but of course Adelard does not give the Arabic as does Hermann. Finally, we should notice that Hermann concludes Book IX (f.64r) with the common Arabic phrase: "Wa delicah me aradene en nebeienne wa hed hedatu," which we could render in Latin: *Et hoc (est) quod voluimus demonstrare et finis explicit*. Some such similar expression terminates almost every proof in the Arabic texts of Euclid.

There can be no doubt that the translator had at least Version II of Adelard before him, for the definitions, axioms, postulates and enunciations are either drawn directly from Version II or are quite close in wording. It may be that Hermann also used Version I, but, if so, he did not copy directly from it, but used it only as an auxiliary. The translator probably was also familiar with the Boethian excerpts, and perhaps even some Greek. Not only does he borrow an occasional Greek term such as *rumbum* from the

---

[21] See Birkenmajer's article "Biblioteka Ryszarda de Fournival," in *Rozprawy* of the Cracow Academy, 60, No. 4 (1922), pp. 49–52, as noted by Haskins, *op. cit.*, p. 50. Cf. *Isis*, 5, p. 215.

[22] L. Delisle, *Le Cabinet des manuscrits de la Bibliothèque Nationale*, 2, 1874, p. 526. See also the Sorbonne Catalogue in Volume 3, p. 68, Item No. 48, where it is marked as a gift of Gerard d'Abbeville.

[23] On Hermann, see Haskins, *op. cit.*, Chap. III.

Boethian text, but he uses the Greek *zeta* to render the Arabic *zay*, apparently assuming that the *zay* originally rendered the *zeta*.

The Hermann translation seems to have been made from the same basic text as that of Version I, possibly the al-Hajjāj text. However it contained the various definitions of the parts of a proof missing in Version I, but present in the Introduction of Version III. Furthermore the extant al-Hajjāj text uses for axioms "al-ʿalūm al-mutaʿarifat" instead of Hermann's "aelmam geme" (i.e., ʿilm jamīʿ?). But the Hermann proofs follow the al-Hajjāj fairly closely, although not so closely as do those of Adelard I.

Finally, it should be observed that the single extant copy of this translation contains only twelve books.

INCIPIT: "Septem sunt omnis discipline. . . ." (See Appendix II (1)).

EXPLICIT: (Book XII) "Omnium sperarum que diametrorum suorum proportio triplicata est . . . reliquantur sperarum que diametrorum proportio triplicata (108r)."

EXTANT MS: Paris, BN latin 16646, 13c, 1r–108r, Books I–XII.

II. The Gerard of Cremona Translation

It has been known for some time that Gerard of Cremona made a translation of Euclid's *Elements*, for we are so informed by his *Vita*.[24] But it was not until Björnbo discovered in 1901 a Vatican MS (Reg. Lat. 1268) which contained Books X–XV in a translation differing from any of the versions he knew that the first extant trace of the Gerard translation was found.[25] Later in 1904 he found other complete copies of the translation in Paris, Boulougne, and Bruges MSS, and certain fragments at Oxford.[26] Although none of the various manuscripts (see the list below) bears the name of Gerard, Björnbo demonstrates, on the basis of characteristic translating vocabulary as exhibited in other translations of Gerard, that this version of the *Elements* is to be attributed to Gerard of Cremona. We are to suppose that Gerard of Cremona did his version from the so-called Ishāq ibn Hunain — Thabit ibn Qurra text rather than the Hajjāj text.[27] At least Gerard's text is quite different from that used by Adelard in Version I, and is in fact a great deal closer to the best Greek tradition. To be sure, we may not be confronted here with the pristine Ishāq-Thabit text, but I suspect it generally follows that text. Presumably, in accordance with the Arabic style, Ishaq and Thabit made their additions in the third person so as to contrast them with other readings. For example, in numerous places the text has some such phrase as the following: (BN Lat. 7216, f. 16r): "Thebit dixit: Inveni in alia greca scriptura aliam probationem. Et ipsa est. . . ." (Cf. 10r, 15v., 20r., 28v. and *passim*.) Similarly in connection with IX.20 and IX.21 we read (f. 54v): "Figuram que invenitur tricesima et figuram que est tricesima prima non inveni in

---

[24] F. Wüstenfeld, "Die Uebersetzungen arabischer Werke in das Lateinische," *Abhandlungen der K. Gesellschaft der Wissenschaften d. Göttingen, 22* (1877), p. 59, where the Euclid translation is the fourth item.

[25] A. A. Björnbo described his discovery of the Vatican MS in the *Abhandlungen zur Gesch. der math. Wissenschaften, 14* (1902), pp. 138–142.

[26] A. A. Björnbo, "Gerhard von Cremonas Uebersetzung von Alkwarizmis Algebra und von Euklids Elementen," *Bibliotheca Mathematica*, Dritte Folge, *6* (1905), 239–248. Cf. M. A. Kugener, "Les versions latines des 'Elements' d'Euclide" etc.," 2ᵐᵉ *Congrès National des Sciences, Comptes Rendus, 1* (1935), pp. 70–72. Kugener mentions that Mlle Baudoux has transcribed Book I of Gerard's version, but so far as I know this has not been published. He also examines another version of the *Elements* in

Bruges 529, which he tends to identify with a rendering of the Ishaq text. It contains only eight books. My guess without having examined it would be that it is another copy of Version I of Adelard.

[27] Two Arabic MSS (Bodlean 279 and 280) contain the Thabit version and I hope in the future to compare these with the Gerard of Cremona text. That the Gerard text is much closer to the best Greek tradition than the Adelard or Hermann translations is quite evident. In Book I the Gerard text contains 48 propositions instead of 47. In specific propositions like II.14 it substitutes a much better tradition than the text used by Adelard. Furthermore, the Gerard text contains the introduction of Hypsikles to Book XIV, and it is clear in the Gerard text that only Books I–XIII are by Euclid.

28

aliqua ex scriptures grecis que reperiantur, sed inveni eas in libro de numeris, dixit thebit, ex eis que prediximus manifestum est quod. . . ." And in Book X (f. 62r): "Refert thebit qui transtulit hunc librum de greco in arabicam linguam se invenisse quod additur ante figuram tricesimam primam huius partis in quibusdam scriptis graecis cuiusdam Joanitii babiloniensis, quod tamen non est de libro." This may be a later editor talking rather than Thabit. Similarly, a later editor appears to be speaking in 64r: "Hic additur quedam figura que non est huius libri neque invenitur in translatione thebit, sed quia est necessaria. . . ." Notice finally the reference to Ishāq: (f. 100v): In scriptis que transtulit ysaach hoc quod sequitur invenit post figuram vicesimam primam libri euclidis in qua ponuntur. . . ." I hardly need point out that *third* person references to Thabit or Ishāq do not necessarily mean that the text is later than Thabit.

I have discovered one interesting reference to the Gerard translation in a fourteenth century MS of the British Museum, Harleian 5266. It is a copy of the Campanus version, but has numerous marginal notes, in some of which he refers to the Gerard translation anonymously as *alia translatio* (see particularly folio 1v where the scribe gives the Cremona rendering of the parallel postulate). The scribe also makes reference to al-Narizi's commentary translated by Gerard of Cremona (e.g., on f. 6r where the scribe says "hanalitius in commento 23 ponit aliam figurationem. . . .").

We should also notice that the introductory material in this translation of Gerard is found inserted in a multiple commentary from the Arabic, a fragment of which is contained in Paris BN 7215, f. 4r.

INCIPIT: The *incipit* (BN lat. 7216, 1r): "Ea a quibus procedit scientia . . . (See Appendix II (2)). Or it is possible that the *incipit* should rather be "Omnium diversarum scientiarum diverse sunt conclusiones . . . ," as contained in BN 7215, for that latter MS contains several lines later the whole introductory passage contained in the Gerard translation.

EXPLICIT: The *explicit* for the text (Bk. XV.5) is as follows: ". . . erit habens duodecim bases penthagonales equilaterum et equiangulorum. Et ilud est quod demonstrare voluimus. Explitus est liber Euclidis simul cum duabus partibus a asiculo editis, cuius partes fuerint quindecim." (BN 7216, 107r) It is obvious from the colophon that it was recognized in the Arabic text used by Gerard that the last two books were not by Euclid, but were believed to be by Hypsicles. Following the main text there are additional proofs which begin (107v): "Demonstratio de composita propositione quam Euclidis in quinti libri prohemio. . . . Et trigonus *gzu* erat equalis trigono *m*. Et trigonus ergo *m* est equalis trigono *fgh*. Et hoc est quod demonstrare voluimus."

EXTANT MSS:

1. Paris, BN lat. 7216, 15c, 1r–108r, Books I–XV and scholia.
2. Boulougne-sur-Mer, Bonien. 196, 14c, 1r–148r, I–XV and scholia.
3. Bruges, Bibl. publ. cod. 521, 14c (See Laude catalogue p. 452–453).
4. Rome, Vatican Reg. lat. 1268, 14c, 92r–142v, X–XV and scolia.
5. Oxford, Bodl. Digby 174 (end of 12c), 162r–173v, XI.2–XIV.3

III. The Translation of John Ocreat

For a possible translation from the Arabic of the *Elements* by Adelard's student John Ocreat, see the discussion of Adelard's Version II above.

CONCLUSION

In addition to the above mentioned translations of the text of the Euclid, we should note finally that portions of the text were translated in the various commentaries translated from the Arabic. These include (*1*) the lengthy commentary of Al-Narizi

(Anaritius) translated by Gerard of Cremona,[28] (*2*) the commentary on Book X believed to be by Muhammad ibn Abdulbāqī al-Baghdādī,[29] probably also translated by Gerard of Cremona, (*3*) and a fragment of the commentary on Book X by Pappus translated in all probability by Gerard of Cremona.[30]

We have now completed our preliminary description of the various translations of Euclid's *Elements* from the Arabic made in the twelfth century. We have left for future presentation a series of Latin commentaries on the *Elements* composed in the thirteenth and fourteenth centuries, including the celebrated version of Campanus of Novarra.[31] We can point out here, however, that the Campanus text is in all likeli-

---

[28] The text has been published by M. Curtze in J. L. Heiberg's Teubner edition of the *Opera omnia* of Euclid as a *Supplementum*, Leipzig, 1899. He used Cracow MS 569, 14c, pp. 7–80. We can cite in addition Vat. Reg. lat. 1268, 13 or 14c, ff. 144–205, and the abridgement of Oxford, Bodl. Digby 168, 14c, 124r–125r. A multiple commentary based quite largely on the Anaritius commentary is found in Paris, BN latin 7215, at least for the definitions of Book I. The introduction takes a paragraph from the introduction to Gerard's translation of the Thabit text. The text of the definitions in BN 7215 is given in the form of Adelard's Version II.

[29] This commentary was published by Curtze as the last part of the al-Narizi commentary in the *Supplementum* cited in Note 28, pp. 252–386. Its probable composition by Muhammad ibn Abdalbāqī has been argued in two articles by Heinrich Suter in the *Bibliotheca Mathematica*, Dritte Folge, 4 (1903), pp. 22 et seq.; 7 (1907), 234–51. The second article contains some corrections to the Curtze edition. This work is identical with the *De numeris et datis*, translated by Gerard of Cremona and published by B. Boncompagni, 1863–4. Furthermore it is contained in the Gerard corpus of Paris, BN latin 9335, 92v–110v.

[30] Appears in Paris, BN Lat. 7377A, ff. 68r–70v. See *Bibliotheca Mathematica*, Dritte Folge, 4, p. 25.

[31] We can at this point list some of the later paraphrases and reworkings of the twelfth century texts. In almost all cases they use the enunciations of the propositions contained in Version II of Adelard.

(1) British Museum, Sloane MS 285. Uses def. post. axioms, enunciations from Version II, but from Books I–VI.4 has different proofs. From VI.4 this is straight Version II. The catalogue calls this 14th century, but it looks more like 13th century to me. For identification purposes I give the proof of I.1: "Facta dispositione figure per secundam petitionem bis sumptam, 'a puncto et cetera,' per primam conceptionem constat propositum, quia 'que uni et eidem et cetera.'" As this example shows, the proofs are abridged in the manner of Version II.

(2) Paris, BN Latin 7374, 13 or 14c. From I. Def. through I.46 this is straight Version II. Then in I.47 it suddenly shifts to a completely different version. For identification, I give the beginning and end of I.47 (*proof:*) "Sit triangulus *abc*, itaque quadratum huius lateris sit tanquam quadrata aliorum duorum laterum. Dico ergo quod iste angulus *bac* sit rectus. Ducatur enim ab *a*. . . . Sed iste angulus *dac* est rectus, et ita habemus quod habere proposuimus." Although the author has used Version II, he has probably

gotten the content of the proofs from Version I, reworking them. This version goes through definitions in Book X, ending ". . . . Quicunque nesciret tibi respondere sic ut daret tibi vacam albam."

(3) Bodleian Library, D'Orville 70, 13 or 14c?, Books I–X.36 (From X.36–X.2 is on Version I). This is the same version which is contained in Vienna, Nat. Bibl. 83, 14c, 39r–65v. Its enunciations are in the form of Version II. For identification, I give the proof of I.1: "Sit *ab* linea data, duo eius capita centra sint circulorum eam occupantium . . . Quare linee *ac* et *bc* sunt equales, per primam animi conceptionem. Constat ergo propositum."

(4) Version of Campanus of Novara. The MSS are legion and their listing must await a separate treatment of the Campanus question. For published editions, see note 9. For the text of a few of the propositions compared with Adelard II, see the article by Weissenborn cited in note 16. For identification, I give the beginning and end of the proof and commentary of I.1 (the enunciation is as are most of them, in the form of Adelard's Version II): "Esto data linea recta *ab*. Volo super ipsam triangulum equilaterum constituere; super alteram cuius extremitatem. . . . Sic igitur super datam lineam rectam, omnes triangulorum species collocavimus." The explicit of the whole work (XV.13) is as follows: ". . . . Quare assignato corpori constat nos speram, quemadmodum propositum erat, inscripsisse." It is worth pointing out that Gerard has probably used Gerard of Cremona's translation of Anartius' commentary on the *Elements*, for he adds two additional demonstrations to I.1 which appear in Latin in that commentary, although they ultimately go back to Hero, and are given in Proclus' commentary in Greek.

(5) I feel sure that we should also place Version IIIB which we have discussed above and several propositions from which we have given in Appendix I (5) in the thirteenth century. The first five books of this are contained in Bodleian Library, Saville 19, 13c. From Book VI.1 on we have a straight Campanus text. The first five books were obviously influenced by Adelard's true Version III. The selections given in the Appendix can be used for identification purposes.

(6) In Oxford, MS C.C.C. 234 of the 15th century, ff. 10r–170r, we have following each enunciation (in Adelard II) two "commenta." The second one is that of Campanus. The first one, constituting still another version, is unknown. The proof for I.1 in this first *commentum* runs: "Detur linea *ab*, est itaque propositum auctoris, super *ab* lineam triangulum equilaterum collocare. Quod sic fiet: Fiat *a*

hood a paraphrase and commentary rather than a new translation from the Arabic. It has always puzzled students of the Euclid question that there is no original Arabic or Greek text even remotely like the Campanus version in actual wording. But the principal reason they considered it a new translation was that although it used the enunciations from Adelard (i.e., from his Version II), the demonstrations were full and complete in the manner of the text of Euclid rather than abbreviated like those of Version II with which it was compared. Hence it was thought that Campanus could not have started from the indications of proofs in Version II and developed completed proofs so much like Euclid in substance if not terminology without having made some kind of translation from the Arabic. But we have shown in this article that numerous versions with the complete proofs circulated long before the time of Campanus (e.g., Versions I & III of Adelard and the translations of Hermann and Gerard). And so it is much more likely that Campanus merely paraphrased one of the earlier complete versions (and as yet I am not sure which one it was), adding much material of his own foreign to any version we know of in Latin, Greek, or Arabic.

Needless to say, it would be helpful to the student of medieval mathematics to have the whole corpus of Euclid translations and commentaries in modern edition. For at the present time there is no edition of any of the medieval versions of Euclid, except the early editions that exist of Campanus' text. A corpus of *Euclides Latinus* is then a pressing desiderandum of medieval scientific history.

APPENDIX I

(1)	(2)
ADELARD OF BATH — VERSION I	**ADELARD OF BATH — VERSION II**
(Oxford, Trinity College MS 47, 139r–180v. See text above for proper order of leaves.) Institutio artis geometrice ab Euclide descripta XV libros continens, per Adelardum Batoniensem ex Arabico in Latinum Sermonem translata incipit.	(Brit. Mus. Royal A. 15. 27, 1r et seq.; Oxford Balliol 257, 2r–v; Oxford, Trinity 47, 104v et seq.; BM Addit. 34018, 1r et seq.; and Oxford, Corp. Christi 251, 22r et seq.) Artis Geometrice per Adelardum Bathoniensem ex arabica lingua in latinam translate liber primus euclidis incipit.

(*I. Def.*) Punctus est illud cui pars non est. Linea est longitudo sine latitudine, cuius extremitates quidem duo puncta. Linea recta est ab uno puncto ad alium extensio, in extremitates suas utrumque eorum recipiens . . . Alia equilatera quidem est, sed rectangula non est, et dicitur elmuain . . . Que cumque vero preter has quas exposuimus quadrilatere fuerint vocabuntur irregulares. Equidistantes linee sunt que in una superficie collocate, et in aliquam partem protracte, non coniungentur, etiam si in infinita protrahantur.	(*I. Def.*) Punctus est cui (*or* cuius) pars non est. Linea est longitudo sine latitudine, cuius extremitates quidem (sunt) duo puncta. Linea recta est ab uno puncto ad alium extensio, in extremitates suas utrumque eorum recipiens . . . Alia (est) elmua(h)in, estque (*or* quod est) equilaterum, sed rectangulum non est . . . preter has autem omnes quadrilatere figure elmunharife nominantur. Equidistantes linee sunt que in eadem superficie collocate atque in alterutram partem protracte non convenient, etiam si in infinitum protrahantur.

punctus centrum et secundum quantitatem *ab* linee describatur circumferentia, ex secunda petitione. . . . Habemus enim triangulum equilaterum supra datam lineam collocatum." (7) Paris, BN Latin 7292, 15c, 188r–267v. Enunciations borrowed from Adelard II. I.1 (*proof:*) "Ad huius propositi demonstrationem, notanda est tertia petitio . . . omnia latera sunt equalia et ita ille triangulus est equilaterus." For further identification, we can cite the proof of I.46: "Fiat triangulus *abc* lineis clausus, et sit *b* rectus angulus. Dico quod quadratum factum . . . super *ab* latus. Et sic illud totum quadratum est equale aliis duobus, quod erat propositum." (8) Still another commentary appears in the margins of the 13th century Bodl. F.5.28, ff.	ii–15r, whose main text is a copy of Adelard's Version II. I would date this before 1250. It is not a full commentary, but sometimes it specifies the proof by giving letters where the proof is only indicated or abridged in Version II. Also it occasionally begins a marginal comment with the expression: "Mens hiuius propositionis est." Thus it has for I.1: "Mens huius propositionis est quod super quamlibet lineam rectam potest collacari triangulus equilaterus . . . et sic sumptur(?) in libro de datis." For further purposes of identification we can note the marginal specification for II.1: "Sic *cdef* est equale: *cgek* et *ghkt* et *hdtf*. Sed *cdef* est id quod fit . . . est equale his que fiunt ex ductu *ce* in quamlibet partem *cd*."

(*I. Post.*) Peticiones quinque. Intendenti autem mensurare, quinque prescire necessarium est: A quolibet puncto in quemlibet punctum rectam extrahere lineam. Assignatamque lineam rectam quantolibet spacio directe protrahere . . . Si linea recta supra duas lineas rectas ceciderit, duoque anguli ex una parte duobus rectis minores fuerint, illas duas lineas ex illa parte protractas proculdubio coniunctum iri. Item duas lineas rectas planum non continere.

(*I. Post.*) Peticiones autem (sunt) quinque. A quolibet puncto in quemlibet punctum rectam lineam ducere, atque lineam definitam in continuum rectumque (*or* et directum) quantumlibet protrahere . . . Item si linea recta supra duas lineas rectas ceciderit, duoque anguli ex una parte duobus rectis minores fuerint, illas suas lineas in eam partem protractes (proculdubio) coniunctum iri. Item duas lineas rectas superficiem non (*or* nullam) concludere.

(*I. Axioms*) Scientia universaliter communis. Si fuerint alique due res alicui equales, unaqueque earum erit equalis alteri . . . Omne totum sua parte maius est. (In marg: Nota quod multas communes scientias pretermisit. Verbi gratia, si due quantitates . . . multitudo enim in infinitum crescit.)

(*I. Axioms*) Communes vero animi conceptiones (*or* communes scientie) sunt hec: Que eidem (*or* Que cunque uni et eidem) equalia sunt et sibi invicem sunt equalia . . . Omne totum sua parte maius est. Nota quod (*or* quia) multas communes scientias pretermisit Euclides, que (*or* quoniam) infinite sunt (et innumerabiles), quarum hec est una, si due quantitates . . . multitudo quippe crescit in infinitum.

(*I.1*) Nunc demonstrandum est quomodo superficiem triangulam equalium laterum super lineam rectam assignate quantitatis faciamus. (*Proof:*) Sit linea assignata *ab*. Ponaturque centrum supra *a* occupando spacium quod est inter *a* et *b* circulo, supra quem *gdb*. Item ponatur supra centrum *b* occupando spacium inter *a* et *b* circulo alio, supra quem *gah*. Exeantque de puncto *g* supra quem incisio circulorum due linee recte ad punctum *a* et ad punctum *b*. Sintque ille *ga* et *gb*. Dico quod ecce fecimus triangulum equalium laterum supra lineam *ab* assignatam. Rationis causa. Quia punctum *a* factum est centrum circuli *gdb*, facta est linea *ag* equalis linee *ab*. Et quia punctum *b* est centrum circuli *gah*, facta est linea *bg* equalis linee *ba*. Sicque unaquaque linearum *ga* et *gb* equalis linee *ab*. Equalium autem uni rei unumquodque equale alteri. Itaque linee tres *ag* et *ab* et *bg* invicem equales. Triangulus igitur equalium laterum *abg* factus est supra lineam *ab* assignatam, et hoc est quod in hac figura demonstrare intendimus.

(*I.1*) Triangulum equilaterum supra datam lineam rectam collocare. (*Proof:*) A duobus terminis date linee ipsam lineam occupando cum circino duos circulos sese invicem secantes describe. Et ab ipsa communi sectione circulorum ad duos terminos linee proposite duas lineas rectas dirige. Deinde igitur ex circuli descriptione argumentum elice.

(*I.47.*) Omnis trianguli a cuius aliquo latere in se ipsum ducto quadratum constitutum duobus quadratis constitutis ex reliquis lateribus in se ipsa ductis equale fuerit, angulum lateri illi oppositum rectum esse necesse est. (*Proof:*) Sit triangulus *abg*. Sitque latus *bg* in se ipsum ductum. Sitque quadratum ex eo constitutum sicut duo quadrata constituta ex duobus reliquis lateribus *ba* et *ag* in se ipsa ductis. Dico quod angulus *bag* rectus. Rationis causa. Extrahatur enim a linea *ag* a puncto *a* supra angulum rectum linea usque ad *d* equalis linee *ab*. Iungaturque *d* cum *g*. Quoniam ergo *ba* sicut *ad* erit quadratum quod fiet ex ductu *ba* in se ipsum sicut quod fiet ex ductu *ad* in se ipsum. Sit autem quadratum quod ex *ag* in se ipso (!) ducto commune. Duo itaque quadrata que fiunt ex lineis *ba* et *ag* ductis in se ipsas duobus quadratis que ex *ad* et *ag* ductis in se ipsas equalia. Atqui duo quadrata que ex *ad* et *ag* ductis in se ipsas sicut quadratum quod ex *gd*

(*I.47.*) Si quod ab uno trigoni (*or* trianguli) latere in se ipsum producetur equum fuerit duobus (quadratis) que a duobus reliquis lateribus describuntur, rectus est (*or* erit) angulus cui (etiam) illud latus (*or* latus illud) opponitur. (*Proof:*) Si enim ab angulo controversie linea recta ortogonaliter ducatur equalis ei cui directe adiacere videtur, et adiacet quidem ut (quadrata) earum equivalentia (*or* equalia) esse convincantur. Deinde ipsi angulo recto basis subtendatur a termino adiecte linee ad angulum dati trigoni. Et deinde quadrato perpendicularis secundum continuum ex premissa propositione et triplici propositione, scilicet: Unum quadratum duobus equum esse. Itemque lineam ab angulo ortogonaliter ductam esse. Itemque ei cui adiecta est equalem esse. Atque ex octava propositione angulus rectus esse convincetur ex necessitate.

32

ducto in se ipsum. Angulus enim *dag* rectus. Duo vero quadrata que ex *ba* et *ag* in se ipsas ductis consituuntur sicut quod ex *bg* in se ipsum. Quare quod ex *bg* in se ipsum equale erit quadrato quod ex *gd* in se ipsum. Sicque latus *bg* equale lateri *gd*; latus quoque *ba* equale erat lateri *ad*. Sitque latus *ag* commune. Erunt quoque *ba* et *ag* sicut *da* et *ag*, unumquodque sicut respiciens se. Sed erit basis *bg* sicut basis *gd*. Angulus ergo *bag* sicut angulus *gad*. Angulus autem *gad* rectus. Angulus itaque *bag* rectus. Sic igitur ostensum est quod omnis trianguli cum ex ductu alicuius suorum laterum in se ipsum quadratum constituitur, si quadratum illud sicut duo quadrata ex ductu reliquorum duum laterum in se ipsa, angulum trianguli a duobus illis lateribus continentum rectum esse necesse est. Et hoc quod demonstrare proposuimus.

(*II.Def.*) Omnem superficiem equidistantium laterum rectangulam necesse est duabus lineis rectum angulum continentibus concludi. Omnis superficiei equidistantium laterum . . . Quod si una earum cum duabus complentibus coniungatur, que ex utraque parte diametri cadunt omnia ista elaalem vocabuntur. (*In marg*: elaalem, arabice; gnomono, grece; vexillum vel signum, latine.)

(*II.Def.*) Omne parallelogramum rectangulum sub duabus lineis angulum rectum ambientibus dicitur contineri (*or* continetur). Nota (quod) parallelogramum idem esse quod superficiem equidistantium laterum. Nota quoque quod nos gnomonem id arabes elaalem (*or* elkalem) dicunt. Nota (quod) de solo rectangulo producto his ubique agit. Omnis parallelogramum spacii. . . . cum duobus supplementis gnomo nominatur.

(*II.1*) Si fuerint due linee quarum una in quotlibet partes dividatur, erit illud quod ex ductu unius in aliam fiet sicut illud quod ex ductu linee indivise in omnes partes linee particulatim divise. (*Proof:*) Sint linee: linea *a* et linea *bg*. Quarum una dividatur, sitque *bg*, supra duo puncta *d* et *h*. Dico quia quod fiet ex ductu linee *a* in partem *bd* et in partem *dh* et in partem *hg* erit sicut illud quod fiet ex ductu *a* in *bg*. Rationis causa. Producatur enim de linea *bg* de puncto *b* linea supra rectum angulum sicut linea *a*, sitque *bz*. Extrahaturque de puncto *z* linea equidistans linee *bg*, sitque *zH*. Extrahaturque de puncto *d* et *h* et *g* linee equidistantes linee *bz*, sintque *dt*, *hk*, *gH*. Manifestum erit quia superficies *zg* sicut tres superficies *zd* et *th* et *kg*. Omnes vero he superficies equidistantium laterum sunt et rectorum angulorum. Atqui superficies *zg* facta est ex ductu linee *a* in lineam *bg*. Continent enim eam linee *bz* et *bg*, lineaque *zb* equalis linee *a*. Atqui superficies *zd* facta est ex ductu *a* in *bd*. Continent enim eam linee *zb* et *bd* et linea *bz* equalis linee *a*. Atqui superficies *th* facta est ex ductu linee *a* in lineam *dh*. Continentque eam linee *td* et *dh*, et linea *td* equalis linee *a*. Atqui superficies *kg* facta est ex ductu linee *a* in lineam *hg*. Continentque eam linee *kh* et *hg*, lineaque *kh* equalis linee *a*. Sic igitur manifestum est quia quod factum est ex ductu linee *a* in lineam *bg* equalis est illis que facta fiunt ex ductu linee *a* in lineas *bd* et *dh* et *hg*. Et hoc est quod demonstrare intendimus.

(*II.1*) Si fuerint due linee quarum una in quotlibet partes dividatur, illud quod ex ductu unius earum in alteram fiet, equum erit hiis que ex ductu linee indivise in unamquamque partem linee particulatim divise rectangula producentur. (*Proof:*) Si enim a terminis linee divise ortogonaliter linee ducantur ei que indivisa erat equales. Atque a punctis sectionum hiis equales et equidistantes per primam descriptionem (que habet dicitur contineri quod propositum est evidentissimum fit) (*or replacing* que . . . fit *is the phrase* quid est sub lineis contineri).

ADELARD OF BATH — VERSION I

(*II.14*) Nunc demonstrandum est quomodo superficies quadrata equalis superficiei trianguli assignati fieri queat. (*Proof:*) Sit triangulus assignatus *abg*, cui cum equalem superficiem quadratam facere voluerimus. Superficiem equidistantium laterum rectangulam faciemus triangulo *abg* equalem, sitque *dH*. Si itaque fuerit *dh* sicut *hH*, facta est superficies quadrata equalis triangulo *abg* sicut voluimus. Quod si ita non fuerit, sit *dh* longius *hH*. Extrahaturque linea *dh* directe usque ad punctum *t*. Sitque *ht* sicut *hH*. Dividatur itaque linea *dt* in duo media supra punctum *k*. Designeturque supra lineam *dt* semicirculus supra quem *dlt*. Protrahaturque linea *Hh* directe ad punctum *l*, iungaturque *l* cum *k*. Sed linea *dt* iam divisa est in duo media supra punctum *k*, atque erit in duas partes inequales supra punctum *h*. Quod ergo ex ductu *dh* in *th* et ductu *hk* in se ipsam sciut quod ex ductu *kt* in se ipsam. Atqui *kt* sicut *kl*. Quod itaque ex *dh* in *ht* et ductu *hk* in se ipsam sicut illud quod ex ductu *kl* in se ipsam. Quod vero ex ductu *kl* in se ipsam sicut quod ex ductu *kh* et *hl* in se ipsas, quoniam angulus *khl* rectus. Quare quod ex ductu *dh* in *ht* et ductu *kh* in se ipsam sicut illud quod ex ductu *kh* et *hl* in se ipsas. Quod autem ex ductu *kh* in se ipsam commune abiciatur. Quare quod ex ductu *dh* in *ht* sicut quod ex ductu *hl* in se ipsam. Quod autem ex ductu *dh* in *ht* est superficies *dH*, quia *ht* sicut *hH*. Quod itaque ex ductu *hl* in se ipsam sicut superficies *dH*. Sed *dH* sicut triangulus *abg*. Quod ergo ex ductu *hl* in se ipsam sicut triangulus. Manifestum igitur est, quia triangulus *abg* sicut superficies quadrata, que est ex ductu *hl* in se ipsam. Et hoc est quod in hac figura demonstrare intendimus.

ADELARD OF BATH — VERSION II

(*II.14*) Dato triangulo equum quadratum describere. (*Proof:*) Dato trigono equum parallelogramum rectangulum designabis per 42 primi. Deinde longiori lateri equale brevioris in directum (*or* directe) adicies; atque tunc circa (*or* super eam) totam semicirculum describes; breviusque latus parallelogrammi ad circumferentiam directe protrahes et a centro ad idem punctum circumferentie ypotenusam ducto (*or* duces); atque ex 5 secundi et penultima primi argumentum elice. Nota quoque quod hinc inveniri potest latus tetragonicum, (quod dicunt elgydar, cuius libet parte altera longioris forme.) (*In MS Trinity 47 for the phrase* quod . . . forme *we have* quod apud arabes caldaice elgidher dicunt.)

---

(3)

ADELARD OF BATH — VERSION III

Introduction (Contained only in Oxford, Bodl. Digby 174, f. 99r–v): Geometrie sicut et reliquarum facultatum usus suum antecessit artificium. Cuius usus apud egiptios inolevit. Horum limitates agrorum enim nili superfluens singlis annis obducens, invenerunt quasdam ceteras mensurandi rationes quibus mensure prioris agros recuperarent. Has itaque rationes philosophi, in declarantes (?) et aprobantes (?) de superiori occasione mensurandi operam, adhibuer(unt?). Sed propter rei difficultatem singuli citra perfectionem operis, subsistentes, quidam in talia, ut pythagoras, th⟨ales⟩, ⟨amer⟩istus, et archytas, alii progressiva (!) ut plato et aristoteles et alexander instruxerunt, quousque tandem Euclides confirmationem et ordinem superadieciens artem positam contexuit et edidit. Et sic invidimenta (?) signam.

Vis in distinctiones ex ordio hec preconsideranda proponimus, scilicet, quid sit ipsa ars, que partes, qui nomen, ⟨que⟩ causa nominis, qui genus, quis artifex, quid officium, quid instru-

34

mentum, que intentio, que materia, que utilitas vel finis, quis modus, quis ordo, quis titulus. Est igitur ars posita magnitudinis immobilis scientia secundum rationabiles figuras contemplatas (!). "Immobilis magnitudo," i.e. quantitas [sine] motu considerata, ad oppositum astronomie que mobilem magnitudinem, i.e. spacium, secundum varia, cum motu considerat. "Secundum rationabiles figuras," addam helmurife et aliarum irrationalium figuras quas geometria nequaque attendit.

Partes huius artis distinguuntur secundum partes sue materie, scilicet, lineam, superficiem, corpus, et numerum; vel secundum xv distinctiones factas ab ipso auctore; factas, inquit, et dictas secundum distinctionem principiorum et dictas modo agendi.

Nomen vero est geometria, a "ge" qui est terra et "metrus" qui est mensura. Causa nominis sicut dicta est in prima instructione, i.e., agrorum mensuratione.

Genus consideratur tripliciter, secundum effectum, secundum continentiam, et secundum naturalem suppositionem. Genus secundum effectum est mensuratio. Est enim eius genus, i.e., effectus, qui reddit suam artificem examinantem et mensurantem. Secundum continentiam sic genus eius est mathematica ad habitus oriendum. Scientia dividitur in sapientiam et eloquentiam. Eloquentia in grammaticam, dialectam, rhetoricam. Sapientia in theoricam et practicam. Practica in ethicam, mathematicam (!), et practicam liberalem. Ethica in monasticam, economicam, et politicam. Mathematica (i.e., mechanica?) dividitur in lanificium, armaturam, agriculturam, navigationem, venationem, medicinam, et theatriam. Practica liberalis dividitur [sicut] in artes liberales, singularis (ars) enim singulatim habet practicas. Theorica dividitur in theologiam, phisicam, et mathematicam. Mathematica in arismeticam, geometriam, astronomiam. Ecce mathematica dividitur in geometriam; quare, genus. Genus denique secundum naturalem suppositionem est magnitudo, circa cuius species posterea versatur.

Artifex vero est tam demonstrator quam exercitator. Officium demonstratoris est ad intellectiam discipline theoremata explicare, ad quod vii sunt necessaria: propositio, exemplum, dispositio, ratio, conclusio, ratiocinatio, et in fine dis[s]olutio. Est autem propositio theorematis expositio. Exemplum est istius specie suppositio. Dispositio est aliarum figurarum ad exemplum applicatio. Ratio est principiorum et suarum conclusionum inductio, i.e. argumentum. Conclusio est illatio propositi. Ratiocinatio est rationis et conclusionis controversio (?), i.e., argumentatio. In fine, dissolutio est cum falsigraphus insistit non sic vel aliter accidere quam geometer affirmat.

Officium exercitatoris est mensurare. Est autem mensuratio circa quantitates asignatio. Instrumentum vero demonstratoris est radius et mensa cum plana. Exercitatoris vero instrumenta sunt mensure geometrice, scilicet, pertica cum palma, digitus, pes, passus, et ulna. Illis vi adde ii cum stadio et miliare. Vel secundum Cerbertum, digitus, uncia, palmus, sextalis dodrans, pes, laterculus, cubitus, gradus, passus, pertica vel decempeda, actus minimus, clima, porta, actus quadratus, agrippennus, iugerum vel iugum vel iugerus, centuria, stadium, miliarium, leuca. Age. Digitus continet iiii grana ordei in longum continuum disposita, dictus digitus geometricus ad differentiam singulorum, qui diversi diversarum sunt quantitatum. Uncia continet digitum et tertiam partem digiti. Palmus quatuor digitos et tres uncias. Sextilis xii digitos et ix uncias et iii palmos. Pes xvi digitos, et cetera. Assignatis enim totalibus mensuris facile est partiales committere et e converso. Laterculus, dictus a latere indeclinabili, habet in latitudinem pedem unum in longum pedem unum et eius deuncem. Cupidus (i.e., cubitus) habet unum pedem et semissem. Gradus habet duos cubitos. Passus habet unum et bisse. Pertica passus duos. Actus minimus superficialis habens in latere pedes iiii, in longum xccl (i.e., xl?). Clima quoque superficialis est habens utrimque pedes lx (xl?). Porta in longum lxxx, in latere xxx. Actus quadratus undique habet perticas duodecim. Iunger duo agrippennis in longitudine, perticas xxiii, in latere, xii perticas, cuius quartum dicitur tabula, continens perticas confractas lxxii. Centuria continet iugera cc, licet etiam solet derogari tam longitudini tam latitudini dum modo summa continetie redintegretur in constrato. Stadium passus cxxv. Miliarium stadia viii; leuca unum miliarium et dimidium.

Sciendum quod omnes mensure in unitatem ducte pertinent ad longitudinem, in se vero semul ad superficiem, in se vero bis (?) ad soliditatem, verbi gratia, semul iiii grana ordei pedis linearem, quot iiii superficialem, quot iiii soliditatem perficiunt. Similiter de reliquis linearibus (?) mensuris.

Intentio auctoris est racionalium figurarum mensurationem explicare. Materia vero sunt rationales figure, in quatuor species distincte; et numerus, in lineas: mediale, binominum, residuum. Sicut autem rationales rationationi huius artis subiecte dicuntur, tam irrationales, i.e., non denominate ab aliqua mensura geometrica secundum aliquem numerum, scilicet, nec monopedalis nec bipedalis; in superficiebus: circulus, triangulus, quadrangulus; in corpore: spera, quadratum (i.e. cubum?), columpna, pyramis, et figure alie.

Utilitas vel finis (rei) mensurandi scire et dimensione mediantis (i.e., mensurantis?) exemplum. Modus agendi is est; agere enim demonstrare. Est autem demonstratio argumentatio, arguens ex primis et veris in illorum conclusionibus. Sic enim ars per accidens contracta est qui (cuius?) sequentia necessario accidunt ex premissis aut principiis deinceps. Est enim scientia demonstrativa que docet demonstrare et demonstrat, ut posteriores a naturalibus quedam demonstrare et non demonstrare; unde geometria.

Duplex est ratio ordinis in hoc opere, ordo

36

scilicet partium ipsius secundum quod a lineis
erat inchoandum, procedendum in superficiem,
et numerum, tandem tractandum etiam corpus.
Secundus ratio demonstrationis derogaret secundum quod ipsa precedentia causarent sequentia, secundum quod proponitur ad illam
partem demonstrationis, de his que ad superficiem vel corpus pertinent. Experientie autem
auctor demonstrationis adquiescit in ordine.

Titulus is est primus liber euclidis philosophi
de arte geometrica incipit, vel incipit ars geometrica ⟨c⟩ ccclxiiii propositiones et proposita
continens, ab euclide in arabico composita et ab
adhelardo bathoniensi in latinum transumpta.
Propositiones vero per indicativum, proposita
per infinitivum explicantur. Item propositiones
ponunt vel aliquod est, vel non est. Proposita
vero ubi aliquod est faciendum, vel non. Cetera
patent. Deinceps littere insisterent incipiendo a
principiis. Hec autem distinguuntur per anxioma,
petitiones, et conceptiones. Anxioma dignitas
inquisitionis. Explicat enim rerum diffinitiones.
Petitiones sunt quibus concessis secundum
ypothesim nullum sequitur inconsequentis. Conceptiones sunt que universaliter humane intel-
ligentie occurrunt in quibus non est exigendum
propter quod. Nec omnia scripta principia.
Principia sunt reputanda secundum ad illorum
similitudinem quibus eque manifesta; ubi quilibet maius equali maiori, maius est equali minori;
principa sine expositione faciunt innotescere.
Triangulum equilaterum (i.e., proposition 1).
Data dicuntur quibus equalia . . . (Similar to
IIIB below for prop. 1, then like IIIA).

(4)	(5)
ADELARDUS OF BATH — VERSION IIIA (Oxford, Balliol Coll. 257, 2r–98v) Primus liber euclidis institutionis artis geometrice incipit xlvii propositiones continens, per Adelardum Bath-oniensem ex arabico in Latinum translatus.	**ADELARD OF BATH — VERSION IIIB** (Oxford, Bodl. Saville No. 19, 1r–39v)
(*I. Def.*) Punctus est illud cui pars non est. Linea est longitudo sine latitudine, cuius ex-tremitates quidem duo puncta sunt. Linea recta est ab uno puncto ad alium extensio, in extremi-tates suas utrumque eorum recipiens . . . Alia elmuain est, quod equilaterum, sed rectangulum non est . . . Equidistantes linee sunt que in eadem superficie collocate et in alterutram partem protracte convenient, etiam si in infini-tum protrahantur.	(*I. Def.*) Like Version II.
(*I. Post.*) The same as Version II.	(*I. Post.*) Like Version II.
(*I. Axioms*) The same as Version II.	(*I. Axioms*) Like Version II.
(*I.1*) Proposition as in Version II. (*Proof:*) Esto exemplum *ab* linea data. Data dicuntur quibus equalia habitudanter invenire possumus (?), scilicet, humane rationi tractabilia, i.e., media inter maxima et parvissima. Dispositio. Supposito itaque centro in *a*, circumferentia vero in *b*, designetur circulus secundum spatium *ab*, iuxta secundam petitioem. Item fixo pede	(*I.1*) Proposition as in Version II. (*Proof:*) Data dicuntur quibus equalia habitudanter in-venire possumus, media, scilicet, inter immensa et brevissima, que, scilicet humana ratio veritatis indaginatione sue rationi supponit. Detur ergo *ab*. Deinde sumatur centrum in *a*, circum-ferentia in *b*, ad circumducendum circulum. Item aliud centrum sumatur in *b*, circum-

pigro circini in *b*, mobili vero in *a*, designetur et alius circulus secundum idem spatium circa *b* centrum, secans priorem in *d* et *c*; hypothenusis deinque erectus ab *ab* in *c* sectionem. Ratiocinatio. Age. *ab*, *ac* linee exeunt ab *a* centro ad circumferentiam. Ergo sunt equales secundum primum (?) anxioma. Item *ba*, *bc* linee exeunt a *b* centro ad circumferentiam. Ergo secundum idem eedem sunt equales. Ergo *ac*, *bc* eidem, scilicet, *ab*, sunt equales; ergo inter se secundum primam conceptionem. Ergo huius trianguli *abc* omnia latera sunt equalia. Sicque super *ab* lineam datam equilaterum triangulum collocando proposito satisfecimus.

ferentia in *a*, circulo similiter, communi sectione notata in *c*. Deinde eringantur due linee a duobus terminis *ab* ad *c*. *ab* et *ac* exeunt ab eodem centro ad circumferentiam; ergo sunt equales. Similiter *ba* et *bc*, quia similiter exeunt. Ergo *ac* et *bc* sunt equales, quia eidem *ab*, secundum secundam (conceptionem). Consequenter tria latera *ab*, *ac*, *bc* sunt equalia in eodem triangulo, quod proposuimus.

(I.47) Proposition as in Version II. (*Proof:*) Esto exemplum *abc*, cuius *ab* lateris quadratum sit equale quadratis reliquorum laterum, scilicet, *ca*, *cb*. Dispositio. Educatur igitur a *c cd* perpendicularis ad *cb*, et equalis *ac*, reflexa linea a *d* in *b*. Ratiocinatio. *c* angulus est rectus. Ergo secundum premissam quadratum *db* est equale quadratis *cd*, *cb*. Sed *cd*, *ca* latera sunt equalia. Ergo quadrata secundum antepremissam. Equalibus ergo commutatis, quadratum *db* est equale quadratis *cb*, *ca*, quibus quadratis *ab* fuerit equale secundum ypothesim. Ergo quadrata *ab*, *bd* sunt equalia. Ergo *ab* et *bd* resunt (!) equalia. Age. Istorum triangulorum *abc*, *bcd*, *ca*, *cb*, et *cb*, *cd* latera sunt equalia, et *ab*, *bd* bases equales. Ergo duo *c* anguli sunt equales. Sed *c* angulus [exterior] rectus; ergo et inferior. Quod proposuimus.

(I.47) Proposition as in Version II. (*Proof:*) Propositum est probare ut quod *abc* angulus est rectus. Hoc modo, fiat angulus rectus super *ab* lineam producta ab puncto *d*, equalis *bc* linee, et illo angulo recto subtendatur *ad* basis hoc modo. Inde sic *abd* angulus est rectus. Ergo per proximam quadratum *ad* linee valet quadrata *ab* et *bd* linearum. Sed *bd* linea est equalis *bc* linee, et *ab* sibi. Ergo quadratum *ad* linee valet quadrata *ab* et *bc* linearum. Sed quadratum *ac* linee ex ypotesi valet quadrata *ab* et *bc* linearum. Ergo per definitionem *ac* linea est equalis *ad* linee. Ex quo sic *ab* et *bc* latera sunt equalia *ab* et *bd* lateribus, et basis basi. Ergo per 7 [primi] et cetera. Sed *abd* angulus respicit *ad* linee, et *abc* angulum respicit *ac* linee, et bases sunt equales. Ergo anguli sunt equales. Sed *abd* est rectus. Ergo *abc* rectus, quod erat propositum.

(II.Def.) Omne parallelogramum rectiangulum sub duabus lineis angulum rectum ambientibus dicitur contineri; specie, non numero, ne videatur derogare ultime petitioni, scilicet, duas lineas rectas superficiem nullam concludere. Ibi enim "duas" apellans numero, hic specie, *ab* namque *cd*, cum sint equales, similiter *ac*, *bd* qua una reputantur. Si quidem *ad* parallelogramum continetur sub *ab*, *bc*, et hinc *ab*, *cd* equalia, inde *dc*, *bd* terminos claudunt. Omnis parallelogrammi spatii. . . . gnomo nominatur *ao*, *od*. "Eorum vero parallelogramorum etc," scilicet, tam *ao* et tam *od* cum *eo* et *ob* supplementis "gnomo nominatur," ut istud.

(II.Def.) As in Version II.

(II.1) Proposition as in Version II. (*Proof:*) Exemplum, *ab*, *ac*. Dispositio. Describatur *ad* parallelogramum ex *ab*, *ac*, et idem distinguatur per partialia parallelograma equidistantibus ad *ac*, *bd*, ductis ab *i* et *l* sectionibus. Ratiocinatio. Age. *ad* constat ex *ae*, *if*, *ld*, nec excedit nec exceditur ab illis; ergo est illis equale. Sed *ae* fit ex ductu *ac* indivise in *ai* portionem *ab*; et *if* ex ductu eiusdem *ac* in *il*, mediante *ie*; *ld* vero eiusdem *ac* in *lb*, mediante *lf*. Sicque colligitur propositum.

(II.1) Proposition as in Version II. (*Proof:*) Sit *ab* prima linea, *op* secunda, et ducatur *ab* in *op*, id est, ab *a* termino *ab* linee ducatur linea equalis *op*, et sit *ac*. A termino [*c*] item *ac* ducatur equidistans *ab*, et sit *cd*. Item a *d* puncto ducatur linea ad *b* equidistans *ac*. Dividatur linea *ab* in duas partes, scilicet, *ag* et *gb*. Et a *g* puncto ducatur *ge* linea equidistans *ac* linee. Constat propositum hoc modo, *ac* valet lineam indivisam, scilicet *op*, et *ag* valet *ce* lineam per 34 [primi] Euclidis. Et similiter *ge* valet *ac*. Ergo *acge* parallelogramum fit ex ductu *ac* linee indivise in *ag*, scilicet, alteram dividentium. Item *gebd* parallelogramum fit ex ductu *ge*, que valet lineam indivisam, in *gb*, scilicet, alteram dividentium. Et sic constat propositum. Eodem modo agendum in quotcumque dividatur *ab*.

(*II.14*) Proposition as in Version II. (*Proof:*) Exemplum, *abc*. Dispositio. Huic trigono *abc* describatur parallelogramum equale et rectangulum secundum 42 primi libri. Deinde *oe* maius latus protrahatur exterius ad equalitatem *oh* minoris, et terminetur in *d*. *de* divisa equaliter in *i*. Postea sumpto centro in *i*, ducatur semicirculus a *d* in *e*. *ho* protracta ad circumferentiam. Linea ducta ab *f* in *i*. Ratiocinatio. Age. *de* in *i* dividitur equaliter, in *o* vero inequaliter. Ergo secundum quintam huius libri, *to* quod fit ex ductu *od*, *oe*, inequalium sectionum, mediante *eh*, cum quadrato *oi*, intraiacentis sectionibus, est equale quadrato *ie*, medietati ergo quadrato *if*, eius equale de ratione centri et circumferentie, ergo et quadratis *fo*, *oi* equalibus secundum anteultimem primi (i.e., I. 46). Dempto ergo [quadrato] *oi* communi, relinquuntur *to* et quadratum *of* equalia. Ergo idem quadratum *of* est equale *abc* trigono, eius equali. Quod proposuimus. Similiter quelibet rectilinea figura quadrari potest, si primo resolvatur in triangulos, trianguli in parallelograma, deinde ut superius ratiocinando. Tota proposita distinctio in duplicem redundat doctrinam, multitudinis, et magnitudinis.

(*II.14*) Proposition as in Version II. (*Proof:*) Sit *t* triangulum cui equum quadratum est describandum. Hoc fit hoc modo. Primo describandum parallelogramum equale triangulo per 41 [primi] Euclidis, et dicitur *abcd*. A *d* puncto ducatur linea equidistans *dc*, equalis *bd*. Deinde dividatur *ce* linea per equa in *p* punctum, et describatur semicirculus de *p* centro. Consequenter protrahatur *bd* linea, cuius terminus in circumferentia dicatur *g*. Et a *p* centro ducatur *pg* linea ad *g*. Inde *pdg* angulus est rectus per 27 (?) [primi] Euclidis. Ergo *pg* et cetera. Sed *pg* est *pe*. Ergo quadratum *pe* valet quadrata *pd* et *dg*. Sed quod fit ex ductu *cd* in *de* valet, cum quadrato *pd*, quadratum *pe*, per 5 huius libri. Ergo a primo quod fit ex ductu *cd* in *de*, cum quadrato *pd*, valet quadratum *pd*, cum quadrato *dg*. Sed quadratum *pd* est commune. Ergo dempto communi, quod fit ex ductu *dc* in *de* valet quadratum *dg*. Ergo e contrario. Sed quod fit ex ductu *cd* in *de* est equale *abcd* parallelogramum. Ergo quadratum *dg* valet parallelogramum *abcd*. Ergo valet *t* triangulum. Quadratum igitur linee *dg* triangulo (?) erit equale, et angulo assignato, quod erat propositum.

APPENDIX II

(1)	(2)
TRANSLATION OF HERMAN OF CARINTHIA	**TRANSLATION OF GERARD OF CREMONA**
Paris BN Latin 16646, 13c	Paris, BN Latin 7216

(*Introduction:*) Septem sunt omnis discipline fundamenta, in quibus omnium rerum ad mathematice studia pertinentium firma essentie conceptio, etiam certusque veritatis intellectus in quadam quasi materia et causa fundata existunt. Sunt autem hec: Preceptum, Exemplum, Alteratio, Collatio, Divisio, Argumentum, Finis. Preceptum est integra sentente quedam et absoluta propositio. Exemplum est precepti in actu et re quedam explanatio. Alteratio est que datum exemplum destruens alioque divertens precepto non convenit, sicque per indirectam ratiocinationem, quod infringere nequid impossibilitate quadam confirmat. Collatio est convenientium coniunctio. Divisio est disputationum (?) disiunctio. Argumentum est ratio ad veritatem precepti. Finis est conveniens omnium conclusio, quo adepto, deinceps extraneum in philosophie disciplinis restet. Atque hoc longius persequi locus non exigit. Nec enim huius singulariter circa negotii verum omnis discipline diffusus atributa. Nunc autem artis elementa quedam et quasi communes loci prestituendi sunt equibus sequentium ratio firmius et evidentius prodita est.

(*Introduction:*) Ea a quibus procedit scientia, ex qua res que scitur comprehenditur sunt septem, videlicet, Propositum, Exemplum, Contrarium, Dispositio, Differentia, Probatio, Conclusio. Propositum autem est id quod antecedit summam scientie ante expositionem. Exemplum vero est corporum et figurarum forma ex proposito intellectorum que ex sua forma significantur super propositi intentionem. Sed contrarium est exempli contrarium et deductio propositi ad impossibile. Dispositio vero est compositionis dispositio conveniens super ordines suos in scientia. Differentia quoque est separatio eius, quod est inter positionem possibilem et impossibilem. Probatio vero est sillogismus super confirmationem propositionis. Conclusio autem est terminus scientie, cum re scita consequens totum quod nominavimus.

(*I. Def.*) (As in Adelard II with an occasional change, e.g., . . . alia elmuaim quam nos rumbum dicimus.)

(*I. Def.*) Punctus est cui pars non est. Linea est longitudo sine latitudine et eius extremitates duo puncta. Linea recta est extensio in oppositione cuiuslibet duorum punctorum que sunt in duabus ipsius extremitatibus unius ad aliud. . . . alia est rombus cuius latera sunt equalia, sed anguli non sunt recti. . . . Linee recte equidistantes sunt que cum in una plana superficie site sint, in utrasque partes usque in infinitum protrahantur, in nulla earum concurrent.

(*I. Post.*) Omnis igitur continua quantitas quinque modis investiganda videtur, ut a quolibet puncto in quemlibet punctum recta linea ducatur, atque linea definita. . . . Item, si linea recta super duas lineas rectas ceciderit duoque anguli ex una parte duobus angulis rectis minores fuerint, illas duas lineas in eam partem protractas coniunctum iri necesse sit. Due vero linee recte superficiem nullam concludant.

(*I. Post.*) Hec sunt peticiones. Cum quibus necesse est convenire, sunt quinque. Ex quibus est, ut linea recta a quolibet puncto ad quodlibet punctum perducatur. Et ut linea recta finita protrahatur super rectitudinem. . . . Et quod si ceciderit linea recta super duas lineas rectas et fecerit in una duarum partium duos angulos interiores minores duobus rectis, ille due linee recte, quando in illam partem protrahentur, coniungantur.

(*I. Axioms*) Est ante omnia complectens quedam sapientia que *aelmam geme* dicitur, eo quod maximas huiusmodi propositiones quas arabice *almukadimas* grece *theoremata* nuncupamus contineat, quibus artis ratio omnino innixa videtur. Que eidem equalia sunt, et sibi invicem sunt equalia . . . Omne totum sua parte maius. His ita constitutis, deinceps operam artificio demus.

(*I. Axioms*) Que eidem rei sunt equalia, sibi invicem sunt equalia. . . . Et totum maius est sua parte. Et due recte non comprehendunt superficiem.

(*I.1*) Primum igitur equilaterum triangulum supra rectam et definite quantitatis lineam collocamus. (*Proof:*) Data siquidem linea recta inter *a* et *b* puncta, acceptaque punctorum distantia, i.e., linee spacio fixo circino supra *a* centrum fiat circulus *bgd*, translato statim equali tenace, eodemque spacio retento, circa *b* centrum fiat alter circulus *agh*. Deinde ab *g* sectionis circulorum puncto descendant recte linee in *a* et *b*. Eritque huiusmodi equilaterus triangulus. Nam a centrum circuli *bgd*, lineam *ag* ii que est *ab* equalem esse cogit. Simili quoque modo *b* centrum circuli *agh* lineam *bg* ei que *ba* (*MS b* ly *a*) coequatur. Centrum enim est punctus a quo omnes linee ad circulum exeuntes sibi invicem equales sunt. Est autem si duo uni sunt equalia, utrumque alteri esse equale. Quoniam ergo duo ea *chateti* (*from Arabic khatt, i.e., linea*) basi sunt adequati, et sibi invicem sunt equales. Est itaque triangulus equilaterus supra rectam lineam collocatus.

(*I.1*) Super rectam lineam definite quantitatis triangulum equilaterum constituere. (*Proof:*) *Exemplum.* Verbi gratia, ponatur linea recta *ab* definite quantitatis, et super centrum *a* secundum quantitatem spacii quod est inter *a* et *b* circumducatur circulus super quem sunt *gdb*. Alius quoque circulus super centrum *b* secundum quantitatem spacii quod est inter *a* et *b* describatur, super quem sunt *gae*. Deinde a puncto *g* in quo unus duorum circulorum alium secuit due recte linee ad duo puncta *a* et *b* protrahantur; sintque linee *ga* et *gb*. Dico igitur quia iam fecimus triangulum equilaterum super lineam *ab* datam. *Probatio.* Huius probatio est. Quia punctum *a* factum est centrum circuli *gdb*, fit linea *ag* equalis linee *ab*; et similiter quia punctum *b* factum est centrum circuli *gae*, fit linea *bg* linee *ba* equalis. Unaqueque harum duarum linearum *ga*, *gb* linee *ab* equalis invenitur. Que autem eiusdem rei equalia sunt sibi, quoque invicem sunt equalia. Tres igitur linee *ag*, *ab*, et *bg* sibi invicem sunt equales. Triangulus igitur *abg* est equalium laterum, qui ut ostensum est super lineam datam *ab* constitus est. Et hoc est quod demonstrare voluimus.

(*I.47*) Si quod ex uno trianguli latere in se ipsum ducto provenit equum fuerit duobus quadratis que ex duobus reliquis lateribus fuerint, angulus cui latus illud opponitur rectus est. (*Proof:*) Ut in triangulo *abg*, scilicet quod ex latere *bg* in se ducto tradetur equum est eis que ex duobus reliquis concurrunt, angulum *gab* rectum esse necesse sit. Quod ita sumatur

(*I.48*) Si quadratum factum ex aliquo latere trianguli duobus quadratis que fiunt ex reliquis duobus lateribus fuerit equale, angulus qui ab illis duobus reliquis lateribus trianguli comprehenditur erit rectus. (*Proof:*) *Exemplum.* Verbi gratia, si triangulus super quem *abg*, et sit quadratum quod fit ex latere eius *bg* duobus quadratis factis ex reliquis duobus lateribus *ba*, ut primum a puncto *a* recto dividentes angulo producamus lineam usque ad notam *d* equalem ei que est *ab*, applicemusque *d* cum *g*. Quoniam itaque *ba* equalis est *ad*, inter ea ex utroque in se ducto provenerunt equalitas erit. Id autem quod *ab* linea in se ducta reddidit commune utrique. Quare(?) vero que ex *ag* et *ab* utroque in se ducto procedunt equalia sunt ei quod *gb* conficit. Eodemque pacto ea que *ga* et *ad* ex se producuntur equalia ei quod *gd* in se ductum generant. Erit latus *gb* equale ei quod est *gd*. Est autem *ba* latus ei quod est *ad* equale. Sitque *ag* commune utrique. Sic itaque *ba* et *ag* equalia erunt eis que sunt *ag* et *ad*. Est autem et basis *bg* basi *gd* equalis. Erit igitur angulus *bag* angulo *gad* equalis. Angulus autem *gad* rectus est. Erit igitur et *bag* de quo differimus rectus angulus, ut angulum oppositum lateri quod in se ductum reliquis adequatur rectum esse constant.

ag equale, dico igitur angulum *bag* rectum fore. Probatio huius. Quia protraham a puncto *a* lineam *ad* erectam super lineam *ag* super rectos angulos, et ponam *ad*, *ab* [equale] et producam lineam *gd*. Et quia quadratum linee *bg* duobus quadratis factis ex duobus lateribus *ba*, *ag* equale existit et linea *ba* linee *ad* equatur, erit quadratum factum ex latere *bg* duobus quadratis que fiunt ex duobus lateribus *ag*, *ad* quadrato facto ex latere *gd* equatur, quoniam angulus *dag* est rectus. Quadratum igitur factum ex latere *dg* quadrato quod fit ex latere *bg* equale existit. Linea ergo *bg* linee *gd* est equalis. Et quoniam linea *ba* linee *ad* equalis existit, linea *ag* ente communi, erunt due linee *ba*, *ag* duabus lineis *da*, *ag* equales, unaqueque sue relative, et basis *bg* basi *gd* equalis est. Ergo angulus *dag* angulo *gab* equalis existit. Angulus autem *gad* est rectus. Ergo angulus *bag* est rectus. Si igitur quadratum quod fit ex latere trianguli duobus quadratis ex reliquis duobus lateribus factis equale fuerit, angulus ab illis duobus reliquis trianguli lateribus comprehensus erit rectus. Et hoc est quod demonstrare voluimus.

(*II. Def.*) Omnis equidistantium superficies rectangula duabus lineis rectum ambientibus angulum continetur. Omnis equidistantium superficiei ea spacia . . . totius comprehensio nominatur *alalem* quod in arabico sermone idem est quod latine vexillum. Visit (?) autem esse *alalem* semicirculus quidem terminans circa diametrum conducens, quod quidam *gnomonem* nuncupati sunt. Nos vero quid gnomonis sit non ignoramus. Itaque propter extremorum quemdam ambitum rectius umbonem appellandum arbitramur.

(*II.Def.*) *Prohemium.* Omnis superficies equidistantium laterum et rectorum angulorum ab hiis duabus rectis lineis dicitur comprehendi que rectum comprehendunt angulum. Omnis figure equidistantium laterum si una ex superficiebus equidistantium laterum que super ipsius dyametrum consistant et quas dyametrus per medium secat duabus superficiebus equidistantium laterum que dicuntur suplementa adiungatur et sunt ab utraque parte dyametri, totum hoc vocatur gnomo.

(*II.1*) Si de duabus lineis altera in quotlibet partes divisa fuerit, illud quod ex ductu alterius in alteram fiet equum est eis que ex ductu linee indivise in omnes partes eius que divisa est producuntur. (*Proof:*) Sint igitur proposite linee due *a* et *bg*, maneatque *a* indivisa. Sed *bg* dividatur ad puncta *d* et *e*. Illud itaque quod ex linea *a* ducta in *bd* et *be* et *eg* concrescunt equum est ei quod *a* in *bg* ducta componit. Primum igitur a puncto *b* linee *bg* recto descernentes angulo producamus lineam equalem *a*, que est *bz*. Deinde ab eiusdem puncto *z* lineam trahimus usque ad *h* equidistantem ei que est *bg*; postremo a singulis partitionum punctis lineas equidistat; educimus ab *d* ad *t*, ab *e* ad *k*, ab *g* ad *h*. Erunt itaque tres superficies, quarum omnium ea que est *gz* que est integritatem equaliter optinet (optimet?) ea vero que est *gz* ex ductu linee *a* in *bg* procedit. Sint enim *bz* et *gh* equalis *a*. Sed *zh* equalis *bg*. Tres autem que infra sunt ex eadem *a* quelibet in suam partem ducta producantur. Nam omnium catheti *a* linee equales sunt, torqusti (?) vero partes *bg* quelibet bases sue adequantur. Erit itaque ductus linee *a* indivise in *bg* integram equaliter ductum eiusdam in omnes partes divise.

(*II.1*) Si fuerint due recte linee quarum una in partes sit divisa, quodcumque fuerint, superficies rectorum angulorum que comprehenditur ab hiis duabus rectis lineis equalis est superficiebus rectorum angulorum que a linea non divisa et ab unaquaque parte linee divise comprehenduntur. (*Proof:*) *Exemplum.* Exempli causa, sint due recte linee super quas fuerit *a*, *bg* et dividatur linea *bg* in divisiones quodcunque sunt super duo puncta *d*, *e*. Dico igitur quod superficies rectorum angulorum que comprehenditur a duabus lineis *a* et *bg* superficiei rectorum angulorum que comprehenditur ab hiis duabus lineis *a* et *bd* et superficiei rectorum angulorum que continetur ab hiis *a*, *de* et superficiei rectorum angulorum etiam que comprehenditur ab hiis *a*, *eg* est equalis. *Probatio.* Probatio huius. Quoniam a puncto *b* linee recte *gb* lineam rectam super rectos angulos protraham, sitque linea *bz*. Et ponam lineam rectam *bz* linee recte *a* equalem, et producam a puncto *z* linea *zh* linee recte *bg* equidistantem. Deinde protraham a punctis *d*, *e*, *g* lineas linee *bz* equidistantes. Sintque linee ille *dt*, *ek*, *gh*. Est itaque unaquaque superficierum *bt*, *dk*, *eh* equidistantium laterum. Superficies quoque *bh* superficiebus *bt*, *dk*, *eh* equatur. Superficies vero *bt* superficiei rectorum angulorum que continetur ab hiis duabus lineis *a*, *bd* equalis est, quoniam *bz* linee *a* est equalis. Et superficies *dk* superficiei rectorum angulorum que comprehenditur ab hiis duabus lineis *a*, *de* equatur, quoniam linea *a* linee *dt* equari invenitur. Superficies quoque *eh* superficiei rectorum angulorum que comprehenditur ab hiis duabus lineis *a*, *eg* equalis existit, quoniam linea *a* linee *ek* est equalis. Superficies ergo rectorum angulorum que comprehenditur ab hiis duabus lineis *a*, *bg* superficiei rectorum angulorum que continetur a duabus lineis *a*, *bd* et superficiei rectorum angulorum que comprehenditur ab hiis duabus lineis *a*, *eg* equalis existit. Si igitur due recte linee fuerint, quarum una fuerit in partes divisa, quodcumque fuerint, superficies rectorum angulorum que comprehenditur ab hiis duabus rectis lineis superficiebus rectorum angulorum que continentur ab illa linea que non dividitur et ab unaquaque ex sectionibus divise linee equalis est. Et hoc est quod demonstrare voluimūs.

(*II.14*) Proposito triangulo equum tetragonum describimus. (*Proof:*) Sit enim propositus trigonus, *abg*; hinc equalem in primis equidistantium superficiem ponimus *dezh*, cuius latera quoniam equalia non sunt protrahimus *de* usque ad *t*, ut sit *et* equalis *eh*; totam itaque lineam per medium secamus sub puncto *k*, ut sit equaliter divisa ad punctum *k*, inequaliter ad notam *e*. Deinde *kt* occupans super *k* centrum fiat semicirculus *dt*. Statimque *he* producatur recte usque ad circumferentiam, atque loco nota *l* signato. Ab eodem puncto ad centrum recta linea descendat. His ita perfectis, quoniam ut supra datum est iuxta equalem et inequalium linee divisionem *de* maior inequalium in *et* minorem, quodque inter minorem et medietatem relinquitur in se ductum medietati totius linee *kt* videlicet adequantur. Est autem *kl* equalis *kt*. Erit igitur et *kl* in se ducta similis itaque ductis equalis eadem vero *kl* eis que sunt *le* et *ek* ductu adequantur. Erunt ergo *de* in *et* et *ek* in se ducta *le* et *ek* in se ductis equales. Cum itaque communis utrique *ek* de medio exierit, remanebit *de* in *et* ducta *le* in se ducte equalis. Ac vero ex *de* in *et* ducta subiecta superficies que dato triangulo equalis posita est. Processit igitur *le* in se ducta eidem trigono equum tetragonum producet.

(*II.14*) Quadratum figure rectorum laterum date equale describere. (*Proof:*) *Exemplum.* Exempli causa. Sit figura rectorum laterum data *a*, et oportet ut faciam quadratum figure rectorum laterum *a* equale. Describam ergo superficiem equidistantium laterum et rectorum angulorum figure rectorum angulorum et laterum *a* equalem. Sitque superficies *bgde*. Erit ergo *be* lateri *ed* equale aut unum eorum altero maius erit. Si ergo fuerint equalia, iam fecimus quod voluimus. Sed si non fuerint equalia, unum ergo eorum altero erit maius. Sit ergo una duarum linearum *be*, *ed* altera maior; sitque *be* maior. Et protraham lineam *ez* in rectitudine linee recte *be*, et ponam *ez* lineam linee *ed* equalem. Deinde dividam lineam *bz* in duo media in puncto *h*, et circumducam super centrum *h* secundum quantitatem longitudinis duarum linearum *hb*, *hz* circuli medietatem *btz*. Et protraham lineam *et* rectam in rectitudine linee *de*, et producam lineam *th*. Dico ergo quadratum equale figure rectorum laterum descriptum fore. *Probatio.* Probatio huius. Et quoniam linea recta *bz* iam est divisa in duas equales sectiones in puncto *h*, et duas inequales in puncto *e*, erit superficies rectorum angulorum que ab hiis duabus comprehenditur lineis *be*, *ez*, cum quadrato quod fit ex linea *eh*, quadrato quod fit ex linea *hz* equalis. Linea vero *hz* linee *ht* est equalis. Superficies ergo rectorum angulorum que ab hiis duabus continetur lineis *be*, *ez*, cum quadrato quod fit ex linea *eh* quadrato quod fit ex linea *ht* est equalis. Sed quadratum quod fit ex linea *ht* duobus quadratis que fiunt lineis *te*, *eh* equatur, quoniam angulus *teh* est rectus. Superficies ergo rectorum angulorum que ab hiis duabus continetur lineis *be*, *ez*, cum quadrato quod fit ex linea *eh*, duobus quadratis que fiunt ex duabus lineis *he*, *et* equalis existit. Remoto itaque quadrato communi quod fit ex linea *eh*, remanet superficies rectorum angulorum que continetur ab hiis duabus lineis *be*, *ez* reliquo quadrato quod fit ex linea *et* equalis. Superficies autem rectorum angulorum que ab hiis duabus continetur lineis *be*, *ed* est equalis, quoniam linea *ze* linee *de* equatur. Superficies ergo *bd* quadrato quod fit ex linea *et* equalis existit. Sed superficies *bd* figure *a* rectilinee fuerit equalis. Figura ergo *a* rectilinea quadrato quod fit ex linea *et* equatur. Iam ergo fecimus quadratum figure rectilinee super quam est *a* equale quod est quadratum quod fit ex linea *et*. Et hoc est quod demonstrare voluimus. Huius preterea theorematis propositum et dispositio aliter inveniuntur secundum quod proponitur ut fiat quadratum data triangule figure equale, et in exemplo ponitur triangulus cui quadratum equale fiat. Cetera vero non mutantur. Explicit liber secundus.

A MEDIEVAL TREATMENT OF HERO'S THEOREM ON THE AREA OF A TRIANGLE IN TERMS OF ITS SIDES

The theorem for the area of a triangle as a function of its sides (namely, $A = \sqrt{s\,(s\text{-}a)\,(s\text{-}b)\,(s\text{-}c)}$, where s is the semiperimeter and a, b, c, are the sides) has a long history since its enunciation by Hero of Alexandria [1]. The most important early study of this history was done by F. Hultsch in 1864 [2]. Much of the recent historical investigation of this theorem was summarized succinctly by S. Gandz [3]. In brief, we can note that this theorem was given without proof by the author of the Mishnat ha-Middot (ca. 150) [4], by one of the *agrimensores* (ca. 400 ?), also without proof [5], by the Indian mathematician Brahmagupta (ca. 628) [6], who extends it to a quadrilateral but has no proof, by the Banū Mūsā accompanied by a proof [7], by that superb polymath al-Bīrūnī (ca. 1000) [8], who assigns the enunciation to Archimedes and takes his proof from one

---

[1] Hero of Alexandria, *Metrica*, I, viii (Edition of H. Schöne in *Heronis Alexandrini opera... omnia*, Vol. 3 [Leipzig 1903] 18-24) ; *Dioptra* xxix (*Ibid.*, 280-84) ; *Geometrica* (Ed. of J. L. Heiberg, *Ibid.*, Vol. 4 [Leipzig 1912] 248). Incidentally, Al-Bīrūnī (see footnote 8 below) assigns the theorem to Archimedes. While we have no antique evidence of this, it certainly makes sense that the theorem is earlier than Hero.

[2] F. Hultsch, 'Der Heronische Lehrsatz über die Fläche des Dreieckes als Function der drei Seiten', *Zeitschrift für Mathematik und Physik*, Vol. 9 (1864) 225-49.

[3] S. Gandz, edit., *Mishnat ha-Middot, Quellen und Studien zur Geschichte der Mathematik, Astronomie und Physik*, Abt. A : *Quellen*, Vol. 2 (1932) 45, n. 40.

[4] *Ibid.*, 45-46.

[5] M. Cantor, *Die römischen Agrimensores und ihre Stellung in der Geschichte der Feldmesskunst* (Leipzig 1875) p. 107. See the text of the theorem by Marcus Junus Nipsus in his so-called *Podismus*, F. Blume, K. Lachmann, and A. Rudorff, *Die Schriften der römischen Feldmesser*, Vol. 1 (Berlin 1848) 300-301.

[6] H. T. Colebrooke, *Algebra with Arithmetic and Mensuration from the Sanscrit of Brahmegupta and Bhascara* (London) 1817) pp. 295-96 (also p. 72). Cf. Hultsch, *op. cit.* in note 2, p. 239. Other later Indian authors took up the theorem.

[7] M. Curtze, 'Verba Filiorum Moysi, Filii Sekir, id est Maumeti, Hameti, et Hasen. Der Liber trium fratrum de Geometria. Nach der Lesart des Codex Basileensis F. II. 33 mit Einleitung und Commentar' *Nova Acta der Ksl. Leop. Carol.*

80

Abū Abdallāh al-Shannī, by al-Karkhī (ca. 1020) (⁹), by Savasorda (12th century) without proof (¹⁰), by Leonardo of Pisa, from the Banū Mūsā (¹¹), by Luca Pacioli in 1494, with a proof from Leonardo (¹²), by Widmann without proof in 1489 (¹³), by Leonardo of Cremona in the same century, again without proof (¹⁴), and in the 16th century by Pierre de la Ramée (¹⁵) (with a proof similar to that of the Banū Mūsā) and no doubt by others in that century. Incidentally, while the theorem itself is not given by Campanus, its possibility of development is suggested by the thirteenth century mathematician in his commentary on the *Elements* (see footnote 24 below).

The first contact of Latin scholars with a *proof* of the theorem came with the translation by Gerard of Cremona of the *Verba filiorum* of the Banū Mūsā (see footnote 7) and it was this proof that was reflected in the subsequent treatments by Leonardo of Pisa, Pacioli, and Pierre de la Ramée. A quite different proof of this theorem also

---

Deutschen Akademie der Naturforscher, Vol. 49 (Halle 1885), Prop. VII, pp. 131-35. This is a most defective text, based as it is on a single manuscript. I have prepared a new text from all the extant manuscripts in my *Archimedes in the Middle Ages*, Vol. 1 (in press, Madison, Wisconsin), Chapter IV.

(⁸) H. SUTER, 'Das Buch der Auffindung der Sehnen im Kreise von Abū 'l-Raiḥān Muḥ el-Bīrūnī', *Bibliotheca Mathematica*, 3. Folge, Vol. 11 (1910-11) 39-40, 70.

(⁹) GANDZ, *op. cit.* in note 3, *loc. cit.*

(¹⁰) "Der 'Liber embadorum' des Savasorda in der Übersetzung des Plato von Tivoli", in M. CURTZE, *Urkunden zur Geschichte der Mathematik im Mittelalter und der Renaissance, Abhandlungen zur Geschichte der mathematischen Wissenschaften, 12.* Heft (Leipzig 1902) 72.

(¹¹) *Practica geometrie*, in *Scritti di Leonardo Pisano*, edit. by B. Boncampagni, Vol. 2 (Rome 1862) 40-42.

(¹²) HULTSCH, *op. cit.* in note 2, pp. 242-46, gives the citation to Pacioli's *Summa*, and he translates and discusses the proof given by Pacioli.

(¹³) See the note by G. ENESTRÖM, *Bibliotheca Mathematica*, 3. Folge, Vol. 5 (1904) 311.

(¹⁴) "Die 'Practica geometriae' des Leonardo Mainardi aus Cremona", in M. CURTZE, *Urkunden etc.* (cited in note 10), *Abhandlungen etc.*, 13. Heft (Leipzig 1902) 386-87.

(¹⁵) P. DE LA RAMÉE. *Scholarum mathematicarum libri unus et triginta* (Francofurti 1599), p. 313. Ramée's proof is substantially the same as that of the Banū Mūsā. He appears to claim that the proof comes from Jordanus and Tartaglia, and that it is lacking in logic. My guess is that he found the proof in a work of Tartaglia, who perhaps got it from Pacioli. I suspect that Tartaglia also saw the different proof that circulated with the *De ratione ponderis* of Jordanus and that perhaps he mentioned such a proof without reproducing it, reproducing rather the proof of the Banū Mūsā as given by Pacioli.

circulated during the Middle Ages, a proof associated with the name of Jordanus. It is this proof, somewhat closer to the proof by Hero, that is the object of our discussion in this paper. It existed in two versions, both of which have been edited here.

The first of these two versions was previously published by Curtze on the basis of a single manuscript (namely, *I*) ([16]). His only suggestion as to authorship he based on a table of contents at the beginning of the manuscript where we read *Theoremata Cratili* ([17]). He thought that our theorem might be one of those included under that title. But as to who Cratilus was, Curtze had no idea. Now this proof discovered by Curtze was associated with the name of Jordanus for the first time by Pierre Duhem for the following reasons ([18]) : (1) The theorem appears in the manuscripts in close proximity to (or even as a part of) works attributed to Jordanus ([19]). (2) There is a reference in the Vatican manuscript (*Xa*) to the effect that the theorem 'is a part of the Phyloteigni and ought to be joined to it' (cf. Version I, variant reading to line 1) ; but Jordanus refers in his *Elementa de ponderibus* to a work of his entitled *Philotegni*, a work which can now confidently be identified with his *De triangulis* ([20]). Still

---

([16]) M. CURTZE, 'Über eine Handschrift der Kœnigl. öffentl. Bibliotek zu Dresden', *Zeitschrift für Mathematik und Physik*, Vol. 28 (1883), Hist. -lit. Abtheilung. 5-6, 78.

([17]) *Ibid.*, 4, 6. The hand giving this table of contents is that of Valentinus Thaus (Thaw) and is dated 1580 (see p. 1). Incidentally, it is Thaus' hand that adds the combined Greek and Latin marginal note opposite the theorem under consideration (see Variant Readings, Version I, line 1).

([18]) P. DUHEM, 'Un ouvrage perdu cité par Jordanus de Nemore : le Philotechnes', *Bibliotheca Mathematica*, 3. Folge, Vol. 5 (1904) 323-25.

([19]) In manuscript *Q* (40r) it comes at the end of the *Liber de ratione ponderis* attributed to Jordanus and actually precedes the explicit of that work (40v) : 'Explicit liber quartus Iordani de ponderibus'. In manuscript *R* it follows two folios after the *Liber de ratione ponderis*, which occupies folios 50r-55v. In manuscript *I* there are a number of works attributed to Jordanus, including the *De triangulis* (50r-61v), an *Arithmetica* (61v-110v), *Elementa de ponderibus* (186r-187v), *De forma spere in plano* (224r-225v), *De numeris datis* (228r-242v), *De ratione ponderis* (243r-249v). Our theorem is not particularly close to any of these, occupying folios 178r-v. But the association of the theorem with Jordanus, or at least with one of the tracts *De ponderibus*, is suggested by a statement in Leonardo de Cremona's *Practica geometriae* (*ed. cit.* in note 14, p. 386) where he notes that he has found the theorem in a book of mechanics (*libro de mechanici*). This could very possibly have been a manuscript of one of the tracts *De ponderibus* attributed to Jordanus, as Curtze asserts.

([20]) See E. A. MOODY and M. CLAGETT, *The Medieval Science of Weights*, 2nd printing (Madison 1960), pp. 130, 134-36, 379, 381. Note further that in MS Bruges,

82

the accuracy of this marginal reference may well be questioned since none of the four copies of the *De triangulis* which I have examined contains the proof in question, while in manuscript *I* our proof is separated from the *De triangulis* by more than 100 folio pages. Furthermore, manuscripts *Xa* and *Q* of Version I of our theorem say that 'this rule is said to have been written in Arabic' (Cf. Version I, line 7), while *Ya*, the unique manuscript of Version II, states more categorically : 'this rule concerning the triangle was written in Arabic.' (Cf. Version II, line 165).

These remarks certainly seem to throw some doubt on Jordanus' authorship of the theorem. But let us examine them more closely, speculating as to their meaning. At least three possibilities suggest themselves, the first of which is incompatible with the idea of Jordanus as the original author of the theorem. The first interpretation would hold that the meaning of the remarks is that the whole theorem (enunciation and proof) was composed in Arabic and merely translated into Latin. In support of such a theory we recognize that the proof, unlike the proof of Banū Mūsā, was fairly close to an Arabic version of Hero's proof that circulated in the Middle Ages [21], although to be sure we can readily see that that Arabic version was not itself the text from which our proof was translated in spite of the general similarity of the two. Supposing this interpretation to be the correct one, then, if Jordanus had any connection at all with the theorem, it was merely to transmit it, or possibly to modify it somewhat. We know that Jordanus did on occasion take certain theorems almost verbatim from other authors, as for example when he drew the theorems and proofs concerning the trisection of an angle and the finding of two mean proportionals from the Banū Mūsā [22].

A second possible interpretation of the remarks on the Arabic writing of this theorem is that they refer only to the enunciation, and not to the proof. This seems to have been Duhem's opinion [23]. In support of this theory is the fact that the comment follows immedi-

---

Stadsbibliotheek 530, iv-8v, the *De triangulis* is specifically entitled *Phylotegni Iordani de triangulis incipit liber primus* (iv ; cf. 8v).

[21] This version of Hero's proof was added to the end of al-Tūsī's edition of the geometry of the Banū Mūsā *Majmūᶜ al-Rasā'il*, Vol. 2 ([Hyderabad 1940]). Cf. H. SUTER, 'Über die Geometrie der Söhne des Mûsâ ben Schâkir', *Bibliotheca Mathematica*, 3. Folge, Vol. 3 (1902) 271-72.

[22] See CLAGETT, *Archimedes in the Middle Ages*, Vol. 1, Appendices V and VI.

[23] DUHEM, *op. cit.* in note 18, pp. 323-24.

ately after the enunciation in manuscripts *Xa* and *Q* and that in manuscript *Ya* it is added only in connection with the full, formal enunciation as it is given at the end. But one could object to this interpretation (and as a matter of fact to the first one as well) by pointing out that the remark in *Xa* and *Q* is by no means an assertive statement. If the author of that remark really *knew* that this rule was written in Arabic (and translated therefrom) why did he use the rather tentative verb, *dicitur*, 'is said'. This objection, of course, does not hold for the statement in *Ya*, which flatly asserts that the rule was written in Arabic. Still there is considerable evidence that Version IJ is merely a rewrite of Version I in the *Xa* and *Q* tradition. Hence the change in tone in the remark as found in *Ya* ought perhaps to have no significance.

A third interpretation would deny the Arabic origin of either enunciation or proof. It would hold that the remark is only a vague, general statement that the same rule can be found in writing of Arabic origin (as for example, in the *Verba filiorum* known to be of Arabic origin), but that both the enunciation and proof as presented here are independent of Arabic sources. That is, they are either Greek in origin or original with Jordanus or some other Latin geometer. There is no sure way to decide between these varying interpretations and I suspect that no decision can be made until further evidence appears.

The line of argument followed in both of the versions is clear enough to demand little additional comment. As I indicated earlier, it resembles the proof of the theorem given in Hero's *Metrica* and *Dioptra* (see footnote 1) more than it does that found in the *Verba filiorum* of the Banū Mūsā. It is evident, however, that the medieval proof is less economical than Hero's proof. This is sharply brought out by contrasting Version II with Hero's proof, since Version II has added geometrical steps lacking in Version I. But even in Version I the author takes an excessive number of steps to draw his obvious conclusions after showing the similarity of triangles PAC and DBF and that of triangles DFH and CPQ. I do not know when the elaboration represented by Version II was composed, but its unique manuscript dates from the fifteenth century. It seems probable to me that it was composed from some copy of Version I that was closer to the tradition of manuscripts *Xa* and *Q* than *I* since many of the variant readings of *Xa* and *Q* have been incorporated in Version II. That Version II postdated Campanus' commentary on the *Elements* of Euclid is possible since in Version II the enunciation is changed to a form similar to that found in Campanus' commentary ([24]). Incidentally, the only author cited in either version is Euclid, and he is cited only once in Version II (line 103), a very general citation to Book XI.

My text of Version I has been constructed on the basis of the three manuscripts listed in the *Sigla* below. It is clear from examining the variant readings that *Xa* and *Q* again and again agree with each other against *I*, but sometimes circumstances of style and meaning have demanded that I follow *I* and sometimes *Xa* and *Q*. As between *Xa* and *Q*, *Xa* is to be preferred since on occasion *Q* omits necessary material (e. g., see variants to lines 49-51 and 65-67) and it often alters correct forms (e. g., see *medietasque* instead of *mediatatisque* in line 2, *superadditus L* instead of *Sed super AC* in line 34, and so on). *Xa* has no figure. The bottom part of the triangle is missing in the figure in *I*, as are the lines R, S, and T ; but solids X and Y are represented as rectangular parallelopipeds. My text of Version II follows *Ya* with only occasional changes, the pristine readings having been included in the variant readings. In both texts I have been free with capitalization and punctuation, capitalizing the letters referring to geometrical magnitudes where small letters are used in the manuscripts and the first letter of words beginning sentences although the manuscripts sometimes use small letters and sometimes capital letters. I have also capitalized the enunciation of Version I to indicate that it is written in larger letters than the proof in manuscript *I*, a custom often followed in medieval mathematical manuscripts. I have adopted modern punctuation as the meaning demands, but have followed the medieval spelling, e. g., *e* for *ae*, *agregatur* in Version II, etc. The marginal folio numbers for Version I refer to manuscript *I*, those for Version II to manuscript *Ya*.

([24]) Thaw points out in a comment on the margin of manuscript *I* opposite the theorem (see Variant Readings, Version I, line 1) that there is another way of proof in Campanus' treatment of Proposition II. 13 of the *Elements*, a remark that puzzled Curtze since he could find no such reference. However, in the edition of Basel, 1546, p. 51, we read after the proof of II.13 : 'Notandum autem per hanc et precedentem et penultimam primi quod cognitis lateribus omnis trianguli, cognoscitur area ipsius, et auxiliantibus tabulis de chorda et arcu, cognoscitur omnis eius angulus'. Compare lines 1-2 of Version II, where the long enunciation of Version I has been abandoned in favor of short problematic statement of the theorem, in the manner of the brief statement of Campanus.

SIGLA OF MANUSCRIPTS

Versio I

I = Dresden Sächs. Landesbibliothek, Db. 86, 178r-v, 14c. (Cf. the text
of M. Curtze, cited in footnote 16).
Xa = Rome, Vat. Reg. Suev. 1261, 57v-58r, ca. 1350-1375.
Q = Paris, BN lat. 7378A, 40r-v, 14c.

Versio II

Ya = Munich, Bay. Staatsbibliothek, cod. 234, 105v-108v, 15c.

[Area trianguli tribus lateribus mensurata]

[Versio I]

178r / SI TRIA TRIANGULI LATERA COACERVENTUR,
MEDIETATISQUE COMPOSITI AD SINGULA LATERA DIF-
FERENTIE SUMANTUR, PRIMAQUE IN SECUNDAM DU-
CATUR ET IN PRODUCTUM TERTIA, ITEMQUE QUOD
5 INDE PROVENIT IN PREDICTAM MEDIETATEM, ILLIUS
ULTIMO PRODUCTI RADIX ERIT AREA TRIANGULI.

[Regula hec in arabico conscripta dicitur, in qua quoniam
tertia multiplicatio, videlicet linee in solidum, que in continuis
non habetur, rationem in ea sumere oportet numerorum].

---

1   *mg. Xa* Hec est pars Phyloteigni et debet ei subiungi
     mg. 16c *m. I* σφάληρον θεώρημα. vide hac de re Campanum Prop. 13,
     lib. 2, Euclidis aliter idem
     tria trianguli *tr. XaQ*
2   medietasque *Q*
4   quod *XaQ om. I*
5   provenit *XaQ* productum *I*
5-6 illius ultimo *XaQ om. I*
6   producti radix *tr. I*
7-9 [Regula...numerorum] *XaQ om. I*

10   Sit itaque datus triangulus ABC [Fig. 1] et medietas coacer-
vati ex lateribus ipsius sit linea KLM, differentieque ipsius sint
ad AB linea T, et ad AC S, et ad BC R. Sitque solidum quod
continetur R, S, T designatum nota Z. Intelligamus itaque in

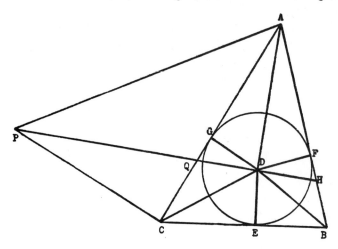

Fig. 1

---

10   itaque *XaQ om. I*
11   — que *I om. XaQ*
12   linea *om. Xa*
        et [1] *om. Q*
        S : scilicet (?) *Xa*
        et [2] *om. Q*
        Sitque *XaQ sed I*
13   continetur *XaQ* continent *I*
        R, S, T *XaQ* idest *I*
        itaque *XaQ* igitur *I*

dato triangulo circulum contingentem latera notis F, G, E, cuius
15 centrum D, a quo prodeant ad puncta contactus linee DF, DE, DG
singulis lateribus perpendiculares, atque inter se equales. Sed et
a D ad tres angulos linee protrahantur, que singulos angulos per
equa partientur, quorum omnium medietates quoniam equantur
uni recto et quia angulus DBF et angulus BDF equantur recto, erit
20 angulus BDF tanquam angulus DAF et angulus GCD. Ducatur
itaque linea PQDH ita ut angulus HDF equatur angulo DAF,
eritque angulus HDA rectus, et similiter angulus QDA, et ob
id etiam angulus GDQ equalis angulo GAD, et reliquus, scilicet
QDC, equalis angulo DBH, quoniam totus angulus GDC est
25 equalis duobus, scilicet GAD et DBE. Erigatur itaque perpendicularis a C donec concurrat cum linea HDQP, que sit CP, et
continuetur P cum A. Deinde, quoniam AF est equalis AG, et
BF equalis BE, atque CE equalis CG, linee enim ab eodem
puncto exeuntes usque ad contactus circuli sunt equales, tunc
30 AF et FB et CE sunt tanquam tres relique. Erunt ergo medietas
trium laterum componentque lineam KLM, sitque K equalis
AF et L equalis FB et M equatur CE. Manifestum est etiam

14	triangulo *XaQ* trigono *I*
	latera *I om. XaQ*
15	*ante* ad *add. XaQ* D
	DG *I* GD *XaQ*
16	lateribus *XaQ* lineis *I*
18	medietas *Q*
19	uni *I* cum *XaQ*
	DBF *I* BDF *XaQ*
	BDF *I* DBS *XaQ*
20	BDF *I* DBF *XaQ*
	GCD *I* DCG *XaQ*
21	itaque *XaQ* ergo *I*
	equatur *IQ* equetur *Xa*
23	id etiam *I* hoc *XaQ*
	GDQ *Xa* QDG *Q* GQD *I*
24	QDC *XaI* QDE *Q*
	GDC *XaQ* QDC *I*
25	duobus scilicet *I* angulo *XaQ*
26	HDQP *I* HDPQ *XaQ*
28-29	enim... exeuntes *I* siquidem exeuntes ab eodem puncto *Xa*
29	contactus *I* contactum *XaQ*
31	— que *XaQ* ergo *I*
32	equatur *I* equetur *XaQ*

88

quod KLM addit super AB quantum est M, quare M equalis
est T. Sed super AC addit L ; quare L equatur S. Atque super
35 BC addit K, itaque K est equalis R. Sint etiam linee N et O
equales perpendicularibus, quarum una DF, solidumque sub
lineis KLM et N et O contentum sit Y. Quia item anguli ADP
et ACQ sunt recti et equales, puncta A, D, C, P in eodem semi-
circulo consistent ; quare anguli PAC et PDC sunt equales.
40 Erunt ergo anguli PAC et DBF equales, itaque trianguli PAC
et DBF sunt similes, / eritque BF ad FD sicut AC ad CP. Sed
etiam FD ad FH sicut PC ad CQ, quoniam trianguli DFH et
CPQ sunt similes. Itaque BF ad FH tanquam AC ad QC. Ergo
coniunctim BF et AC ad FH et CQ sicut BF ad FH. Sed FH
45 est tanquam QG propter triangulos similes et quia FD equalis
est DG, [et itaque FH et QC sunt tanquam GC]. Itemque BF
et AC sunt tanquam KLM. Est ergo KLM ad GC sicut BF ad
FH. Sed proportio BF ad FH aggregatur ex proportione BF ad

33-34	equalis est I est equale XaQ
34	Sed super AC Xa superadditus L Q sed et similiter AC I
	quare I itaque XaQ
	L I B XaQ
	super XaQ similiter I
35	itaque I ergo XaQ
	equalis I equale XaQ
	R I Y XaQ
	O I D XaQ
37	item Xa idem I tunc Q
	anguli I trianguli XaQ
38	et equales I om. XaQ
39	consistent I consistunt XaQ
	quare I et XaQ
40	DBF XaQ FBD I
40-41	PAC et DBF XaQ FBD et ACP I
42	PC XaQ PO (?) I
43	CPQ I PQC Xa PAC Q
	QC I CQ XaQ
44	FH ^{2} XaQ HF I
	FH ^{3} *correxi ex* BF IXa *et* BH BF Q
45	*post* similes *add.* XaQ FDB, DGQ
46	DG I GD XaQ
	[et... GC] *supplevi*
	Itemque : tuncque Q
47	sunt I om. XaQ

DF et ex proportione DF ad FH. Sed AF ad FD tanquam FD
50 ad FH. Proportio ergo BF ad FH constat ex proportionibus
BF ad DF et AF ad DF. Sed proportio BF ad DF tanquam
linee S ad N, et AF ad DF sicut differentie R ad O proportio.
Itaque KLM ad T aggregatur ex proportione S ad N et R ad O.
Quia igitur solidi Y ad solidum Z aggregatur proportio ex pro-
55 portionibus KLM ad T et N ad S et O ad R, et proportiones
N ad S et O ad R faciunt proportionem T ad KLM, erit solidum
Y equale solido Z. Producit autem KLM in O superficiem equalem triangulo dato. Divisus est enim triangulus ABC in tres
triangulos ADB et BDC et ADC. Sed ex dimidio AB in DF fit
60 equale triangulo ADB et ex dimidio AC in DG equum est trigono
ADC et quod ex dimidio BC in DE est equale triangulo BDC.
Et quia omnes perpendiculares sunt equales linee O, ideo quod
fit ex dimidio omnium laterum, et ipsum est KLM, in O est
tanquam area trianguli ABC. Cum ergo solidum Y habeat tria
65 latera N, O, KLM, et ex O in KLM fiat area trianguli, tunc

---

49	ex *I om. XaQ*
49-51	tanquam... DF [2] *om. Q*
50	Proportio ergo *tr. Xa*
	proportionibus : proportione *Xa*
	DF [1] : FD *Xa*
54	aggregatur proportio *tr. XaQ*
57-58	equalem... dato *I* dato triangulo equalem *XaQ*
58	Divisus *I* quia divisus *XaQ*
	enim *I om. XaQ*
59	ADB *I* scilicet ABD *XaQ*
	et [1], [2] *I om. XaQ*
	fit *I* sit *XaQ*
60	triangulo *I* trigono *XaQ*
	ADB : ABD *Q*
	est *I om. XaQ*
61	est equale *I* equum est *XaQ*
	BDC *I* DBC *XaQ*
62	sunt... O *I* linee O sunt equales *XaQ*
63	fit *I* fit P *XaQ*
	ipsum *I* ipse *XaQ*
	in O *tr. I ante* et
64	ergo *I* igitur *XaQ*
	habeat tria *XaQ* habet *I*
65	N, O *XaQ* O, N *I*
65-67	tunc... trianguli *om. Q*

N ducta in eandem aream perficit solidum Y. Quare cum O sit equalis N, erit area trianguli inter lineam KLM et solidum Y [medium] proportionale ; quare inter eandem et solidum Z. Si ergo, ut proponitur, KLM ducatur in Z, radix producti erit
70 area trianguli.

[Versio II]

105v / Cognitis tribus lateribus cuiusque trianguli eius aream invenire.

Propositum est cognitis tribus lateribus trianguli invenire
106r quanta sit superficies que ab eis ambitur. Ad huius itaque /
5 exemplum ponam ut sint latera trianguli ABG propositi nota [Fig. 2], et per hoc inveniam eius aream. Inscribam autem intra propositum triangulum circulum DFE supra centrum D et a centro ducam lineas tres ad tria puncta in quibus circulus contingit latera trianguli, que sint DC, DF, DE, et ab eodem centro ducam
10 lineas tres ad tres angulos eius, que sunt DA, DB, DG. Ex hoc igitur declaratur nobis quod quilibet illorum angulorum divisus est in duos angulos equales, propter hoc quod latera trianguli ADE sunt equalia lateribus trianguli ADC, et latera trianguli BDC lateribus trianguli BDF, et similiter latera trianguli GDF
15 lateribus trianguli GDE. Cum igitur tres anguli trianguli ABG sint equales duobus rectis, tres medietates eorum, que sunt anguli DAE, DBF, et DGE, erunt equales uni recto. Set angulus DGE cum angulo EDG valet unum rectum, quoniam tertius, scilicet GED, est rectus. Ergo duo anguli DGE et EDG sunt
20 equales tribus angulis qui sunt DAE, DBF, et DGE. Ergo remoto angulo DGE remanebit angulus EDG equalis duobus angulis DAE et DBF. Ergo ipse erit maior angulo DAE. Faciam itaque ipsum sibi equalem et protraham lineam HD que cum ED faciat

---

66   perficit *I* producit *Xa*
67   equalis *Xa* equale *I*
68   [medium] *supplevi*
69   proponitur *I* proponatur *XaQ*
70   trianguli : trianguli explicit *Xa*

17   anguli *corr. ex* angulus
18   valet *corr. ex* valent

angulum equale angulo DAE, quam etiam producam in con-
25 tinuum usque ad P, et ponam notam Q ubi intersecabit lineam
AB. Et quoniam medietas trium laterum cuiusque trianguli
est maior quolibet latere ipsius, erit tota linea AG cum BF maior

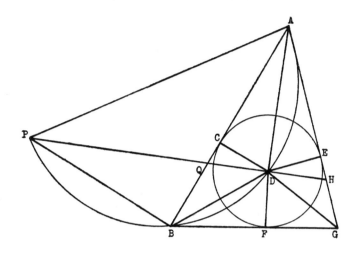

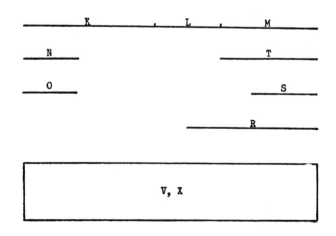

Fig. 2

<hr />

24   producam *corr. ex* productam

quolibet latere propositi trianguli, nam AE est medietas AE et
AC, et EG est medietas EG et GF, et FB est medietas BF et
30 BC. Ergo AG et FB simul sunt medietates omnium laterum
106v / predicti trianguli. Sit itaque linea KLM, quod agregatur ex
AG et FB, et sumatur differentia eius ad quodlibet trium laterum. Et sit eius differentia ad latus quidem AG T et ad latus
AB sit S et ad latus BG sit R. Manifestum est igitur quod T
35 erit equalis CB, eo quod medietas trium laterum excedit AG
in CB, et S erit equalis GE quoniam medietas trium laterum
excedit AB in GE, et R erit equalis AE quoniam medietas trium
laterum excedit BG in AE. Sit itaque solidum quod fit ex T,
S, et R solidum Z. Sumantur etiam due linee equales perpendi-
40 culari DE, que sunt N et O, et fiat ex eis et linea KLM solidum
Y. Dico igitur tunc quod solidum Y est equale solido Z, quod
sic ostendam.

Angulus HDE est equalis angulo DAE. Set angulus DAE
cum angulo EDA valet unum rectum quoniam tertius est rectus.
45 Ergo angulus HDE cum angulo EDA valet unum rectum. Ergo
totus angulus ADH est rectus. Ergo etiam reliquus ADP erit
rectus, propter hoc quod linea HDP est linea una. Ergo ipsi
sunt ambo equales. Si igitur ab eis demantur duo anguli ADC
et ADE, qui propter hoc quod reliqui duo sunt equales reliquis
50 duobus remanebit angulus EDH equalis angulo CDQ. Et iterum
quoniam anguli DAC et DBC et DGE sunt equales uni recto,
eo quod ipsi sunt medietas trium angulorum trianguli, et angulus QDC est equalis angulo DAC, erunt tres anguli QDC, DBC,
et DGE equales uni recto. Set etiam anguli CDB et CBD equantur
55 uni recto propter quod alter est rectus. Ergo tres anguli QDC,
CBD et DGE equantur duobus qui sunt CDB et CBD. Quare
demptis duobus qui sunt QDC et CBD remanebit QDB equalis
DGE. Producam itaque a puncto B perpendicularem a linea
107r AB / et protraham ipsam quousque concurrat cum HQP et
60 ponam P in loco concursus. Concurrent autem quoniam angulus
BQD est maior recto, eo quod ipse est extrinsecus ad QCD, qui
est rectus. Ergo reliquus BQP est minor recto et PBQ est rectus.

---

30   BC corr. ex FC
39   S corr. ex scilicet
54   anguli corr. ex angulus
57   QDB corr. ex QBD
58   producam corr. ex productam

Ergo isti duo simul sunt minores duobus rectis. Igitur in eandem
partem protracte concurrent. Deinde a puncto P ducam lineam
65 ad punctum A super eum. Igitur describam semicirculum supra
quem sunt A, D, B, P ; et quoniam ipsa opponitur angulo recto
qui est ABP et angulo recto qui est ADP, erit uterque illorum
in circumferentia semicirculi descripta supra eam. Ergo puncta
A, D, B, et P sunt in circumferentia eiusdem circuli. Propter
70 hoc igitur erit angulus PDB equalis angulo PAB, eo quod ambo
sunt super unam portionem circuli que est PB. Quapropter
angulus PAB est equalis angulo DGE, erunt igitur duo trianguli
APB et DGE equianguli et eorum latera proportionalia. Set
et similiter erunt duo trianguli PQB et DQC, quoniam angulus
75 DQC est equalis angulo PQB quia sunt contra se positi et reliquus utriusque est rectus ; et quia triangulus QDC fuit equian-
gulus et equilaterus cum triangulo HDE, erit HDE similis QPB
et eorum latera proportionalia. Erit itaque proportio AB ad
BP sicut proportio GE ad ED. Set etiam proportio PB ad BQ
80 est sicut proportio DE ad EH. Ergo a primo proportio AB ad
BQ est sicut proportio GE ad HE. Ergo coniunctim proportio
AB et GE simul ad BQ et HE simul est sicut proportio GE ad
HE. Set HE est equalis QC, ut ostensum est prius. Ergo proportio AB et GE simul ad BQ et CQ simul ita quod CQ acci-
85 piatur loco HE est sicut proportio GE ad HE. Set AB et GE
simul est equalis KLM, eo quod ipsa est medietas trium laterum ;
107v / et BC est equalis T, quoniam T fuit differentia medietatis
trium laterum ad latus AG. Ergo proportio KLM ad T est sicut
proportio EG ad HE. Proportio autem EG ad HE est agregata
90 ex duplici proportione, scilicet GE ad ED et DE ad EH. Set
proportio DE ad EH est sicut proportio AE ad ED, propter
hoc quod ab angulo ADH recto cadit DE perpendiculariter super
latus quod opponitur recto angulo. Ergo facit duos triangulos
similes sibi invicem. Et totalis ergo proportio GE ad EH est
95 agregata ex proportione GE ad ED et AE ad ED. Sed iam fuit
proportio KLM ad T sicut GE ad EH. Ergo proportio KLM
ad T est agregata ex proportione GE ad ED et AE ad ED. Set
GE est equalis S et ED est equalis N, et iterum AE est equalis
R, et ED est equalis O. Ergo proportio KLM ad T est agregata

70   PDB *corr. ex* PDP
      PAB *corr. ex* PAP

100 ex proportione S ad N et R ad O. Ergo everse proportio T ad
KLM est agregata ex proportione N ad S et O ad R. Set rursus
proportio solidi Y ad solidum Z est agregata ex proportione
laterum suorum, sicut patet per undecimum Euclidis. Ex proportione itaque KLM ad T et N ad S et O ad R fit proportio
105 solidi Y ad solidum Z. Set iam fuit proportio N ad S ducta in
proportionem O ad R sicut proportio T ad KLM. Ergo proportio solidi Y ad solidum Z fit ex proportione KLM ad T et T
[ad] KLM. Cum itaque fiat ex proportione primi ad secundum
et secundi ad primum, ipsa erit proportio equalitatis. Ergo
110 solidum Y est equale solido Z. Quare cum sit solidum Z notum
108r quia omnia eius latera sunt / nota, erit propter hoc solidum Y
notum.

Set ex ductu medietatis laterum in perpendicularem fit area
trianguli, quoniam ex ductu AE in ED fit superficies quedam
115 dupla ad triangulum AED propter quod ipse est orthogonius.
Ergo ipsa superficies est equalis toti superficiei ACDE. Et similiter ex ductu GE in DE fit quedam superficies equalis GEDF.
Et ex ductu BF in FD, que est equalis CD, fit superficies equalis
BFDC. Ergo ex ductu medietatis trium laterum in perpendi-
120 cularem fit area trianguli. Ergo ex ductu KLM in N fit area
trianguli. Set ex ductu O in hoc productum fit solidum Y. Ergo
ex ductu perpendicularis in aream trianguli fit solidum Y ; ergo
et solidum Z. Ponam itaque ut area trianguli sit V et accipiam
aliam sibi equalem que sit X et accipiam [N] in loco O propter
125 hoc quod ipse sunt equales. Dico ergo quod ex ductu N in KLM
fit V, quod est ipsa area trianguli, et ex ductu N, quod est O,
in X, quod etiam est area trianguli, fit solidum Y. Hic una et
eadem quantitas, scilicet N, multiplicat duas, videlicet KLM
et X. Ergo eadem est proportio multiplicatorum et productorum.
130 Ergo que est proportio KLM ad X eadem V ad Y. Sed que est
KLM ad X eadem est KLM ad V, eo quod X et V sunt equalia.
Ergo que est proportio KLM ad V eadem est V ad Y. Ergo V
est medio loco proportionalis inter lineam KLM et solidum Y.

---

100   *mg. c. 2 107v add.* Ya Esto memor ponere ut sit unum atus 21 et
         aliud 20 et tertium 19 et tunc sit area trianguli fere 172 partes
         quadrate quarum longitudo et latitudo erit secundum magni-
         tudinis unius partium positarum.
105   Y *corr. ex* X
120-122 Ergo... trianguli *mg.* Ya

Set solidum Y est equale solido Z. Ergo area trianguli est medio
135 loco proportionalis inter KLM et Z, quod est, inter medietatem
trium suorum laterum et solidum quod fit ex tribus differentiis
medietatis trium laterum ad t ia latera. Ergo quod fit ex ductu
medietatis laterum [in] illud solidum valet illud quod fit ex
ductu aree in se. Si igitur coacervemus tria latera trianguli et
140 agregati sumpserimus medietatem et illius ad tria eius latera
sumpserimus differentias et multiplicaverimus primam in se-
108v cundam et tertiam in productum fiet solidum / illud inter quod
et medietatem laterum est area trianguli proportionalis, et si
in ipsum multiplicaverimus medietatem illorum trium laterum
145 quod provenerit erit equale ei quod fit ex ductu aree in se. Si
igitur illius sumpserimus radicem, ipsa area trianguli et hoc
est quod voluimus demo[n]strare.

Est etiam alter modus inveniendi per hanc eandem demo[n]-
strationem in cognitione aree trianguli suppositis notis late-
150 ribus. Demo[n]stratum enim est quod solidum quod fit ex tribus
differentiis que sunt medietatis laterum ad tria latera valet
solidum quod fit ex medietate trium laterum in perpendicularem
vel in semidiametrum circuli inscripti et ex ductu eiusdem semi-
diametri in productum. Item demo[n]stratum est quod illud
155 quod fit ex semidiametro in medietatem trium laterum est
equale aree trianguli. Si ergo semidiameter circuli esset nobis
nota cum medietas trium laterum sit nota quia omnia latera
nota, multiplicaremus semidiametrum in illam medietatem laterum et fieret productum area trianguli. Set semidiameter sic
160 fiet nota. Solidum quod fit ex tribus differentiis in se divide per
medietatem laterum et exibit superficies que fit ex semidiametro
in se. Quare ergo illius radicem [sume], et ipsa erit semidiameter,
quam multiplica in medietatem laterum et habebis aream. Et
iste modus est omnino idem cum priori.
165 Hec autem regula de triangulo fuit scripta in arabico : Si
tria latera trianguli coacerventur, medietatisque compositi ad
singula latera differentie sumantur, primaque in secundam ducatur et in productum tertia, itemque quod inde provenerit in predictam medietatem, illius ultimo producti radix erit area trianguli.

---

155   medietatem *corr. ex* medietate
162   [sume] *supplevi*
168   *post* productum *habet* Ya *et delevi* in
169   producti *corr. ex* producta

X

A MEDIEVAL ARCHIMEDEAN-TYPE PROOF
OF THE LAW OF THE LEVER

I know of no more appropriate way to express my appreciation for the scholarly work of André Combes than to present here a short, hitherto-unedited text. This text consists of a proof of the law of the lever which I have recently discovered in a Parisian manuscript (BN lat. 7377 B, 93v-94r). The section of the manuscript containing this proof was probably written in the fourteenth century at Paris. Its significance lies in its substitution of a statical-symmetrical proof of the Archimedean kind for the dynamic proof presented by the celebrated mathematician Jordanus de Nemore in his *Elementa de ponderibus*. In fact, the anonymous author appears to be the only known medieval Latin author to present such a proof.

It has been surmised that Jordanus' *Elementa*, written in the early thirteenth century, was composed to provide a theoretical background or justification for a brilliant little tract on the material balance of unequal arm lengths entitled *De canonio* [1]. The *De canonio*, containing four propositions, was translated from the Greek and appears to have had wide circulation in the later middle ages [2]. Jordanus' *Elementa*, then, presents a series of propositions that lead up to a dynamic proof of the law of the lever (Proposition 8) [3], followed by a last

---

[1] For the general suggestion that the medieval *auctores de ponderibus* undertook to supply the missing theoretical underpinning for the *De canonio*, see E. A. MOODY and M. CLAGETT, *The Medieval Science of Weights* (Madison, 1952; 2nd printing, 1960), 61. For an excellent elaboration of this theme with special attention to Jordanus' *Elementa*, see JOSEPH BROWN, *The « Scientia de ponderibus » in the Later Middle Ages* (Dissertation, University of Wisconsin, 1967).

[2] Father Brown in his above-mentioned dissertation (pp. 10-11) lists 26 manuscripts of the *De canonio*, the largest number of manuscripts for any single work of the *corpus de ponderibus*.

[3] MOODY and CLAGETT, *The Medieval Science of Weights*, 128-40.

theorem (Proposition 9) that forms a bridge between the analysis involving weightless beams and that of the *De canonio* involving material
beams. It is this last proposition that appears to have stimulated our
anonymous author to try a new way of introducing the *De canonio*
and so I quote it in full ([4]):

> *E. 9* IF TWO OBLONG BODIES, WHOLLY SIMILAR
> AND EQUAL IN SIZE AND WEIGHT, ARE SUSPENDED
> ON A BALANCE BEAM IN SUCH MANNER THAT ONE
> IS FIXED HORIZONTALLY TO THE END OF ONE ARM,
> AND THE OTHER IS HUNG VERTICALLY, AND
> SO THAT THE DISTANCE FROM THE AXIS OF SUP-
> PORT, TO THE POINT FROM WHICH THE VERTICAL-
> LY SUSPENDED BODY HANGS, IS THE SAME AS THE
> DISTANCE FROM THE AXIS OF SUPPORT TO THE
> MID-POINT OF THE OTHER BODY, THEN THE SUS-
> PENDED WEIGHTS WILL BE OF EQUAL POSITIONAL
> GRAVITY.

Let A and B be the ends of the beam, C the axis, and
let the weight fixed horizontally to the beam be ADE, its
mid-point D; and let the other weight, which hangs, be BG.
And let BC be equal to CAD. I say that the weights ADE
and BG, in this position, will be of equal heaviness.

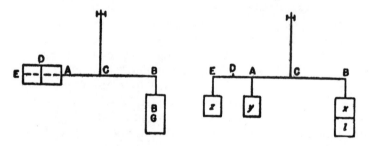

To make this evident, we say that if the beam, on the
side of A, were as long as CE, and if there were suspended
at A and E two equal weights z and y, and if a weight double

([4]) *Ibid.*, 141. I shall not repeat the Latin text here. I have used Moody's
translation except that I have restored the errors concerning proportions (« as *EC*
is to *CB* » and « as *AC* is to *CB* ») found in all of the manuscripts, errors which
Moody had corrected in his translation without any comment. See the next footnote.
I have also substituted « ratio » as the translation of *proportio* for Moody's
« proportion ».

each one of these, *xl*, were suspended from B, then in this
position also *xl* would be equal in heaviness to *z* and *y*. For
let its halves be *x* and *l;* then the weight *x* will be to the
weight *z*, as EC is to CB; and the weight *l* will be to the
weight *y*, in this position, as AC is to CB. Hence the
ratio of *xl* to *z* plus *y*, in this position, will be the same
as the ratio of EC plus AC to twice BC. But because
twice BC is equal to EC plus AC, *xl*. will likewise be equal
in positional weight to *z* plus *y*. For this reason, therefore,
since all the parts of GB are of equal weight, in this position,
and since any two parts of ADE, equidistant from D, are equal
in weight to two equal parts of BG, it follows that the whole
of ADE is equal to the whole of BG (in positional weight).
And this is what needed to be proved.

Now there is a multiple error of ratios in the proof (« as EC
is to CB » and « as AC is to CB ») in all of the manuscripts of *Elementa*
that caused later commentators to change the proof ([5]) and perhaps
provided one of the reasons why our author decided to take a different
tack. At any rate, from this proposition one then passes to Proposition
I of the *De canonio* (although in fact Proposition 9 of the *Elementa*
concerns a balance between equal weights and *De canonio*, Proposition
I, a balance between unequal weights) ([6]):

I. IF THERE IS A BEAM OF UNIFORM MAGNITUDE
LENGTH AND OF THE SAME SUBSTANCE, AND IF IT IS
DIVIDED INTO TWO UNEQUAL PARTS, AND IF AT
THE END OF THE SHORTER SEGMENT THERE IS
SUSPENDED A WEIGHT WHICH HOLDS THE BEAM
PARALLEL TO THE PLANE OF THE HORIZON, THEN
THE RATIO OF THAT WEIGHT TO THE EXCESS OF
THE WEIGHT OF THE LONGER SEGMENT OF THE
BEAM OVER THE WEIGHT OF THE SHORTER SEG-
MENT IS AS THE RATIO OF THE LENGTH OF THE
WHOLE BEAM TO TWICE THE LENGTH OF THE
SHORTER SEGMENT.

([5]) Father Brown has shown the significance of these errors in the preparation
of later commentaries, *op. cit.*, 4, 7-8, 51, 60-61, 317, 627-30.
([6]) MOODY and CLAGETT, *The Medieval Science of Weights*, 64-66.

Let the beam AB, then, be divided into two unequal segments at the point G, and let the shorter segment be AG, and the longer one GB. And let z be the weight suspended from the point A. And let GD, equal in length to AG, be marked off. It is then evident that the remainder, DB, is the excess of the segment GB over the segment AG, both in length and in weight. I say therefore that the ratio of the weight z to the weight of the segment DB, is as the ratio of the length of AB to the length of AD.

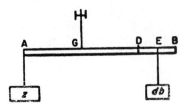

The proof of this is as follows: A beam of the length AD, if understood to be suspended from the mid-point G, without any weight being attached at either of its ends, will undoubtedly be parallel to the plane of the horizon. And when, from the point A, the weight z is suspended, the beam DA will fall on the side of A, by reason of the gravity of the weight z. Consequently, that which holds the beam AD parallel to the horizon, is the weight of DB. And it has been demonstrated in the books which speak of these matters, that it makes no difference whether the weight of DB is equally distributed along the whole line DB, or whether it is suspended from the mid-point of that segment. Let the weight db, then, be suspended from the point E. Since the line AE is divided into two unequal parts at the point G, and since the weight z is suspended from the point A and the weight db from the point E, and since the beam AE is then parallel to the horizon, the ratio of the segment EG to the segment GA will be as the ratio of the weight z to the weight db; as has been proved by Euclid, Archimedes, and others. And this is the foundation on which all [the propositions] depend. Now AB is twice EG, and AD is twice AG; therefore the ratio of the weight z to the weight of DB, is as the ratio of the line AB to the line AD.

It will be noticed that this proof depends on a series of specified assumptions: 1) a material beam without pendent weights but supported from its mid-point will be in equilibrium; 2) if a weight is suspended from one end of such a beam it will decline at that end; 3) a material beam segment may be replaced by an equal pendent weight hung from the mid-point of the length of the beam segment without disturbing equilibrium; and 4) the law of the lever, i. e., unequal weights suspended at distances inversely proportional to the weghts will be in equilibrium. It is only in connection with the last assumption, the general law of the lever, that the author specifies the previous mathematicians responsible for its proof: « Euclid, Archimedes, and others ». Presumably the first three suppositions emerged from a consideration of symmetry and center of gravity in the manner of Archimedes' *On the Equilibrium of Planes*. Indeed, the third assumption is merely the inverse of the assumption that Archimedes makes in his lever proof (Proposition 6) when he distributes a weight along a weightless beam without altering the center of gravity.

Having examined Proposition I of the *De canonio* we can briefly note that Proposition II is the inverse of Proposition I (and thus constitutes a form of the lever law as applied to a material beam balance of unequal arm lengths). We give its enunciation without proof (which proof is an indirect one based on Proposition I) ([7]):

II. IF THE RATIO OF THE WEIGHT SUSPENDED AT THE END OF THE SHORTER SEGMENT, TO THE EXCESS OF THE WEIGHT OF THE LONGER SEGMENT OVER THE WEIGHT OF THE SHORTER ONE, IS AS THE RATIO OF THE LENGTH OF THE WHOLE BEAM TO TWICE THE LENGTH OF THE SHORTER ARM, THEN THE BEAM WILL HOLD PARALLEL TO THE PLANE OF THE HORIZON.

We are now prepared to consider the anonymous medieval proof from the Parisian manuscript. It was perhaps undertaken by someone disgruntled with the efficacy and soudness of Jordanus' *Elementa* as an introduction to the *De canonio*. Clearly, the author had read Proposition 9 of Jordanus' *Elementa,* for he uses the same term for a material beam segment: *oblongum,* a term not used in any of the other tracts on weights. He also uses the same expression for equi-

---

([7]) *Ibid.,* 67.

librium: *equidistare orizonti* (although to be sure this expression was occasionally used in other tracts) ([8]). In considering Proposition 9, he perhaps was disturbed by the above-noted errors in ratios that are found in Proposition 9 and possibly even more disturbed by the fact that one does not really need Proposition 9 at all in the passage to Proposition I of the *De canonio*. In fact, all that one really needs is Proposition 8. But presumably our author was also dissatisfied by the dynamic proof of Proposition 8. Hence, he seems to have decided to give an entirely new entrance into the problems of the *De canonio* by starting from the special assumption of the equilibrium of equal weights at equal distances and then proceeding by strictly symmetrical considerations to a modified form of the lever law where the basic beam is still weightless but where a material beam segment is balanced against a pendent weight. Hence, the whole of the *Elementa* is scrapped except a part of the actual wording of the lever law (*brachia libre fuerint proportionalia ponderibus ita ut breviori gravius appendatur*) which is straight out of Jordanus' eighth proposition.

Thus, if my interpretation of the proof is correct, it is Archimedean. But it is a direct proof, while that of Archimedes in Proposition 6 of *On the Equilibrium of Planes* is indirect since Archimedes reduces the situation involving unequal weights hung at inversely proportional distances to the special case. Before discussing the details of the anonymous proof itself it should be observed that the author specifies at the end of the proof four assumptions beyond his assumption in the body of the proof of the special case of the lever law. These assumptions have been bracketed in the translation below. Assumption [2], which holds that a uniform material beam is in equilibrium when hung from its mid-point, is identical with assumption 1) of the proof in *De canonio*, Proposition I. Both authors use it to authorize the elimination of the effects of equal beam segments about the center of movement, i. e., to make these equal material segments as if they were weightless. In addition, the medieval author seems to use this assumption in the very beginning of his proof when he hangs his material beam segment from its mid-point and thus it affords a kind of substitute for a center of gravity concept. Assumption [3], which holds that if a material beam is not hung from the mid-point it declines on the side of the longer arm, is not directly used by the medieval author and probably is merely an effort on his part to complete assumption [2] by assuming that it is only in the mid-point that a uniform beam may be supported and

---

([8]) *Ibid.*, 154, 174, 282. And see M. CLAGETT, *The Science of Mechanics in the Middle Ages* (Madison, 1958; 2nd printing, 1961), 69, footnote 1.

preserve equilibrium. The author's assumption [4], which in effect establishes that the rules of addition and subtraction apply to quantities of weights, allows him to assume the direct proportionality of the lengths of uniform material beams and the weights of those lengths. Finally, assumption [1] is merely a geometric restatement of one of the basic steps in the proof. I have explained it in footnote 2 of the translation.

Turning from the assumptions to the proof itself, we note that the most interesting ploy of the author occurs when he moves the center of movement from point H in the weightless beam down to point C in the material beam segment. This is effectively the same as distributing the weight of the material segment along the weightless beam. In the fact that the material beam segment AE extends beyond the center of movement, we have a situation similar to that in Proposition 6 of Archimedes' *On the Equilibrium of Planes*, where in distributing the longer of two unequal weights along the weightless beam some of the distributed weight extends beyond the center of the whole weightless beam. Whether or not the author drew this technique from Archimedes I cannot say. All I can say is that if the author was composing this proof in Paris in the fourteenth century he could have had access to Moerbeke's translation of *On the Equilibrium of Planes* in the same manuscript of the translation used by Johannes de Muris, Nicole Oresme and perhaps others.

In order to understand better the intention of the author in his proof, I would like to add some further possible considerations that remain unspecified by the author. Let us suppose for a moment that not only is the initial material beam segment AE distributed over the weightless beam (as the author proposes) but also that weight f, its equal, is formed into a like beam and similarly distributed (this the author does not do). But lest the beams interpenetrate, let us have two abutting beams hanging as illustrated in Fig. A. 1. Equilibrium remains even though the beams are not in line so long as they are affixed one to the other and the system is properly suspended. Now suppose, with the author, that we remove the equal segments m and no from the nearer beam and at the same time bring l and $A'E'$ into line. The result is that we have a straight beam composed of l and $A'E'$, which is in equilibrium about C. Then, following the technique of *On the Equilibrium of Planes*, Proposition 6, we could say that this single beam is equivalent to having two weights, one l with its center at Q and the remaining segment $A'E'$ with its center at D. And then if we leave l where it is and take the weight f, equal to the remaining segment $A'E'$, and hang it from D, we do not, in the

Archimedean procedure, disturb the equilibrium and in fact we have the condition of equilibrium specified by the author of this anonymous proof. Then since it can be proved that $\dfrac{f}{l} = \dfrac{QC}{CD}$, and QC and CD are the distances at which f and l act and preserve equilibrium,

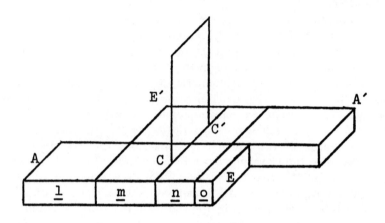

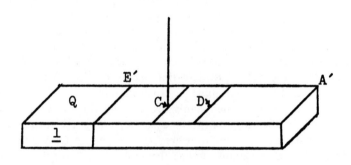

FIG. A. 1

we have thus shown that an equilibrium exists when the weights are inversely as the distances.

One last observation concerning our anonymous proof is worth making. The form of the lever law that is here under consideration

416

is but a step from that given in *De canonio,* Proposition II. In the latter the whole balance beam is material, while in the anonymous proposition the basic beam is still weightless with one of the weights being a material beam segment.

In the text below, the figures are as given in the manuscript except that the proportion of the lines has been somewhat altered to agree with the ratios specified in the text. All the letters marking the quantities in the manuscript text and figures are lower-case letters. But following the practice of *The Medieval Science of Weights,* I have used capital letters for line lengths and points, small letters for weights.

AN ANONYMOUS LEVER LAW PROOF

[*Anonymi demonstratio*]

/ Pono pondus totius oblongi *ABCDE* esse equale ponderi
f et sunt brachia libre *GH, HI* equalia [Fig. A. 2]. Et pono dictum
oblongum dependere < a > medio puncto sui, hoc est a puncto *B*, per
perpendicularem *BG*, et pondus *f* dependere per perpendicularem *AK*
5 (*! IK*). Et quoniam puncta *G*, *I* equaliter distant a puncto *H*, quod est
centrum motus, et ab ipsis dependent duo pondera equalia-- scilicet
oblongum predictum et pondus *f*-- propter hoc equidistant *GH, HI*,
que sunt brachia libre, orizonti. Tunc abscindam ex linea *AC* lineam
CP equalem *CE*. Et faciam descendere *H*, quod est centrum motus,
10 usque ad *C*. Deinde tollam ab utroque lateri centri motus ex dicto
oblongo duas partes equales, que sunt *m* et *no* (et tunc procedit ad
secundum [figuram A. 3]). Et propter hoc remanebit *l* pars oblongi dependens a medio puncto sui, hoc est a puncto *Q*, equalis ponderis in illo
situ ponderi *f*. Modo intendo probare quod proportio brachii *QC* ad
15 brachium *CD* est tanquam proportio ponderis *f* ad pondus *l*, que est
pars oblongi. Dico in primis quod proportio linee *QC*, que est medietas
totius linee *AE*, ad lineam *CD*, que est medietas linee *BD*, est tanquam
proportio totius linee *AE* ad lineam *BD*. Sed proportio linee *AE* ad
lineam *BD*, vel ad lineam *AP* sibi equalem, est tanquam proportio
20 ponderis totius oblongi ad pondus *l* partis oblongi. Sed pondus *f*
est equale ponderi totius oblongi. Ergo a primo proportio linee *QC*
ad lineam *CD* tanquam ponderis *f* ad ad (*!*) pondus *l*, que est pars
oblongi totius. Et per hoc patet brachia libre fuerint proportionalia
ponderibus suis ita ut breviori gravius appendatur [et] lingula examinis
25 in neutram partem nutum faciet.

/ Si linea in duo equalia duoque inequalia secetur, et a tota
dematur duplum illius quod a minori sectione et a dimidio linee
interiacentis sectionibus aggregatur, id quod de tota linea remanet
equale est ei quod utrique sectioni interiacet.
30   Quando aliquod oblongum equalis grossitiei et ponderis in omni
parte dependet a medio puncto sui, tunc equedistet orizonti.
    Si vero non a medio puncto sui dependeat, tunc ex parte longiori
declinat.
    Omnis corporis pondus equale est omnibus ponderibus omnium
35 partium suarum aggregatis.

[*A Proof by an Anonymous Author*] ([1])

I posit that the weight of the whole oblong *ABCDE* is equal to weight *f* and that the arms of the balance *GH* and *HI* are equal [see Fig. A. 2]. And I posit that the aforementioned oblong hangs from its mid-point, that is, from point *B*, by the perpendicular *BG*, and that the weight *f* hangs by the perpendicular *IK*. And since points *G* and *I* are equally distant from point *H*, which is the center of motion, and two equal weights hang from them--namely, the aforementioned oblong and weight *f*--for that reason *GH* and *HI*, which are the balance

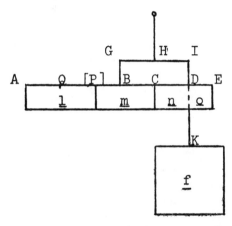

Fig. A. 2

arms, are parallel to the horizon. Then from the line *AC* I shall mark off *CP* equal to *CE*. And I shall cause *H*, which is the center of movement, to descend to *C*. Then on each side of the center of movement I shall take from the aforementioned oblong two equal parts, which are *m* and *no* (and then [the demonstration] advances to the second [figure, Fig. A. 3]). And for that reason part *l* of the oblong hung from its mid-point, that is from point *Q*, will remain of the same weight in that position as weight *f* [, that is, it will remain in equilibrium]. Now I intend to prove that the ratio of arm *QC* to arm *CD* is the same as the ratio of the weight of *f* to the weight of *l* which is part of the oblong. First of all, I say that the ratio of line *QC*, which is half of the whole line *AE*, to line *CD*, which is half

---

([1]) See Commentary, lines 1-25.

of line *BD,* is as the ratio of the whole line *AE* to line *BD.* But the ratio of line *AE* to line *BD,* or to its equal, line *AP,* is as the ratio of the weight of the whole oblong to the weight of *l,* a part of the oblong. But the weight of *f* is equal to the weight of the whole oblong. Therefore, from above the ratio of *QC* to line *CD* is as that

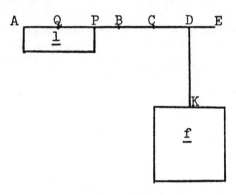

Fig. A. 3

of weight *f* to the weight of *l,* which is part of the whole oblong. And by that reasoning it is evident that the arms of the balance are proportional to their weight in such a way that the heavier weight is hung from the shorter arm while the tongue of the balance makes no movement on either side.

[1] If a line is cut into equal halves and the two halves are cut unequally [in the same way], and there is subtracted from the whole line a magnitude double that of the combined length of the lesser segment and half of the length of the segment lying between [the two lesser segments], that which will remain of the whole line is equal to the segment lying between each of the cutoff segments (^{2}).

[2] When some oblong body of equal denseness and weight in every part hangs from its mid-point then it is parallel to the horizon.

[3] If it does not hang from its mid-point then it declines on the side of the longer [segment].

[4] The weight of every heavy body is equal to the aggregate of the weights of all its parts.

(^{2}) This is more easily followed when expressed in specific quantities: (1) $AB = BE$; (2) $AB = (AP + PB)$ and $BE = (BD + DE)$; and (3) by subtraction, $AE - 2 (PB + BC) = BD = AP.$

COMMENTARY

1-25 « Pono... faciet ». The proof can be represented succinctly as follows: (1) Since $ae = f$ and $GH = HI$, ae and f are in equilibrium. (Note: ae is the weight of segment AE). (2) With the center of motion moved perpendicularly to C, the beam AE is effectively moved to the balance beam. Then remove from each side of C the equal material segments m and no (the weights respectively of segments CP and CE). Hence l remains in equilibrium with f (see the introduction for a discussion of this point). (3) Now prove that $\dfrac{QC}{CD} = \dfrac{f}{l}$ as follows:

1) $\dfrac{QC}{CD} = \dfrac{AE}{BD}$ since $QC = \dfrac{1}{2} AE$ and $CD = \dfrac{1}{2} BD$; 2) $\dfrac{QC}{CD} = \dfrac{AE}{AP}$

since $AP = BD$ by assumption [1] appearing at end of text; 3) but

$\dfrac{AE}{AB} = \dfrac{ae}{l} = \dfrac{f}{l}$ by assumption [4] at the end of the text; 4) there-

fore, $\dfrac{QC}{CD} = \dfrac{f}{l}$. (4) Hence l and f are in equilibrium when they are

inversely proportional to the arm lengths.

ARCHIMEDES AND SCHOLASTIC GEOMETRY

Much attention has been given to the efforts of schoolmen in the fourteenth century to apply mathematics to physical problems — in particular to their attempts to treat qualities and velocities quantitatively. Considerably less attention had been paid to a reverse current of « scholasticizing » geometry by techniques followed in natural philosophy at about the same time. This current could be illustrated well by describing and analysing some particular geometrical work like the elementary geometrical treatise of Thomas of Bradwardine[1]. But it is not my intention in this paper to treat the problem of the development of scholastic geometry by an analysis of a full geometrical tract. Rather I think it will be instructive to take a particular geometrical problem like that of a proof for the quadrature of the circle as given in Archimedes' *Measurement of the Circle* and follow the various changes that take place in the presentation and proof during the period of the late thirteenth and fourteenth centuries. It is my hope that some rather general considerations will emerge from this technique, as they have in so

1. An analysis of Thomas BRADWARDINE's *Geometria speculativa* would everywhere show the conflation of philosophical and mathematical arguments. A case in point occurs in his treatment of quadrature (edition of Paris, 1495, 14ᵛ-15ʳ(where in addition to Archimedes' *De mensura circuli* Aristotle's *Prior Analytics* is drawn into the argument, and the « proof » is presented in the form of a syllogism. See my *Archimedes in the Middle Ages*, vol. I (Madison, in press), chap. II, sect. 2, footnote 11. (This work will be hereafter abbreviated as *Archimedes*.)

many of Prof. Koyré's investigations when he used a similar technique.

Attention of Latin schoolmen was first drawn to the quadrature proof of Archimedes by the translation of the *Measurement of the Circle* twice during the course of the twelfth century[2]. By far the more important of these translations is the one
that we can with considerable surety ascribe to Gerard of Cremona. This accurate and complete translation was often repro-
duced in the next three centuries and quoted by many medieval
geometers[3]. In this translation the essential features of the Archimedean quadrature proof as given in Proposition I are preserved, but with some elaboration[4]. Archimedes' aim was first to
show that if a circle is said to exceed a right triangle whose sides
including the right angles are respectively equal to the circumference and radius of the circle, there will result some inscribed
polygon which is at the same time greater than and less than the
given triangle, an obvious contradiction. A similar contradiction
concerning some inscribed regular polygon ensues if we assume
the circle to be less than the given triangle.

In the course of the thirteenth and fourteenth centuries the treatment of Archimedes' proof as rendered in translation by Gerard
of Cremona was subjected to much paraphrasing and elaboration. At least ten different variant forms of Proposition I were
prepared[5] : The Cambridge Version (13th century), the Naples
Version (perhaps by Gerard of Brussels in the thirteenth century), two Florence Versions — F. IA and F. IB (both probably

2. The first translation was possibly made by Plato of Tivoli, the second
almost certainly by Gerard of Cremona. I published the texts of both of these
translations in *Osiris* 10 (1952), 599-606. Both of these texts have been revised,
republished, translated into English, and analysed in Chapter Two of my
Archimedes.

3. For the list of the manuscripts of this translation and the numerous citations of the text in the middle ages, see *Archimedes*, chap. II, sect. 2.

4. The elaboration present in the Gerard translation has been treated in
detail in my commentary to the text in *Archimedes*, chap. II, sect. 2. The
most interesting addition is a corollary to Prop. I relative to the area of a
sector of a circle, which, while missing in the extant Greek text, is assigned
to Archimedes by Hero.

5. The names of these versions are largely drawn from their principal manuscripts. All of these texts, with English translations, have been edited in
Chapters III and V of my *Archimedes*.

from the early fourteenth century), the Gordanus Version (late thirteenth or early fourteenth century), the Corpus Christi Version (early fourteenth century), the Munich Version (fourteenth century), the Pseudo-Bradwardine Version (fourteenth century), a *Versio abbreviata* (end of fourteenth century ?), and the *Questio Alberti de Saxiona de quadratura circuli* (ca. 1360 ?). The different ways in which the basic proof of Archimedes is handled in these various versions will illustrate, I maintain, some of the tendencies present in the scholastic treatment of geometry.

In the first two versions, namely those of Cambridge and Naples, the mathematical form of the proof was well preserved and only a modicum of elaboration was present. This elaboration was concerned largely with writing out the steps implied by Archimedes and citing the proper propositions of Euclid's *Elements* that would apply to these steps (there are eight citations to the *Elements* in the Cambridge proof and ten in that of Naples). More important, these early paraphrases set the custom of specifying Proposition X.1 of the *Elements* as the authority for the procedure whereby, starting with an inscribed regular polygon and then successively doubling the number of sides of inscribed regular polygons and thus at each step of doubling exhausting more than half of the area between the polygon and the circumference, one can ultimately produce a polygon such that the difference between it and the circle is less than any assigned quantity [6]. All but two of the ten versions cite Proposition X.1 as the point of departure for the extraction procedure. Incidentally, we can also note the tendency, started in these early versions and continued in most of the succeeding ones, to feel it necessary to specify by some letter (and in some cases by some particular geometrical figure) the assigned quantity by which the circle is said to exceed or be exceeded by the trian-

6. Most of the paraphrases and versions were acquainted with Proposition X. 1 in the form of a translation of the *Elements* which is now commonly called Adelard II. It reads (MS Brit. Museum Add. 34018, 38^v) : « Si duabus quantitatibus inequalibus positis maius dimidio [a] maiori detrahatur itemque de reliquo maius dimidio dematur deincepsque eodem modo, necesse est ut tandem minore positarum minor quantitas relinquitur. »

gle [7]. Furthermore, it should be noted in connection with the custom of citing the *Elements* inaugurated in these early versions that such citations multiply exceedingly in the scholastic versions (e.g., the *Elements* is cited thirty-six times in the Munich Version).

Another important addition found in the Cambridge Version is the prefixing of postulates that appear fundamental to the proof. The Cambridge postulates have no particular scholastic character, but, as we shall see below, they constitute a kind of first step in the growing self-consciousness concerning the logic of the proof which is very characteristic of the scholastic elaborations. The first postulate assumes that there can be found a straight line equal to a curve and vice versa [8]. This is, of course, crucial to the whole treatment of quadrature and must be at least tacitly assumed by all who adopt the Archimedean proposition and its proof. One might suggest that the author borrowed this postulate from the *De curvis superficiebus* of Johannes de Tinemue, translated in the thirteenth century [9]. Or one might say that the occasion for such a postulate was to recognize and at the outset to set aside the doubt expressed by Aristotle as to the equation of circular and straight lines, a doubt based on what was felt to be a basic specific difference between curves and straight lines (*Physics*, VII, 4, 249 *a*).

The second postulate of the Cambridge Version assumed that an arc is greater than its chord (see footnote 8). Such an assumption is, of course, embraced by the more general postulate given

7. For example, in the Cambridge Version the assigned excess is specified as D and it is noted that it does not make any difference as to the geometric form of that magnitude : « Sit ille excessus D, et non refert cuius forme sit », MS Cambridge, Gon. & Caius College 504/271, 108ᵛ, c. 1; cf. *Osiris, 10,* (1951), 610.

8. MS cit. in note 7, *ibid.* : « Cuilibet recte linee aliquam curvam esse equalem et cuilibet curve aliquam rectam. Item cordam quamlibet esse minorem arcu. Item ambitum cuiuslibet figure includentis esse maiorem ambitu figure incluse. » Cf. *Osiris, 10,* 609.

9. MS Oxford, Bodl. Library, Auct. F. 5 28, 111ʳ : « Presentis demonstrationes ypothesis et tota sequentium theorematum series lineam rectam curve et superficiem rectam curve esse equalem sibi postulat admitti. » This text also has been re-edited and translated into English in my *Archimedes*, chap. VI, sect. 2; but for this passage; cf. *Osiris, 11* (1954), 302, lines 15-17.

44

by Archimedes in his *Sphere and Cylinder* [10] : « Of all lines which have the same extremities the straight line is the least. » Similarly the Cambridge Version's third postulate, namely, that the perimeter of an including figure is greater than the perimeter of the included figure (see footnote 8), obviously has the same purpose as the conclusion drawn by Archimedes from his postulates in the *Sphere and Cylinder* [11] : « If a polygon is inscribed in a circle, it is evident that the perimeter of the inscribed polygon is less than the circumference of the circle », together with the first proposition of that work : « If a polygon is circumscribed about a circle, the perimeter of the circumscribed polygon is greater than the perimeter of the circle. » While the Cambridge postulate looks to be more general than the statements of Archimedes, its chief defect is that it does not make precise what constitutes « including » and « included ».

The actual proof of the proposition in the Cambridge Version demands no further comment, but it is of interest that both the Cambridge and Naples Versions add, as do most later versions, a supplement to the proof, namely, the conversion of the right triangle to a square by the use of Proposition II. 14 of the *Elements* [12].

While I have tended to treat the Cambridge and Naples tracts together, there are two special features of the Naples Version that are of concern to the problem at hand. First, it is of some interest that the author of the proof immediately tells us that it is to be a proof *per impossibile*, that is, by reduction to absurdity [13]. This seems to be another one of the early steps in the growing compulsion on the part of schoolmen to explain the logical form of the argument. The second feature is somewhat trivial but is also illustrative of a concern to explain why various steps are taken : in describing the procedure of continuously

10. ARCHIMEDES, *Opera omnia*, 1 (Leipzig, 1910), 8.
11. *Ibid.*, 10.
12. For example, in the Cambridge Version (*MS cit.*, 109ʳ, c. 2; *Osiris, 10*, 613, c. 2, lines 151-154) : « Sed dato trigono equum quadratum invenire per ultimam secundi; ergo dato circulo quadratum; et hoc erat propositum. »
13. MS Naples, Bibl. Naz. VIII, c. 22, 65ᵛ. Cf. *Osiris, 10*, 607, line 4 : « Hoc (i. e. Prop. I) probatur per impossibile. »

doubling the sides of the inscribed regular polygon, our author stops with an inscribed octagon. He tells us that he does this for the sake of brevity (*causa compendii*) [14]. Other versions say the same sort of thing and one (that of Gordanus [15]) adds that he does this « lest the oppression (*pressura*) of many lines impede the demonstration ».

The two Florence versions of Proposition I of the *De mensura circuli* — which I have elsewhere labelled as F.IA and F.IB — demand no special attention within the context of our study of changes in form and content. They still preserve a strictly mathematical form, they cite Euclid's *Elements* making Proposition X. 1 the heart of the proof, they explain why they are stopping with an inscribed octagon in the first part of the proof (F.IA says *causa brevitatis* and F.IB indicates in a perceptive way that we can posit that segments between the octagon and the circle are less than the posited difference between circle and triangle, since in such a problem one does not have to proceed to infinity « non est procedere in infinitum — » [16]), and F.IA points out in the course of the argument where the contradiction lies. Quite incidentally I can add that the Florence version of Proposition III which appears with F.IA and F.IB is of considerable interest, and I have treated it at length in my *Archimedes in the Middle Ages* (Chapter III, Section 3), since it contains a detailed elaboration of Archimedes' calculations for π different in some respects from that of Eutocius. One very interesting comment by the scholar who copied this version of Proposition III is worth mentioning, for it shows in an extreme way how devotion to a text and to authority can be an obsession of a schoolman. In one place our scholarly scribe realizes that the original author has made some serious calculating errors and he accordingly gives the corrected calculations in the mar-

14. *Ibid.*, 66^r; *Osiris, 10,* 608, lines 23-24.
15. Vat. lat. 1389, 109^r; cf. the text in my *Archimedes*, chap. III, sect. 4 : « Sed ad presens gratia compendii, ne pressura multarum linearum demonstrationem impediat, concedantur ille 8 portiuncule esse minores quantitate H. »
16. These two Florence Versions of Proposition I of the *De mensura circuli* are both contained in MS Florence, Bibl. Naz. Con. Soppr.. J. V. 30. F. IA is on folio 12^r, F. IB on 12^v. Cf. *Osiris, 10,* 609-613 for F. IA and 615-616 for F. IB.

46

gin, leaving the text as it was; but he introduces the correct calculations with the comment [17] : « It seems to me that here the
commentator has erred in multiplying, for, as it seems to me,
it is based on a false root. Accordingly, I have placed a [correct-
ed] reading [in the margin]... » Then following the long marginal correction he adds : « Note that I have interposed this
reading because the comment seemed to be based on a false root.
However, *I have not dared to correct* [the text]... » (*Italics
mine*). While the mathematician might scorn our scribe's lack
of courage, no doubt the philologist will applaud his scrupulous
preservation of the original text.

Turning to the Gordanus Version, which comprises Chapter 23
(« To Square a Circle Demonstratively ») of Part VIII of a mathematical and astronomical *Compilatio* attributed to one Gordanus, we find ourselves confronted with an elaboration of far
greater detail than any of those so-far mentioned. The author
took with great seriousness his avowed objective « to treat of and
explain further the proof of Archimedes [18] ». Euclid is cited
thirty-three times, every step being proved by an appropriate
citation to the *Elements*. A case in point occurs where the author
proves in a detailed and obvious manner that the four triangles
formed by the corner angles of the circumscribed square and the
appropriate sides of a circumscribed regular octagon constitute
an area greater than half of the area included between the circumference of the circle and the sides of the square [19]. The
author's attempt earlier to reduce a geometrical argument to
numbers is another example of his tendency to elaborate the obvious, and perhaps is indicative of a tendency of some authors to
consider numerical arguments as clearer and simpler than geo-

17. MS cit. in note 16, 10^r, marg. : « Hic videtur mihi quod commentator
erraverit in multiplicando, fundat enim se in falsa radice, ut mihi videtur,
propter, quod apposui lecturam... » And on 10^v, marg. : « Nota quod istam
lecturam interposui, quia commentum videbatur mihi fundare se super fal-
sam radicem. Non tamen fui ausus corrigere... »
18. MS cit. in footnote 15, 108^r : « Itaque probationem ipsius Archimenidis
petractare et magis declarare volentes, ostendimus dictum triangulum memo-
rato circulo nec maiorem esse nec minorem et ita propositum reputabimus
nos habere. »
19. *Ibid.*, 110^v-111^r.

metrical ones[20] Further, Gordanus is careful to remind the reader of the structure of the proof when he tells us toward the end of the first half of the proof that the argument has arrived at a conclusion which is « false and impossible » and hence also false and impossible is the original assumption from which it follows[21].

Similar detail and elaboration is found in the version I have called by the name of Corpus Christi. In this version the author presents three postulates or *petita*[22], which he says are « in reality principles (*principia*) and are known per se ». The first postulate is the same as the second one of the Cambridge Version, namely that « an arc is greater than its chord ». The second postulate, that « a curved line is equal to a straight line », resembles the first postulate of the Cambridge Version. The physical justification of the postulate is of interest :

We postulate this, although it is a principle known per se, for if a hair or silk thread is bent around circumference-wise in a plane surface and then afterwards is extended in a straight line in the same plane, who will doubt — unless he is hare-brained — that the hair or thread is the same whether it is bent circumference-wise or extended in a straight line and is just as long the one time as the other.

20. *Ibid.*, 109ʳ : « Hec autem illatio patet magis in numeris... »
21. *Ibid.*, 110⁻ : « Hoc autem est falsum et impossibile; ergo et illud ex quo sequitur... »
22. MS Oxford, Corpus Christi 234, 170ʳ : « Tria petimus que ad demonstracionem quadrature circuli... Licet aut petamus tria, sciendum tamen est quod revera sunt principia et per se nota... primum est, quod arcus sit maior corda. Hoc petimus, licet per se notum sit et cuilibet sane mentis indubitabile... Secundum petitorum est, quod linea curva sit equalis recte. Hoc petimus, licet sit principium et per se notum et cuilibet sani capitis cognitum; si enim capillus vel filum sericum in plana superficie circumferencialiter circumflectatur, deinde idem in directum in eodem plano extendatur, quis nisi cerebrosus dubitet eundem esse capillum et idem filum sive circumflectatur sive in directum extendatur et tantum esse quantum prius. Rursus, si circulus super planam superficiem circumrotetur circumferencia planam superficiem tangente et in directum circumvoluta super planam superficiem ab uno sui puncto donec perveniat ad idem punctum circumferencie quibus nisi mente capitis dubitabile est quin circumferencia descripserit lineam rectam equalem circumferencie. Tercium petitorum tale est : Quelibet linea curva duobus terminis arcus circumferencialis coterminata ex parte convexitatis arcus arcum ambiens maior est illo arcu. »

I think our author might be surprised as to how many « hare-brained » mathematicians there had been and were to be. At any rate, the author adds a further kinematic justification :

Again, if a circle is rolled on a plane, and within a plane, with its circumference tangent to the plane surface, and it is rolled in a straight line from one point of the circumference until it arrives at the same point of the circumference, it can scarcely be doubted by anyone with a sound mind in his head that the circumference will have described a straight line equal to the circumference.

The third *petitum* assumes that « any curved line sharing two termini of a circumferential arc and including it in the direction of the convexity of the arc is greater than the arc ». This should be compared with the second assumption of Archimedes' *Sphere and Cylinder* [23] :

Of other lines in a plane and having the same extremities, [any two] such are unequal whenever they are both concave in the same direction and one of them is either wholly included between the other and a straight line which has the same extremities with it, or is partly included by, and is partly common with, the other; and that [line] which is included is the lesser [of the two].

The general similarity of the two statements is evident. Furthermore, the Corpus Christi author goes on to indicate that « curved lines » include bent lines, and this expression is similar to Archimedes' *kampulai grammai*, rendered, I might add, by William Moerbeke in his translation of 1269 of the *Sphere and Cylinder as curve linee* [24].

Like a number of the versions of the fourteenth century the Corpus Christi text adds some additional propositions before presenting the essential part of the Archimedean proof. Proposition I asserts that « the perimeter of any polygon inscribed in a circle... is less than the circumference », while Proposition II

23. ARCHIMEDES, *Opera omnia*, 1, 8; T. L. HEATH, *Works of Archimedes* (Cambridge, 1887), 4.
24. MS Vatican, Ottob. lat. 1850, 23ᵛ, c. 1.

tells us that the « perimeter of a polygon circumscribed about
a circle is greater than the circumference »[25]. Again, Archimedes'
Sphere and Cylinder appears to have been the ultimate source
of these added propositions[26]. The author then passes on to
still another auxiliary proposition to the effect that the product
of the perimeter of an inscribed or circumscribed regular polygon and the line drawn from the center of the circle to the
middle point of one of the sides of the polygon is double the
area of the polygon[27]. Since this theorem represents an important step in the proof of Proposition I of the Archimedes' *De
mensura circuli*, it is not surprising that many of the authors of
the emended versions thought it necessary to include the auxiliary proposition with some proof of it. The only distinctive feature of the Corpus Christi treatment is that the theorem is presented and proved as a separate proposition prior to the com-
mencement of the proof of the main Archimedean proposition.
Such a theorem, although in slightly different form, is also presented as a preliminary proposition in Albert of Saxony's *Questio de quadratura circuli*[28]. Still one other addition is made to
the Corpus Christi Version, which as I have noted appears in
other versions. This was to make the final objective of Archimedes' proof not just the equation of the right triangle and circle
but rather the finding of a square equal to the triangle after the

25. MS cit. in note 22, 170r-v : « Ex primo principio est manifestum quod
cuiuslibet poligonii circulo inscripti linea curva ambitus poligonii, i. e. linea
curvea ambiens poligonium, minor est circumferencia, cum omnes parciales
arcus sint maiores suis cordis... Ex tercia peticione manifestum est quod
omnis ambitus poligonii circulo circumscripti maior est circumferencia,
cum parciales linee recte exteriores maiores sint arcubus circumferencie. »
26. See footnote 11.
27. MS cit. in note 22, 170v : « Dicimus ergo quod si aliquod poligonium
circulo inscribatur et a centro illius circuli ducatur perpendicularis ad unum
laterum poligonii sive ad medium punctum eiusdem lateris, quod idem est
ex tercia tercii, quod id quod fit ex ductu illius perpendicularis in totalem
lineam continentem sive ambientem poligonium est duplum ad poligonium.
Prorsus eodem modo [si] alicui circulo circumscribatur poligonium et a
centro illius circuli ad medium punctum unius laterum illius poligonii
ducatur linea, dicimus quod id quod fit ex ductu illius linee in totalem
ambientem poligonium est duplum ad ipsum poligonium. »
28. See footnote 50 below.

50

triangle has been shown to be equal to the circle[29]. This is accomplished in the Corpus Christi text by the use of the porism to Proposition VI. 8 of the *Elements*, rather than by the use of Proposition II. 14 as in the other versions.

Earlier I suggested that certain of the fourteenth century versions of the quadrature proof show self-conscious attention to the proof's logical structure. This is particularly true of the Munich Version. It begins by laying out the two propositions which are to be proved :[30] (1) the specified right triangle is equal to the given circle, and (2) a square can be found which is equal to the right triangle. The conclusion from these two propositions, together with Axiom I of the *Elements*, is that a square can be found which is equal to the circle. The author then proceeds to prove the first proposition. Again he presents the steps needed : (a) if the triangle is less than the circle, then the circle (on the authority of Aristotle's *Physics*, IV, 8, 216 *b*) can be divided into a « quantity which is exceeded » and a « quantity by which it (the circle) exceeds ». The excess of the circle over the triangle is designated as P. (b) If P is assumed, then it would be possible to inscribe a regular polygon within the circle which would be greater than the triangle. But this is, in fact, impossible. Therefore, the assumption in (a) is impossible. Having shown the form of the argument, the author then goes on to argue (b) in detail by using Proposition X. 1 of the *Elements*. The second half of the proof involving the assumption of a triangle greater than the circle is treated in a similar manner.

While the author of the Munich text does not, as in the case of the Cambridge and Corpus Christi texts, place any axioms or postulates before the proof, he does on a number of occasions in the course of his proof call attention to basic axioms, which he labels as *communis sciencia, communis animi concepcio,* or simply *communis concepcio*[31]. These axioms are either drawn

29. MS cit., 172^{r-v}.
30. Munich, Bay. Staatsbibl. cod. 56, 182^r; cf. M. CURTZE's text in *Bibliotheca mathematica*, 3. Folge., 2 (1901), 48-49, and the new text in my *Archimedes*, chap. III, sect. 6.
31. For an example of each of the three phrases, see MS cit., 182^r; Curtze text, p. 49 : « ... et tunc erunt circuli et huius quadrati superficies equales uni et eidem triangulo A. Ergo erunt inter se equales per communem animi

from (1) the *Elements*, directly or indirectly, or (2) as we have
seen, from Aristotle in the case of what constitutes « greater
than » in connection with comparable quantities, or (3) indirectly from Archimedes in the case of the axiom that « the
straight line which is the chord is the least which can be drawn
between... two points [32] ».

Turning to another fourteenth century text, the Pseudo-Bradwardine Version that is added to a Vatican manuscript of Bradwardine's *Geometry*, we first notice that it appears to have been
the direct source for still another version, which I have called
the *Versio abbreviata*. The Pseudo-Bradwardine Version also has
a very basic similarity with the text of Albert of Saxony's
Questio de quadratura circuli, which we shall examine shortly.
The basic similarity lies in using what John Murdoch has called
a « betweenness postulate of continuity [33] » instead of directly
using Proposition X. 1 of the *Elements*. Thus instead of the successive polygon procedure commonly used, the author merely
asserts that a polygon greater than the right triangle can be
found because of the continuous divisibility of the postulated
excess of the circle over the triangle. To be sure, such an
assumption might well rest for the author on the unexpressed

concepcionem ». *Ibid.* : « Si autem idem triangulus sit minor circulo dato,
tunc possibile est, quantitatem circuli dividi in quantitatem qua (!que)
excellitur, et in eam quia excellit, que est communis concepcio quam ponit
Aristoteles 4° physicorum capitulo de vacuo. » *Ibid.*, 183ʳ : « Demonstratur
autem sic, sumpta ista communi sciencia, quod [quando] quecunque dupla
sunt maiora ad aliud iposrum, media sunt maiora ad medium alterius. » The
first of these passages is the first axiom of the first book of the *Elements*,
while the third is the extension of one added in the Adelard and Campanus
versions of the *Elements*; the second, is, as noted above, drawn from the
Physics of Aristotle.
32. *Ibid.*; cf. Curtze text, p. 52 : « ... quelibet corda et arcus ducuntur
ab eodem puncto ad eundem, recta autem linea que est corda brevior que
esse potest intra illa duo puncta ».
33. Mr. MURDOCH discusses the « betweenness » postulates in his « The
Medieval Language of Proportions », to appear in the published papers
of the Symposium on the History of Science, Oxford University, 9ᵗʰ-15ᵗʰ
July, 1961.
34. MS Paris BN lat. 9335, 56ʳ, c. 1; cf. my *Archimedes*, chap. ɪᴠ, prop.
III : « Et faciam in circulo ABG figuram lateratam et angulosam non contingentem circulum DZE. Tunc latera figure facte agregata sunt longius
EDZ. »

52

use of Proposition X. 1 of the *Elements*. A somewhat similar betweenness theorem for constructing the perimeter of a regular polygon which is greater than a given line but less than a circumference of a circle which is itself greater than the given line was used by the Banu Musa in their *Verba filiorum*[34] for their proof of the quadrature proposition. Their proof rests on the silent use of Proposition XII. 16 of the *Elements*, which in turn itself depends on the corollary to Proposition X. 1.

The title of the Pseudo-Bradwardine text is of some interest : « To conclude demonstratively the quadrature of a circle[35] ». It resembles the chapter title prefaced to the so-called Gordanus Version. The use of « demonstratively » perhaps reflects the same kind of distinction made in the *Question* of Albert of Saxony between quadrature *ad intellectum* (« finding a square and demonstratively proving that it is equal to the circle ») and quadrature *ad sensum* (« finding a square such that the sense reveals no difference between it and some circle[36] »). Following the title, the author of the Pseudo-Bradwardine Version asserts that in order to reach the objective expressed in the title, he must first prove five other conclusions which will stand as premises for the proof of the main conclusion. The first conclusion equates a regular polygon to a right triangle whose right angle is included by lines equal respectively to the perimeter of the polygon and a line drawn from the center of the polygon to the middle of one of the sides[37]. As in the *Versio abbreviata* the particular polygon employed by Pseudo-Bradwardine is a regular pentagon (while that used by Albert

35. MS Vat. lat. 3102, 111^v; cf. my text in *Archimedes*, chap. v, sect. 1, line 1 : « Quadraturam circuli demonstrative concludere. »

36. MS Bern, Bügerbibliothek A. 50, 169^v; cf. *Archimedes*, chap. v, sect. 3 : « Sic similiter quadratura circuli 5° modo dicta, quedam est ad intellectum, ut invenire unum quadratum et hoc demonstrative probare esse equale circulo; et quedam est ad sensum, ut facere et invenire unum quadratum inter quod et aliquem circulum sensus nequit ponere differentiam... »

37. MS cit. in note 35 *ibid.* : « Omnis figura poligonia equilatera est equalis triangulo orthogonio, cuius unum laterum rectum angulum ambientium <est perpendiculare> a centro poligonie ad punctum medium unius laterius poligonie protractum, et alterum est equale omnibus lateribus illius poligonie simul sumptis. »

in a similar proposition is a square). Pseudo-Bradwardine's second proposition holds that the perimeter of any polygon circumscribed about a circle is longer than the circumference of the circle transformed into a straight line [38]. In the proof of Proposition II Pseudo-Bradwardine makes use of Ptolemy's *Almagest*. Incidentally, the only other authority cited by our author is Euclid, whose *Elements* are cited three times, twice in the first proposition and once in the last. The third and fourth propositions of the Pseudo-Brawardine tract assert that when we have a given surface that is less than a given circle (Proposition III [39]) or a surface greater than a given circle (Proposition IV [40]) we can in the first case inscribe a regular polygon in the given circle which is greater than the given surface (Proposition III) or in the second case circumscribe a regular polygon about the given circle which is less than the proposed surface (Proposition IV). The proof of these propositions is based upon the assumption that the latitude of excess between the given circle and the given surface is always divisible [41], which as I have already noted earlier is the feature distinguishing this version. The fifth proposition merely holds that, if a given surface is neither greater than nor less than a circle, it must be equal to the circle [42]. This is said to be self-evident.

With these five propositions or conclusions assumed, the main quadrature proposition is then provable. For if the circle is said to be greater than the right triangle composed of circumference and radius at the right angle, a regular polygon which is greater than the triangle can be inscribed in the circle by Proposi-

38. *Ibid.*, 112ʳ : « Latera cuiuslibet figure poligonie circulo circumscripte simul sumpta faciunt lineam longiorem circumferentia circuli predicti rectificata. »
39. *Ibid.* : « Dacta (!) aliqua superficie que sit minor dacto (!) circulo est dabilis aliqua figura poligonia inscriptibilis eidem circulo proposito maior superficie data. »
40. *Ibid.* : « Dacta aliqua superficie que sit maior circulo dacto dabilis est poligonia circumscriptibilis circulo que sit maior circulo dato et minor superficie proposita. »
41. *Ibid.* : « Probatur quia ex quo ille circulus est maior dacta superficie et ipsam divisibiliter excedat, ergo infra latitudinem excessus locabilis est et per consequens inscriptibilis est poligonia. »
42. *Ibid.* : « Si aliquam dactam superficiem repugnat esse maiorem et etiam minorem circulo proposito, ipsa erit ei necessario equalis. »

54

tion III. But by Proposition I it is shown that such a polygon must be equal to a right triangle which is less than the proposed triangle composed of circumference and radius at the right angle. This then contradicts the inference drawn from the hypothesis that the circle is greater than the proposed right triangle; hence that hypothesis must be false. In precisely the same way, by the use of Propositions IV and I and II, the hypothesis that the circle is less than the proposed right triangle can be shown to be false. Thus by Proposition V the main proposition follows.

While having the main lines of proof found in the Pseudo-Bradwardine Version, its cognate, the *Versio abbreviata*[43], has some interesting distinguishing features. For example, the author of this abbreviated version omits most of the scholastic elaboration, e.g., omitting the constant reiteration of consequence, antecedent, and the like, found in the longer treatise. The six propositions of the longer tract have been reduced to three. His first and third propositions are drawn from Propositions I and VI of the Pseudo-Bradwardine tract. He omits entirely Propositions II and V of the longer text, while he makes a single proposition (his second) out of Propositions III and IV, although the proof he gives for this second proposition completely differs from the proof of Propositions III and IV. It does not depend on the statement that a polygon can be inscribed or circumscribed in the excess between the triangle and circle because of the infinite divisibility of that excess. But rather he returns to the conventional procedures, using Proposition X.1 of the *Elements* directly. Thus the text is a kind of conflation of the Pseudo-Bradwardine Version and the more common tradition stemming out of Gerard of Gremona's translation of the *De mensura circuli*.

It is finally with Albert of Saxony's *Questio de quadratura circuli* that we arrive at the full scholastic elaboration of Archimedes' proof[44]. As a scholastic *questio*, Albert's small tract has all the common features of the fourteenth century disputative

---

43. The text is in MS Florence, Bibl. Naz. Conv. Soppr. J. IX, 26, 49ᵛ-50ᵛ; cf. my *Archimedes,* chap. v, sect. 2.
44. For the text see the MS cit. in note 36 and my *Archimedes* as cited there; it was printed earlier by H. Suter in *Zeitschrift für Mathematik und Physik, 29* (1884), Hist.-lit. Abt., 81-101.

form : opening affirmative arguments (*rationes quod sic*), negative arguments (*rationes quod non*), clarifying distinctions (*distinctiones*), conclusions (*conclusiones*), followed and terminated
by considerations of the opening arguments. The scholastic technique is also illustrated by the close attention of the author to
the logical form of the argument, where common logical terms
are employed : *consequentia* (implication), *antecedens* (antecedent part of the *consequentia*), *consequens* (conclusion of the
consequentia), major, minor, etc. As in the case of the Munich
Version, Albert tends to outline the structure of his proofs before
proceeding to the detailed proofs.

The opening affirmative arguments are three in number and
are of some interest : (1) Following the views of Antiphon
and Bryson as reported by Aristotle : *Physica* and *Elenchi*, one
could argue that if there can be given a square « greater than »
and a square « less than » a circle, there can be given a square
« equal » to the circle [45]. (2) If a square equal to the circle
does not exist, then it would be possible to pass from « greater » to « lesser » without going through « equal [46] », which
is held at this point to be false. But Albert notes later that the
general argument « If a greater and a lesser, then an equal »,
while true in the case of squares, is not universally true, for
there can be given a rectilinear angle greater than an angle of
a segment (i.e. a mixed angle composed of an arc and a chord),
and also one less than it, yet there is no rectilinear angle equal
to the angle of the segment [47]. (3) The relationship of « squar-

45. MS cit., 172ʳ; Suter text, p. 87 : « Arguitur primo quod sic auctoritate
Antifontis et Brissonis, qui, ut dicit philosophus primo elencorum et primo
physicorum, circulum quadrare sunt conati. Probatur etiam ratione, quia
cuicunque est dare quadratum maius et quadratum minus, sibi est dare quadratum equale; sed circulo est dare quadratum maius, scilicet quadratum
circulo circumscriptum, et quadratum minus, scilicet quadratum circulo ins-
criptum; igitur etiam est dare quadratum circulo equale. »
46. *Ibid.* : « 2° sic : si non esset dare quadratum circulo equale, sequeretur
quod fieret transitus de maiore ad minus, sive de extremo ad extremum
transeundo per omnia media, et tamen nunquam perveniretur ad equale
vel ad medium. Sed hoc est falsum. »
47. *Ibid.*, 171ᵛ; Suter text, p. 93 : « ... quamvis etiam ratio Brissonis bene
probet quadraturam circuli... cum hoc tamen stat et verum est quod maior

ing » to a circle is like that of « cubing » to a sphere. But
a sphere can be cubed since the contents of a spherical vase
can be poured into a cubical vase. Therefore, a circle can be
squared.

The opening negative arguments are of no particular interest. It
ought to be noted that in general the opening arguments or
« principal reasons » are not taken very seriously in a scholastic
disputation, and so often it is only after clarifying distinctions are
drawn that an author presents his own conclusions. Such is the
case with this *questio* of Albert[48].

The first distinction notes there are five ways in which the expression « quadrature of a circle » can be understood. The
first way rests on the ultimate semantic identification of the word
quadrature with « dividing into four » and merely means dividing the circle into four quadrants by two orthogonally inter-
secting diameters; this common usage Albert ascribes to Campanus. The second way of understanding quadrature is as a
procedure of fashioning a square-like figure out of a circle by
rearranging its parts, perhaps as in the accompanying figure   . The expression is also used in a third way to mean
the finding of a square whose perimenter is equal to the circumference of a circle but whose area is not equal to the area of
the circle; this usage he once more ascribes to Campanus. The
fourth way of understanding quadrature — labelled later as
erroneous — is as the finding of a square whose perimeter is
equal to the circumference of a circle and which at the same
time is equal in area to the circle. Finally, the fifth meaning
assigned to quadrature is simply the finding of a square equal to
a circle.

In the second of his distinctions Albert notes that quadrature of
a circle in the third and fifth meanings can be considered « with
respect to sense », or « with respect to intellect » (i.e., demon-

illius rationis non est universaliter vera, quia angulo portionis datur bene
angulus rectilineus maior, sicut est angulus rectus, et angulus rectilineus
minor; non tamen datur angulo portionis angulus rectilineus equalis. »
48. My succeeding discussion paraphrases the text of Albertus, the Latin
of which I have not quoted in detail because of its ready availability in
Suter's text. I shall quote the text only to single out certain of the conclusions and postulates.

stratively), or « with respect to both sense and intellect ». « With
respect to sense » in the third meaning of quadrature merely
means that a square can be found whose perimeter does not
sensibly differ from the circumference of the circle; while in connection with the fifth meaning it means that a square can be
found whose area does not sensibly differ from that of the
circle. « With respect to intellect » in connection with the third
way of understanding quadrature means that one can demonstratively prove the equality of the perimeter of a square and the
circumference of the circle, and in connection with the fifth way
it means that one can demonstratively prove that the one area is
equal to the other.
From these distinctions Albert goes on to his conclusions.
(1) Quadrature understood in the first way as quartering a
circle is, of course, possible and need not be demonstrated.
(2) As understood in the second way, quadrature is possible only
so long as we do not take quadrature in this sense to mean that
the « square-like » figure is equal to the circle. (3) Quadrature in
its third connotation is possible both « with respect to sense »
and « with respect to intellect ». Campanus, Albert says, has
shown that this is so « with respect to sense » when he made the
assumption that the circumference contains the diameter three and
one-seventh times. As for quadrature in the third connotation
« with respect to intellect », our author believes that it is in such
a connotation that Aristotle understands quadrature when he says
in his *Categories* (chap. 7, 7 *b*) : « quadrature of a circle is
knowable but not yet known. » Albert believes that Aristotle
said this because no one had yet demonstrated that the circumference is 3 1/7 the diameter, nor in fact that any straight line
is equal to the circumference. « Nevertheless », Albert says [49]
« it (i.e., quadrature in the third sense) is demonstrable to the
intellect, although such is difficult ». (4) Quadrature in the
fourth sense—where not only is the perimeter of the square equal
to the circumference of the circle but also the area of the same
square is equal to the area of the circle—is quite properly said to
be impossible by Albert. The reason given is that the circle is
the « most capacious » of isoperimetric figures.

49. MS cit. 170r; SUTER text, p. 90 : « ... nihilominus est demonstrabile
ad intellectum quamvis difficile. »

58

Albert has now cleared the way to talk of quadrature in the fifth sense and thus to prove demonstratively that there is a square equal to a circle. To do this, he first proves in the fifth conclusion that every regular polygon is equal to a right triangle one of whose two sides including the right angle is equal to the perimeter of the polygon while the other is equal to the line drawn perpendicularly from the center of the polygon to the midpoint of one of its sides[50]. The wording of this conclusion is obviously drawn from the same tradition as that of the Pseudo-Bradwardine Version. It is easily proved by resolving the regular polygon into triangles, as was done in other emended versions of the *De mensura circuli* (e.g., in Propositions IIIA and IIIB in the Corpus Christi Version).

It is the sixth conclusion[51] of Albert's *Questio* that is the first proposition of the *De mensura circuli*. Albert has taken the wording of this proposition from one of the versions based on the Gerard of Cremona translation. Its wording is closest to that found in the Naples Version. In initiating his proof of the sixth conclusion, Albert first makes three suppositions[52]. The first is in two parts : (1) one figure inscribed in another is less than it, and (2) the perimeter of the inscribed figure is less than that of the figure in which it is inscribed. The second part is similar to the third postulate of the Cambridge Version and

---

50. *Ibid.* : « Omnis figura rectilinea equiangula et equilatera est equalis triangulo orthogono, cuius alterum laterum rectum angulum continentium est equale linee recte quam omnia latera illius figure simul iuncta constituunt et reliquum laterum angulum rectum constituentium equale linee a centro eiusdem figure ad aliquod suorum laterum perpendiculariter ducte. »

51. *Ibid.*, 170^v-171^r; Suter text, p. 91 : « Omnis circulus est equalis triangulo orthogono, cuius alterum laterum rectum angulum continentium est equale circumferentie in rectum extense et reliquum latus (laterum ?) rectum angulum continentium est equale semidyametro eiusdem circuli. »

52. *Ibid.* : « ... suppono primo omnem figuram alteri inscriptam illi cui inscribitur minorem esse, eiusque latera simul iuncta lateribus illius cui inscribitur breviora. 2° suppono omnium triangulorum orthogonorum illum maiorem esse cuius ambo latera angulum rectum constituentia sunt maiora reliquis lateribus reliqui trianguli orthogoni angulum rectum constituentibus, aut alterum reliquo unius (! maius?), et alterm reliquo ipsum respiciente equale... 3° suppono propositis duabus quantitatibus continuis a maiore maiorem minore posse resecare. Ista patet ex eo quod quilibet excessus quo una quantitas aliam excedit est divisibilis. »

in a general way to Propositions I and II of the Corpus Christi
Version and to the second conclusion of the Pseudo-Brawardine
Version. The first part is like the statement in the Pseudo-
Bradwardine Version. The second of Albert's suppositions states
that in the case of two right triangles the one is greater whose
two sides including the right angle are greater, or one of whose
two sides is greater while the other side is equal. The third supposition holds that with two unequal, continuous and, we ought
to add, comparable magnitudes given, it is possible to take from
the greater a quantity greater than lesser. This is said to follow
from the fact that any excess by which one quantity exceeds
another is divisible. I have already indicated the importance of
Albert of Saxony's supposition for proving the main proposition,
particularly when compared with the proof based directly on Proposition X. I of the *Elements* of Euclid.

Albert's proof of the sixth conclusion essentially shows that if
the given right triangle is assumed to be less than the circle, by
Supposition III there can be inscribed in the circle a regular
polygon which is greater than the triangle. This polygon, however, is equal to a right triangle which is then both greater
than the given right triangle (by Supposition III) and less than
it (by Conclusion V and Suppositions I and II), an obvious contradiction. If the given right triangle is said to be greater than
the circle a similar contradiction ensues. Hence the given right
triangle must be equal to the circle. Then in the seventh conclusion the right triangle is converted to a square by reference
to Proposition II. 14 of the *Elements*. Following the proof of
the seventh conclusion Albert considers the initial arguments
with which he opened the question.

With this discussion of Albert's *Questio* my examination of the
various quadrature proofs is completed. We have seen that, starting with the basic proof of Archimedes, there developed a progressive tendency to elaborate the proof. The elaboration was
not only strictly geometric, as in the incrasing citation of the
Elements of Euclid, but also lay in specifying and explaining the
postulates that were believed to lie at the base of the proof.
Further there was an evident tendency to stress and point out
specifically the logical form of the proof, much in the manner
of the adherence to logic found in philosophical *questiones* of the

60

fourteenth century. Conjoined with this was an effort to introduce physical and philosophical arguments into mathematics, an
effort that reached its culmination in the *questio* of Albert of
Saxony. It is not an exaggeration to say that in this case of
a crucial problem in plane geometry the juncture of the philosophical and mathematical was complete, however immature and
tentative the results of that juncture might have been.

XII

Archimedes in the Late Middle Ages

IN SOME RESPECTS the title of my paper is not correct, for in emphasizing the course of Archimedean studies in the late Middle Ages, I have found it necessary to summarize in the beginning the knowledge of Archimedes among the Arabs and the Latin schoolmen in the high Middle Ages to provide some points of contrast and continuity. Furthermore, I have also made numerous references to the fate of the medieval Archimedes in the Renaissance.

Needless to say, the whole paper reflects both my already published first volume on *Archimedes in the Middle Ages: The Arabo-Latin Tradition* and the two subsequent volumes now in preparation. These later volumes will present the full text of the translations of Archimedes made from the Greek by the thirteenth-century Flemish Dominican William of Moerbeke and the step by step use of these translations and the earlier translations from the Arabic down to 1565. Of course, I shall treat here quite lightly many of the problems that I have undertaken to discuss in detail in the three volumes.

This paper also reflects much of the second half of my article on Archimedes that appears in Volume I of the *Dictionary of Scientific Biography*, although I have here often expanded the compendious treatment presented in that article.

Unlike the *Elements* of Euclid, the works of Archimedes were not widely known in antiquity. Our present knowledge of his works depends largely on the interest taken in them at Constantinople from the sixth through the tenth century. It is true that before that time individual works of Archimedes were obviously studied at Alexandria, since Archimedes was often quoted by three eminent mathematicians of Alexandria: Hero, Pappus and Theon. But it is with the activity of Eutocius of Ascalon, who was born toward the end of the fifth century and studied at Alexandria, that the textual history of a collected edition of Archimedes properly begins.

Eutocius composed commentaries on three of Archimedes' works: *On the Sphere and the Cylinder, On the Measurement of the Circle,* and *On the Equilibrium of Planes.* These were no doubt the most popular of Archimedes' works at that time.

The Commentary on the Sphere and the Cylinder is a rich work for historical references to Greek geometry. For example, in an extended comment to Book II, Proposition 1, Eutocius presented manifold solutions by earlier geometers to the problem of finding two mean proportionals between two given lines. This commentary also contains the solution of a subsidiary problem, promised by Archimedes for the end of Proposition 4 of Book II of *On the Sphere and the Cylinder* but which was missing from all copies of Archimedes. Eutocius notes that after "unremitting and extensive research" he found such a solution that still retained vestiges of the Doric dialect and the old terminology for conic sections, and which he therefore thought to be Archimedes' solution.[1] Incidentally, this commentary *On the Sphere and the Cylinder* stimulated the composition of a number of tracts and chapters on the problem of finding two mean proportionals in the Middle Ages and the Renaissance, e.g., in the works of Johannes de Muris, Nicholas of Cusa, Leonardo da Vinci, Johann Werner, Francesco Maurolico, Nicholas Tartaglia, and others.

Eutocius' *Commentary on the Measurement of the Circle* is of interest in its detailed expansion of Archimedes' calculation of π. The works of Archimedes and the commentaries of Eutocius were studied and taught by Isidore of Miletus and Anthemius of Tralles, Justinian's architects of *Sancta Sophia* in Constantinople. It was apparently Isidore who was responsible for the first collected edition of at least the three works commented on by Eutocius and his commentaries. Later Byzantine authors seem gradually to have added other works to this first collected edition until the ninth century, when the educational reformer Leon of Thessalonica produced the compilation represented by Greek manuscript A (adopting the designation used by the modern editor, J. L. Heiberg).[2]

Manuscript A contained all of the Greek works of Archimedes now known excepting *On Floating Bodies, On the Method, Stomachion,* and *The Cattle Problem.* This was one of the two Greek manuscripts available to William of Moerbeke when he made his Latin translations in 1269. It was the source, directly or indirectly, of all of the Renaissance copies of Archimedes.

A second Byzantine manuscript, designated as B, included only the mechanical works: *On the Equilibrium of Planes, On the Quadrature of the Parabola*, and *On Floating Bodies* (and possibly *On Spirals*). It too was available to Moerbeke. But it drops out of history after a reference to it in the early fourteenth century (in a Vatican catalogue of 1311).

Finally, we can mention a third Byzantine manuscript, C, a palimpsest whose Archimedean parts are in a hand of the tenth century. It was not available to the Latin West in the Middle Ages, or indeed in modern times until its identification by Heiberg in 1906 at Constantinople (where it had been brought from Jerusalem). It contains large parts of *On the Sphere and the Cylinder*, almost all of *On Spirals*, and some parts of *On the Measurement of the Circle* and *On the Equilibrium of Planes*, and a part of the *Stomachion*. More important, it contains most of the Greek text of *On Floating Bodies* (a text unavailable in Greek since the disappearance of manuscript B) and a great part of the *On the Method of Mechanical Theorems*, hitherto known only by hearsay (Hero mentions it in his *Metrica* and the Byzantine lexicographer Suidas declares that Theodosius wrote a commentary on it).

At about the same time that Archimedes was being studied in Byzantium, he was also finding a place among the Arabs. The Arabic Archimedes has been studied in only a preliminary fashion, but it seems unlikely that the Arabs possessed any manuscript of his works as complete as manuscript A.[3] Still, they often brilliantly exploited the methods of Archimedes and brought to bear their fine knowledge of conic sections on Archimedean problems. The Arabic Archimedes consisted of the following works:

(1) *On the Sphere and the Cylinder* and at least a part of Eutocius' commentary on it. This work seems to have existed in a poor, early translation, revised in the late ninth century, first by Ishāq ibn Hunain and then by Thābit ibn Qurra. It was re-edited by Nasīr al-Dīn al-Tūsī in the thirteenth century and was on occasion paraphrased and commented on by other Arabic authors (see the index of Suter's *"Die Mathematiker und Astronomen"* under *Archimedes*).

(2) *On the Measurement of the Circle*, translated by Thābit ibn Qurra and re-edited by al-Tūsī. Perhaps the commentary on it by Eutocius was also translated, for the extended calculation of π found in the ninth-century geometrical tract of the Banū Mūsā bears some resemblance to that present in the commentary of Eutocius.

(3) A fragment of *On Floating Bodies*, consisting of a definition of specific gravity not present in the Greek text, a better version of its

basic postulate than exists in the Greek text, the enunciations without proofs of seven of the nine propositions of Book I and the first proposition of Book II.

(4) Perhaps *On the Quadrature of the Parabola*—at least this problem received the attention of Thābit ibn Qurra.

(5) Some indirect material from *On the Equilibrium of Planes* found in other mechanical works translated into Arabic (such as Hero's *Mechanics*, the so-called Euclid tract *On the Balance*, the *Liber karastonis*, etc.).

(6) In addition, various other works attributed to Archimedes by the Arabs for which there is no extant Greek text: *The Lemmata* or *Liber assumptorum, On Water Clocks, On Touching Circles, On Parallel Lines, On Triangles, On the Properties of the Right Triangle, On Data*, and *On the Division of the Circle into Seven Equal Parts*. Manuscripts of all but two of these works have been noted (and perhaps manuscripts of these two works will also turn up as we study Arabic mathematics in more detail).[4]

Of those additional works, we can single out the *Lemmata* (*Liber assumptorum*), for, although it can not have come directly from Archimedes in its present form, since the name of Archimedes is cited in the proofs, in the opinion of experts several of its propositions are Archimedean in character. One such proposition was Lemma 8, which reduced the problem of the trisection of an angle to a *neusis* or "verging" construction like those used by Archimedes in *On Spiral Lines*.[5]

Special mention should also be made of the *Book on the Division of the Circle into Seven Equal Parts* for its remarkable construction of a regular heptagon that may be originally from Archimedes (its Propositions 16 and 17 lead to that construction).[6] This work stimulated a whole series of Arabic studies of this problem, including one by the famous Alhazen (Ibn al-Haitham).

The key to the whole procedure is an unusual *neusis* presented in Proposition 16 that would allow us to find a straight line divided at two crucial points, that is, a straight line whose divisions allow in Proposition 17 the construction of a circle about the line and the division of its circumference into seven equal arcs. The way in which the *neusis* was solved by Archimedes (or whoever was the author of this tract) is not known. Alhazen, in his later treatment of the heptagon, mentions the Archimedean *neusis* but then goes on to show that one does not need it. Rather he shows that the two crucial points in Proposition 17 can be found by the intersection of a parabola and a hyperbola.[7]

It should be observed that all but two of Propositions 1–13 in this tract concern right triangles, and those two are ones necessary for propositions concerning right triangles. It seems probable, therefore, that Propositions 1–13 comprise the so-called *On the Properties of the Right Triangle* attributed in the *Fihrist* to Archimedes (although at least some of these propositions are Arabic interpolations). Incidentally, Propositions 7–10 have as their objective the formulation $A=(s-a)\cdot(s-c)$, where A is the area and a and c are the sides including the right angle and s is the semiperimeter, and Proposition 13 has as its objective $A=s(s-b)$, where b is the hypotenuse. Hence, if we multiply the two formulations, we have

$$A^2 = s(s-a)\cdot(s-b)\cdot(s-c)$$

or

$$A=\sqrt{s(s-a)\cdot(s-b)\cdot(s-c)},$$

Hero's formula for the area of a triangle in terms of its sides—at least in the case of a right triangle. Interestingly, the Arab scholar al-Bīrūnī attributed the general Heronian formula to Archimedes.

Propositions 14 and 15 of the tract make no reference to Propositions 1–13 and concern chords. Each leads to a formulation in terms of chords equivalent to

$$\sin a/2=\sqrt{(1-\cos a)/2}.$$

Thus Propositions 14–15 seem to be from some other work (and at least Proposition 15 is an Arabic interpolation). If Proposition 14 was in the Greek text translated by Thābit ibn Qurra and does go back to Archimedes, then we would have to conclude that this formula was his discovery rather than Ptolemy's, as it is usually assumed to be.

The Latin West received its knowledge of Archimedes from both the sources just described: Byzantium and Islam. There is no trace of the earlier translations imputed by Cassiodorus to Boethius. Such knowledge that was had in the West before the twelfth century consisted of some rather general hydrostatic information that may have indirectly had its source in Archimedes.

It was in the twelfth century that the translation of Archimedean texts from the Arabic first began.[8] The small tract *On the Measurement of the Circle* was twice translated from the Arabic. The first translation was a rather defective one and was possibly executed by Plato of Tivoli. There are many numerical errors in the extant copies of it and the second half of Proposition 3 on the calculation of π is missing.

The second translation was almost certainly done by the twelfth century's foremost translator, Gerard of Cremona. The Arabic text from which he translated (without doubt the text of Thābit ibn Qurra) included a corollary on the area of a sector of a circle attributed by Hero to Archimedes but missing from our extant Greek text. Not only was Gerard's translation widely quoted by medieval geometers such as Gerard of Brussels, Roger Bacon, Thomas Bradwardine, and others, but it served as the point of departure for a whole series of emended versions and paraphrases of the tract in the course of the thirteenth and fourteenth centuries.

Among these are the so-called Naples, Cambridge, Florence and Gordanus versions of the thirteenth century; and the Corpus Christi, Munich and Albert of Saxony versions of the fourteenth. These versions were expanded by including pertinent references to Euclid and the spelling-out of the geometrical steps only implied in the Archimedean text. In addition, we see attempts to specify the postulates which underlie the proof of Proposition I. For example, in the Cambridge version three postulates (*petitiones*) introduce the text:[9] "[1] There is some curved line equal to any straight line and some straight line to any curved line. [2] Any chord is less than [its] arc. [3] The perimeter of any including figure is greater than the perimeter of the included figure." Furthermore, self-conscious attention was given in some versions to the logical nature of the proof of Proposition I. Thus, the Naples version immediately announced that the proof was to be *per impossibile*, i.e., by reduction to absurdity.

In the Gordanus, Corpus Christi and Munich versions we see a tendency to elaborate the proofs in the manner of scholastic tracts. The culmination of this kind of elaboration appeared in the *Questio de quadratura circuli* of Albert of Saxony, composed some time in the third quarter of the fourteenth century. The Hellenistic mathematical form of the original text was submerged in an intricate scholastic structure which included multiple terminological distinctions and the argument and counter-argument technique represented by initial arguments ("principal reasons") and their final refutations.

Another trend in the later versions was the introduction of rather foolish physical justifications for postulates. In the Corpus Christi version, the second postulate to the effect that a straight line may be equal to a curved line is supported by the statement that "if a hair or silk thread is bent around circumference-wise in a plane surface and then afterwards is extended in a straight line in the same plane, who will doubt—unless he is hare-brained—that the hair or thread is the same whether it is bent circumference-wise or extended in a straight line and is just as long the one time as the other."[10]

Similarly, Albert of Saxony in his *Questio* declared that a sphere can be "cubed" since the contents of a spherical vase can be poured into a cubical vase. A somewhat similar "pouring technique" involving a cylinder and a cube was employed later by Nicholas of Cusa and Francesco Maurolico for the problem of the quadrature of a circle.[11] Incidentally, Albert based his proof of the quadrature of the circle not directly on Proposition X.1 of the *Elements*, as was the case in the other medieval versions of *On the Measurement of the Circle*, but rather on a "betweenness" postulate: "I suppose that with two continuous [and comparable] quantities proposed, a magnitude greater than the 'lesser' can be cut from the 'greater.' "[12] A similar postulate was employed in still another fourteenth-century version of the *De mensura circuli* called the Pseudo-Bradwardine version.

Finally, in regard to the manifold medieval versions of *On the Measurement of the Circle*, it can be noted that the Florence version of Proposition 3 (dateable close to 1400) contained a detailed elaboration of the calculation of π. One might have supposed that the author had consulted Eutocius' commentary, except that his arithmetical procedures differed widely from those used by Eutocius. Furthermore, no translation of Eutocius' commentary appears to have been made before 1450, and the Florence version certainly must be dated before that time. The influence of Gerard's translation was still vividly apparent in a version copied by Regiomontanus in the fifteenth century and in a special work written by Francesco Maurolico in the sixteenth.[13]

In addition to his translation of *On the Measurement of the Circle*, Gerard of Cremona also translated the geometrical *Discourse of the Sons of Moses* (*Verba filiorum*) composed by the ninth-century Arabic mathematicians, the Banū Mūsā. This Latin translation was of particular importance for the introduction of Archimedes into the West. We can single out these contributions of the treatise:

(1) A proof of Proposition I of *On the Measurement of the Circle* somewhat different from that of Archimedes but still fundamentally based on the exhaustion method.

(2) A determination of the value of π drawn from Proposition 3 of the same treatise but with further calculations similar to those found in the commentary of Eutocius.

(3) Hero's theorem for the area of a triangle in terms of its sides (noted above), with the first demonstration of that theorem in Latin (the enunciation of this theorem had already appeared in the writings of the *agrimensores* and in Plato of Tivoli's translation of the *Liber embadorum* of Savasorda).

(4) Theorems for the area and volume of a cone, again with demonstrations.

(5) Theorems for the area and volume of a sphere with demonstrations of an Archimedean character.

(6) A use of the formula for the area of a circle equivalent to $A = \pi r^2$ in addition to the more common Archimedean form, $A = \frac{1}{2}(cr)$. Instead of the modern symbol π the authors used the expression "the quantity which when multiplied by the diameter produces the circumference."

(7) The introduction into the West of the problem of finding two mean proportionals between two given lines. In this treatise we find two solutions: (a) one attributed by the Banū Mūsā to Menelaus and by Eutocius to Archytas, (b) the other presented by the Banū Mūsā as their own but similar to the solution attributed by Eutocius to Plato.

(8) The first solution in Latin of the problem of the trisection of an angle.

(9) A method of approximating cube roots to any desired limit.

The *Verba filiorum* was, then, rich fare for the geometers of the twelfth century when compared with the simplistic geometry of the Roman *agrimensores* or the geometry of Gerbert at the end of the tenth century.

The *Verba filiorum* was quite widely cited in the thirteenth and fourteenth centuries (and indeed it was known to Regiomontanus in the fifteenth century). In the thirteenth century the eminent mathematicians Jordanus de Nemore and Leonardo Fibonacci made use of it. For example, the latter in his *Practica geometrie* excerpted both of the solutions of the mean proportionals problem given by the Banū Mūsā, while the former in his *De triangulis* (if indeed he is the author of this part of the tract) presented one of them together with an entirely different solution, namely that one assigned by Eutocius to Philo of Byzantium. Similarly, Jordanus (or possibly a somewhat later continuator) extracted the solution of the trisection of an angle from the *Verba filiorum* but in addition made the remarkably perspicacious suggestion that the *neusis* can be solved by the use of a proposition from Alhazen's *Optics* which solves a similar *neusis* by conic sections.

The expanded *De triangulis* of Jordanus also contains a solution of the problem of the construction of a regular heptagon which appears to be based on Arabic sources.[14] And incidentally, in view of our interest in the use of these tracts later, we can note that Regiomontanus read the *De triangulis* of Jordanus[15] and that much of

Leonardo Fibonacci's *Practica* reappears in the *Summa de arithme-tica etc.* of Luca Pacioli (Venice, 1494), a work of some influence in the Renaissance.[16]

Some of the results and techniques of *On the Sphere and the Cyl-inder* also became known through a treatise entitled *De curvis super-ficiebus Archimenidis* and said to be by Johannes de Tinemue. This seems to have been translated from the Greek in the early thirteenth century or at least composed on the basis of a Greek tract. The *De curvis superficiebus* contained ten propositions with several corol-laries and was concerned for the most part with the surfaces and volumes of cones, cylinders and spheres. This work is essentially Archimedean, employing a version of the method of exhaustion. It simplifies the proof of Archimedes' main conclusions by assuming that to any plane surface there exists an equal conical, cylindrical or spherical surface; and that with two surfaces given there is a surface of the same kind as and symmetrically akin to one of the given surfaces and equal to the other.

These assumptions are coupled with the principle that an "included figure" cannot be greater than an "including figure." The including figure for this author always surrounds the included and in no way touches it (cf. Euclid, *Elements*, Proposition XII.16). The method that the author used was probably suggested to him by the proof of Proposition XII.18 of the *Elements* of Euclid.[17]

The *De curvis superficiebus* was a very popular work and was often cited by later authors. Like Gerard of Cremona's translation of *On the Measurement of the Circle*, the *De curvis superficiebus* was emended by Latin authors, two original propositions being added to one version (represented by manuscript *D* of the *De curvis super-ficiebus*) and three quite different propositions being added to an-other (represented by manuscript *M* of the *De curvis*). In the first of the additions to the latter version, the Latin author applied the ex-haustion method to a problem involving the surface of a segment of a sphere, showing that at least this author had made the method his own. The techniques and propositions of the *De curvis superficiebus* were taken over completely by Francesco Maurolico and integrated beautifully with Archimedes' *On the Sphere and the Cylinder* in 1534.[18]

About the same time as the appearance of the *De curvis superficie-bus* in the early thirteenth century, the geometer, Gerard of Brussels, in his *De motu* used Archimedean *reductio* techniques in a highly original manner. He compared two rotating figures by comparing the motions of two corresponding line elements of the figures. This has, as I have pointed out, some similarity with the method used by Archimedes of balancing corresponding line or surface elements of two figures.[19] But Gerard's methods, while owing something to Archimedean geometry, appear to be essentially his own.

In 1269, some decades after the appearance of the *De curvis superficiebus*, the next important step was taken in the passage of Archimedes to the West when much of the Byzantine corpus was translated from the Greek by the Flemish Dominican, William of Moerbeke. In this translation Moerbeke employed Greek manuscripts A and B which had passed to the Pope's library in 1266 from the collection of the Norman Kings of the Two Sicilies that Charles of Anjou gave to the Pope after Manfred's defeat at Benevento. All the works included in manuscripts A and B except for *The Sandreckoner* and Eutocius' *Commentary on the Measurement of the Circle* were rendered into Latin by William. Needless to say, *On the Method*, *The Cattle-Problem*, and the *Stomachion*, all absent from manuscripts A and B, were not among William's translations. Although William's translations are not without error (and indeed some of the errors are serious),[20] the translations, on the whole, present the Archimedean works in an understandable if literal way.

We possess the original holograph of Moerbeke's translations in manuscript Vatican, Ottobonianus latinus 1850. This manuscript was not widely copied. The translation of *On Spirals* was copied from it in the fourteenth century (MS Vat. Reg. lat 1253, 14r–33r); several works were copied from it in the fifteenth century in an Italian manuscript now at Madrid (Bibl. Nac. 9119), and one work (*On Floating Bodies*) was copied from it in the sixteenth century (Vat. Barb. lat. 304, 124r–41v, 160v–61v).[21]

But, in fact, the Moerbeke translations were utilized more than one would expect from the paucity of manuscripts. They were used by several schoolmen at the University of Paris toward the middle of the fourteenth century. Chief among them was the astronomer and mathematician Johannes de Muris, who appears to have been the compositor of a hybrid tract in 1340 entitled *Circuli quadratura*.[22]

This tract consisted of fourteen propositions. The first thirteen were drawn from Moerbeke's translation of *On Spirals* and were just those propositions necessary for the proof of Proposition 18 of *On Spirals*: "If a straight line is tangent to the extremity of a spiral described in the first revolution, and if from the point of origin of the spiral one erects a perpendicular on the initial line of revolution, the perpendicular will meet the tangent so that the line intercepted between the tangent and the origin of the spiral will be equal to the circumference of the first circle." The fourteenth proposition of the hybrid tract was Proposition 1 from Moerbeke's translation of *On the Measurement of the Circle*. Thus this author realized that by the use of Proposition 18 from *On Spirals*, he had achieved the necessary rectification of the circumference of a circle preparatory to the final quadrature of the circle accomplished in *On the Measurement of the Circle*, Proposition 1.

Incidentally, the hybrid tract did not merely use the Moerbeke translations verbatim but also included considerable commentary. In fact, this medieval Latin tract was the first known commentary on Archimedes' *On Spirals*. That the commentary was at times quite perceptive is indicated by the fact that the author suggested that the *neusis* introduced by Archimedes in Proposition 7 of *On Spirals* could be solved by means of an *instrumentum conchoydeale*.[23] The only place in which a medieval Latin commentator could have learned of such an instrument would have been in that section of Eutocius' *Commentary on the Sphere and the Cylinder*, Book II, Proposition 1, where Eutocius describes Nicomedes' solution of the problem of finding two mean proportionals.

We have further evidence that Johannes de Muris knew of Eutocius' *Commentary* in the Moerbeke translation when he used sections from this commentary in his *De arte mensurandi* (Chapter VIII Proposition 16) where three of the solutions of the mean proportionals problem given by Eutocius are presented, namely the solutions attributed to Plato, Hero, and Philo.[24] Not only did Johannes incorporate the whole hybrid tract *Circuli quadratura* into Chapter VIII of his *De arte mensurandi* (composed, it seems, shortly after 1343) but in Chapter X of the *De arte* he quoted verbatim many propositions from Moerbeke's translations of *On the Sphere and the Cylinder* and *On Conoids and Spheroids* (which latter he misapplied to problems concerning solids generated by the rotation of circular segments).

Incidentally, the treatment of spiral lines found in the *Circuli quadratura* seems to have influenced the accounts of spirals found in the *De trigono balistario* completed by Giovanni da Fontana in 1440 (MS Bodleian, Canon. Misc. 47, 216v–19v) and in the *Quadratura circuli* of Nicholas of Cusa, dated December, 1450.[25]

Within the next decade or so after Johannes de Muris, his colleague at the University of Paris, Nicole Oresme, in his *De configurationibus qualitatum et motuum* (Part I, Chapter 21) revealed knowledge of *On Spirals*, at least in the form of the hybrid *Circuli quadratura*.[26] His description of the spiral as an example of uniformly difform curvature is reflected in a similar description given by Leonardo da Vinci (Institut MS *E*, 34v). Further, Oresme in his *Questiones super de celo et mundo* quoted at length from Moerbeke's translation of *On Floating Bodies*, while Henry of Hesse, Oresme's junior contemporary at Paris, quoted briefly therefrom.[27]

Before this time, the only knowledge of *On Floating Bodies* had come in a thirteenth-century treatise entitled *De ponderibus Archimenidis sive de incidentibus in humidum*, a pseudo-Archimedean treatise prepared on the basis of Arabic sources. Its first proposition expressed the basic conclusion of the "principle of Archimedes": "The weight of any body in air exceeds its weight in water by the weight of a volume of water equal to its own volume."[28]

The references by Oresme to the genuine *On Floating Bodies* comprise citations to Propositions 3–7 of Book I and Proposition 1 of Book II. The whole section is particularly interesting because Oresme joins the dynamic definition of specific weight (perhaps derived from the *Liber de ponderoso et levi*, a text that appeared in the thirteenth century in Latin)[29] with the Archimedean considerations of *On Floating Bodies*. It is just such a juxtaposition that appears in Tartaglia's Italian translation and commentary on Book I of *On Floating Bodies*,[30] which may have led Benedetti to his modified form of the Peripatetic law, namely that bodies fall with a speed proportional to the excess in specific weight of the falling body over the medium.[31]

Incidentally, most scholars do not realize that Leon Battista Alberti made observations much like Tartaglia's a century earlier. Such observations were not tied directly to the text of *On Floating Bodies*,[32] although the substance of some of the Archimedean propositions is given.

Returning to the fourteenth century, we can conclude from our previous discussion that incontrovertible evidence shows that at the University of Paris in the mid-fourteenth century six of the nine Archimedean translations of William of Moerbeke were known and used: *On Spirals, On the Measurement of the Circle, On the Sphere and the Cylinder, On Conoids and Spheroids, On Floating Bodies*, and Eutocius' *Commentary on the Sphere and the Cylinder*. While no direct evidence exists of the use of the remaining three translations, there has been recently discovered in a manuscript written at Paris in the fourteenth century (BN lat. 7377B, 93v–94r) an Archimedean-type proof of the law of the lever that might have been inspired by Archimedes' *On the Equilibrium of Planes*.[33] But other than this, the influence of Archimedes on medieval statics was entirely indirect.

The anonymous *De canonio*, translated from the Greek in the early thirteenth century, and Thābit ibn Qurra's *Liber karastonis*, translated from the Arabic by Gerard of Cremona, passed on this indirect influence of Archimedes in three respects.[34]

(1) Both tracts illustrated the Archimedean type of geometrical demonstrations of statical theorems and the geometrical form implied in weightless beams and weights that were really only geometrical magnitudes.

(2) They gave specific reference in geometrical language to the law of the lever (and in the *De canonio* the law of the lever is connected directly to Archimedes).

(3) They indirectly reflected the centers-of-gravity doctrine so important to Archimedes, in that both treatises employed the practice of substituting for a material beam segment a weight equal in weight to the material segment but hung from the middle point of the weightless segment used to replace the material segment.

Needless to say, these two tracts played an important role in stimulating the rather impressive statics associated with the name of Jordanus de Nemore.

In the fifteenth century, knowledge of Archimedes in Europe began to expand. While the medieval texts continued to influence various authors of the fifteenth century, as I have indicated above,[35] a new source for Archimedes appeared, namely the Latin translation made by Jacobus Cremonensis in about 1450 by order of Pope Nicholas V. Since this translation was made exclusively from Greek manuscript A, the translation failed to include *On Floating Bodies*, but it did include the two treatises in A omitted by Moerbeke, namely *The Sandreckoner* and Eutocius' *Commentary on the Measurement of the Circle*.

It appears that this new translation was made with an eye on Moerbeke's translations.[36] Not long after its completion, a copy of the new translation was sent by the Pope to Nicholas of Cusa, who made some use of it in his *De mathematicis complementis*, composed in 1453–54; although to be sure his earlier works show some smattering of knowledge from medieval sources of *On Spiral Lines*, *On the Measurement of the Circle*, and *On the Sphere and the Cylinder*, as well as of Eutocius' commentary on the last of these works. There are at least nine extant manuscripts of this new translation, one of which was corrected by Regiomontanus and brought to Germany about 1468 (the Latin translation published with the *editio princeps* of the Greek text in 1544 was taken from this copy). Leonardo da Vinci appears to have seen copies of both the Moerbeke and the Cremonensis translations, and both translations can be shown to have exerted some influence on him.[37]

Incidentally, the fate of Moerbeke's holograph copy of his translations in the fourteenth and fifteenth centuries is not known with any exactness, although we can speculate on its possible transfer to France in the fourteenth and its reappearance in Italy, in the fifteenth century, possibly in the hands of Paolo Toscanelli. As I have said, it would seem to have been seen by Jacobus Cremonensis when he made his new translation. At some unspecified time it appears to have passed into the possession of Pietro Barozzi, Bishop of Padua from 1488 to 1507.[38] At any rate, in the latter date it was acquired by Andreas Coner in Padua and he emended it thoroughly both by comparing it to a Greek manuscript and paying close attention to mathematical sense. From the time of Coner we can easily trace its history until its acquisition by the Vatican Library as part of the Ottobonian collection.

Greek manuscript A itself was copied a number of times. Cardinal Bessarion had one copy prepared between 1449 and 1468 (MS E). Another (MS D) was made from A when it was in the possession of the well-known humanist George Valla. The fate of A and its various copies has been traced skillfully by J. L. Heiberg in his edition of Archimedes' *Opera*. The last known use of manuscript A occurred in 1544, after which time it seems to have disappeared.

The first printed Archimedean materials were in fact merely Latin excerpts of Book II, Propositions 1 and 4, of *On the Sphere and the Cylinder* and Eutocius' lengthy comments on those propositions, that appeared in George Valla's *De expetendis et fugiendis rebus opus* (Venice, 1501) and were based on his reading of manuscript A, which was then in his possession.[39] But the earliest actual printed texts of Archimedes were the Moerbeke translations of *On the Measurement of the Circle* and *On the Quadrature of the Parabola*, published on the basis of the Madrid manuscript in Venice, 1503, by L. Gaurico (*Tetragonismus, id est circuli quadratura etc.*).

In 1543, also at Venice, N. Tartaglia republished the same two translations directly from Gaurico's work, and, in addition, from the same Madrid manuscript, the Moerbeke translations of *On the Equilibrium of Planes* and Book I of *On Floating Bodies* (leaving the erroneous impression that he had made these translations from a Greek manuscript, which he had not since he merely repeated the texts of the Madrid manuscript with virtually all their errors). Incidentally, Curtius Troianus, in Venice, 1565, published from the legacy of Tartaglia both books of *On Floating Bodies* in Moerbeke's translation.

The key event, however, in the further spread of Archimedes was the aforementioned *editio princeps* of the Greek text with the accompanying Latin translation of Jacobus Cremonensis at Basel in 1544. It was printed from a Nürnberg manuscript which had been copied from Greek MS A and contains corrections introduced from the Latin translation of Moerbeke. Since the Greek text rested ultimately on manuscript A, the *On Floating Bodies* was not included.

A further Latin translation of some of the Archimedean texts was published by the perceptive mathematician Federigo Commandino in Bologna in 1558, which the translator supplemented with a skillful mathematical emendation of Moerbeke's translation of *On Floating Bodies* (Bologna, 1565), without any knowledge of the long-lost Greek text. Already in the period 1534–49, a paraphrase of Archimedean texts with some attention to medieval sources had been made by Francesco Maurolico, as I have already noted. This was published in Palermo in 1685 on the basis of an incomplete edition of some years earlier that had been abandoned before general publication.

One other Latin translation by Antonius de Albertis, completed in the early sixteenth century, remains in manuscript only and appears to have exerted no influence on sixteenth-century mathematics and science.

After 1544 the publications on Archimedes and the use of his works began to multiply markedly. His works presented quadrature problems and propositions that mathematicians sought to solve and demonstrate not only with his methods, but also with a developing geometry of infinitesimals that was to anticipate in some respect the infinitesimal calculus of Newton and Leibniz. His hydrostatic conceptions were used to modify Aristotelian mechanics. Archimedes' influence on mechanics and mathematics can be seen in the works of such authors as Commandino, Guido Ubaldi del Monte, Benedetti, Simon Stevin, Luca Valerio, Kepler, Galileo, Cavalieri, Torricelli, and numerous others. For example, Galileo mentions Archimedes more than a hundred times, and the limited inertial doctrine used in his analysis of the parabolic path of a projectile is presented as an Archimedean-type abstraction.

Archimedes began to appear in the vernacular languages. Tartaglia had already rendered into Italian Book I of *On Floating Bodies*, Book I of *On the Sphere and the Cylinder* (apparently on the basis of the translation of Moerbeke) and the section on proportional means from Eutocius' *Commentary on the Sphere and the Cylinder*.

Book I of *On the Equilibrium of Planes* was translated into French in 1565 by Pierre Forcadel. In the same year he published a translation of the medieval *De ponderibus Archimenidis* (or *De incidentibus in humidum*).

It was, however, not until 1670 that a more or less complete translation of Archimedes' works was made into German by J. C. Sturm on the basis of the influential Greek and Latin edition of David Rivault, Paris, 1615. Also notable for its influence was the new Latin edition of Isaac Barrow (London, 1675).

Of the many editions prior to the modern edition of Heiberg, the most important was that of Joseph Torelli, published at Oxford in 1792. By this time, of course, Archimedes' works had been almost completely absorbed into European mathematics and had exerted their not inconsiderable influence on early modern science.

1. Archimedes, *Opera omnia*, ed. J. L. Heiberg, Vol. 3, 130–32.

2. *Ibid.*, xxii. The reference to Leon is given by Heiberg from Greek MS G and he reasons that it was taken from MS A. My account of the Greek manuscripts is based almost entirely on Heiberg's Prolegomena to Vol. 3 of his edition of Archimedes.

3. I have noted the principal literature and known manuscripts of the Arabic Archimedes in the bibliography to my article on Archimedes in the *Dictionary of Scientific Biography*.

4. The only works of which I have found no reference in Arabic manuscripts are those *On Parallel Lines* and *On Data* but I have not made any extensive search of Arabic collections which may well contain these works.

5. For an English translation and discussion of this proposition of the *Lemmata*, see my *Archimedes*, Vol. 1, 667–68.

6. A German translation of these propositions has been given by C. Schoy, *Die trigonometrischen Lehren des persischen Astronomen Abu 'l-Raihân Muh. ibn Ahmad al-Bîrûnî* (Hanover, 1927), 74–84. The whole work has been analyzed in modern fashion by J. Tropfke in *Osiris*, Vol. 1 (1936), 636–51. I have given an English translation of Propositions 16 and 17 in my article on Archimedes in the *DSB*.

7. Schoy, *op. cit.* in note 6, pp. 85–91.

8. Much of the succeeding account of the Arabo-Latin Archimedes is taken from my *Archimedes in the Middle Ages*, Vol. 1: *The Arabo-Latin Tradition* (Madison, 1964).

9. *Ibid.*, 69.

10. *Ibid.*, 170–71.

11. Strictly speaking, the method of Cusa is not a pouring technique but a weighing technique. His method, as described in the *De staticis experimentis* (Part IV of the *Idiota*), which I have read in the edition of L. Baur (Leipzig, 1937), 138, is to take a cylinder (whose base is the circle to be squared) with a given height and then a cube of the same height (with a base equal to the square of the diameter). Each of these vessels is filled with water and weighed. Hence the ratio of the circle to the square is as the ratio of the weights.

Maurolico's procedure does involve pouring. He takes a cylinder with height equal to the diameter of its base and a cube whose edge is equal to same diameter. (*Admirandi Archimedis Syracusani monumenta omnia mathematica, quae extant . . . ex traditione . . . D. Francisci Marolici*, Palermo, 1685, p. 39; and see Paris MS, Bibl. Nat. 7464, f. 25r). Maurolico then suggests filling the cylinder to the top. Afterwards, he pours the contents of the cylinder into the cube and notes the height to which it rises. This permits him to show by geometry that the circle is equal to a rectangle comprised by the height of the liquid in the cube and the diameter of the cylindrical base. He then suggests converting that rectangle into a square by finding the mean proportional between these quantities. The resulting square will be equal to the circle.

Somewhat later in the century John Dee in his preface to H. Billingsley's English translation of Euclid's *Elements* (London, 1570), sig. c i verso, suggests making a sphere and a cube with edge equal to the diameter of the sphere of the same uniform material and then weighing them to show that the ratio of their volumes is 11 to 21. He notes that the ratios of other volumes to each other can be determined in the same way. John Dee was very widely acquainted with medieval mathematical texts.

12. Clagett, *Archimedes*, Vol. 1, 418–19.

13. The version of Regiomontanus occurs in Vienna, Nat.-bibl. 5203, 131v-33r, written by Regiomontanus between 1454 and 1462. It is a paraphrase of the Gordanus Version which I published in *Archimedes*, Vol. 1, 142–65. For Maurolico's version, see his *Archimedis de circuli dimensione libellus*, published in the collected edition given in footnote 11. I have prepared the text and an English translation of this from the holograph of Maurolico (Paris, Bibl. Nat. lat. 7465, 21v–28v).

14. *De triangulis libri IV*, ed. of M. Curtze (Thorn, 1887), 42–44. I have collected films of all of the known manuscripts of the *De triangulis* and there is some evidence that there was an earlier edition without Propositions IV.12–IV.28. These latter propositions, which contain among others the interesting propositions on the finding of two mean proportional, the trisection of an angle, and the construction of a regular heptagon, appear to have circulated separately. I intend to study the text in detail in a later volume.

15. See E. Zinner, *Leben und Wirken des Johannes Müller von Königsberg, genannt Regiomontanus*, 2nd ed. (Osnabrück, 1968), 318.

16. The pertinent parts of Pacioli's *Summa* will be discussed in Vol. 3 of my *Archimedes*.

17. As I have indicated below, the techniques and propositions of the *De curvis superficiebus* were used by Francesco Maurolico in his version of *On the Sphere and the Cylinder* (*ed. cit.* in footnote 11, pp. 40–85). Without naming this treatise, Francesco calls the method "an easier way." (*Ibid.*, 2)

Incidentally, Legendre adapts the method of Proposition XII.18 of the *Elements* to the proof of Proposition XII.2 in the same way as the author of *De curvis superficiebus* does in his somewhat similar Proposition III. For Legendre's proof, see Euclid, *The Elements*, translated with introduction and commentary by Thomas Heath, Vol. 3 (Annapolis, 1947), 377–78, 434–37. For the proof in the *De curvis superficiebus*, see my *Archimedes*, Vol. 1, 462–67. Furthermore, the Italian mathematician, V. Flauti, without realizing that Maurolico or Johannes de Tinemue had employed this method, used it in preparing his *Corso di geometria elementare* in 1808. He followed it for the proof of other propositions of Book XII of the *Elements* as well as for theorems of *On the Sphere and the Cylinder* (V. Flauti, "Sull' Archimede e l'Apollonio di Maurolico," *Memorie della Reale Accademia delle scienze dal 1852 in avanti ripartite nelle tre classi di matematiche, scienze naturali, e scienze morale*, Vol. II (Napoli, 1857), p. XCIII.)

As Flauti notes, he was much later to discover Maurolico's use of this method. The *De curvis superficiebus* was unknown to Flauti and so he did not realize that Maurolico had a predecessor in applying this method to *On the Sphere and the Cylinder*.

18. I shall republish Maurolico's text of *On the Sphere and the Cylinder* with a close analysis of it and a comparison of it with the *De curvis superficiebus* in Vol. 3 of my *Archimedes*. Although Maurolico's Archimedean works were not published until 1685, their composition had begun as early as 1534 and was completed in 1550; the dates of completion of the various works are noted in the 1685 edition.

19. See my *Archimedes*, Vol. 1, 9–10.

20. A partial list of William's errors and misunderstandings of the Greek text was given by Heiberg, *Archimedis opera omnia*, Vol. 3, li–lii. To these I shall add a number of other examples in Vol. 2 of my *Archimedes in the Middle Ages*. It is of interest that in *On the Equilibrium of Planes* Moerbeke often mistranslates τόμος by *sector*, when the meaning is rather that of frustum. A possible explanation for this is that several times the Greek text mistakenly has τομεύς which indeed ought to be rendered by *sector*. Of course the context should have suggested even in these cases that *sector* was not the proper rendering.

Interestingly enough, when he later came to translate *On Conoids and Spheroids*, Moerbeke abandoned the erroneous translation and merely retained the word *tomos* in his Latin text. Moerbeke was also troubled by the Greek word *helix* (i.e., spiral) when he began to translate *On Spirals* and he used words like *volutio* and *revolutio* until settling down with *elix* or the adjectival form *elicus*. The reason for this is not hard to understand since *helix* was commonly used in Latin for ivy or a twisting vine and certainly Moerbeke had no experience with a mathematical *helix* before seeing Archimedes' treatise.

Moerbeke was also greatly puzzled by the signs for fractions and myriads used in Proposition 3 of *On the Measurement of the Circle* and he bungles the large numbers and fractions used in the calculation of π. One would suppose, therefore, that he was unacquainted with the various versions in the Arabo-Latin tradition of *On the Measurement of the Circle*.

21. Heiberg did not know of this last manuscript. Hence he concluded that, since Commandino in making his emended version of *On Floating Bodies* had not used the Madrid manuscript, he must have used Moerbeke's holograph. However, I am reasonably sure that it was Vat. Barb. lat. 304 which Commandino used. This is made virtually certain by the fact that the manuscript includes as the only other item the *De analemmate* of Ptolemy in Moerbeke's translation which Commandino also revised.

22. This treatise will be edited and published for the first time in my *Archimedes*, Vol. 3. I shall also demonstrate the likelihood of Johannes de Muris' authorship.

23. Also perceptive was the realization on the part of the author of this commentary that Archimedes' intention in presenting these *neuseis* in the *On Spirals* was merely to assert the existence of a solution rather than to suggest what the solution was.

24. M. Clagett, "Johannes de Muris and the Problem of Proportional Means," *Medicine, Science and Culture: Historical Essays in Honor of Owsei Temkin*, ed. by L. G. Stevenson and R. Multhauf (Baltimore, 1968), 35–49. In my *Archimedes*, Vol. 3, I shall present the full texts and translations of all of the Archimedean sections of Johannes de Muris' *De arte mensurandi*. I shall also show that Johannes de Muris took an incomplete version of the *De arte mensurandi* (containing only five chapters) and expanded it in an impressive way to include the full range of medieval geometric knowledge.

25. Published with Regiomontanus' *De triangulis* (Norimbergae, 1533), p. 5 separate pagination.

26. M. Clagett, *Nicole Oresme and the Medieval Geometry of Qualities and Motions* (Madison, Wisc., 1968), 220–23. Oresme made use of the spiral in an exceedingly fertile section on the measure of curvature. He represented its curvature by a right triangle after plotting the radius lengths against the values of the angle of rotation (*Ibid.*, 450).

27. The citations to Archimedes' *On Floating Bodies* by Oresme occur in his *Questiones super de celo et mundo*, ed. and transl. by C. Kren (Thesis, Univ. of Wisconsin, 1965), 847–64. The reference by Henry of Hesse is found in his *Questiones super communem perspectivam*, MS Erfurt, Stadtbibl. Amplon. F. 380, 30v, col. 1. The pertinent passages from both of these works will be given in my *Archimedes*, Vol. 3.

28. M. Clagett, *The Science of Mechanics in the Middle Ages* (Madison, Wisc., 1959), 95.

29. *Ibid.*, 435. Albert of Saxony, Oresme's junior contemporary at Paris, gives the dynamic definition of specific weight neatly (*Ibid.*, 137): "With two solid bodies given, it is possible without weighing [them] in a balance to find out [1] whether they are of the same or of different specific weights and [2] which of them is heavier. For let these two bodies be *a* and *b* and let equal volumes of *a* and *b* be taken . . . ; and let these portions be released so that they fall in the same water. Then if they descend equally fast toward the bottom of the water, or if just as large a part of one as of the other [is submerged in the water], say that the said bodies are of equal specific weight. If, however, they descend to the bottom unequally fast, or more of one of these portions is submerged in the water and less of the other, say that the one is heavier according to species whose portion descends more quickly or whose portion is submerged further in the water."

This passage is reflective of the much longer treatment by Oresme referred to in footnote 27, but there is no specific reference here to Archimedes' *On Floating Bodies*.

30. See Tartaglia's *Ragionamenti . . . sopra la sua travagliata inventione* (Venice, 1551) with pagination, but see the second page: "*Et che quelli corpi solidi che sono poi di natura piu gravi di l'acqua posti che siano in acqua, subito se fanno dar loco alla detta acqua, e che non solamente intrano totalmente in quella, ma vanno discendendo continuamente per fin al fondo, e che tanto piu velocemente vanno discendendo quanto che sono piu gravi dell' acqua.*"

31. G. B. Benedetti in his preface to the *Resolutio omnium Euclidis prob-*

lematum etc. (Venice, 1553), sig. ** verso, notes that Tartaglia taught him the first four books of Euclid. He also gives the first exposition of his views of the variation of velocity with specific weight. The main conclusion (sig. *** recto) is: *"Modo dico quod si fuerint duo corpora, eiusdem formae, eiusdemque speciei, aequalia invicem, vel inaequalia, per aequale spacium, in eodem medio, in aequali tempore ferentur."* He had earlier mentioned Archimedes. Cf. my *The Science of Mechanics*, p. 665.

It may be noted that John Dee in his preface to H. Billingsley's translation: *The Elements of Geometrie of . . . Euclide of Megara* (London, 1570), sig. b iiii verso to c i recto, gives a translation of some of the propositions from Archimedes' *On Floating Bodies* and notes that by these propositions "great Errors may be reformed, in Opinion of the Naturall Motion of things, Light and Heavy, Which errors, are in Naturall Philosophie (almost) of all men allowed: too much trusting to Authority: and false Suppositions. As, 'Of any two bodyes, the heavyer, to move downward faster than the lighter.' This error, is not first by me Noted: but by one Iohn Baptist de Benedictis. The chief of his propositions, is this: which seemeth a Paradox. 'If there be two bodyes of one forme, and of one kynde, aequall in quantitie or unaequall, they will move by aequall space, in aequall tyme: So that both theyr movynges be in ayre, or both in water: or in any one Middle [i.e. Medium].' " [*Single quotation marks mine.*]

32. Alberti in his *De' Ludi matematici*, Chap. XX (*Opera volgari*, ed. of A. Bonucci, Vol. 4 [Florence, 1847], 438–39), takes up the crown problem, gives the essential content of Propositions 5–7 of Book I of *On Floating Bodies* and concludes with the dynamic definition: *"E quelli corpi che in sè pesano più che l'acqua staranno sotto; e quanto più peseranno tanto più veloci descenderanno e meno occuperanno dell' acqua sendo tutti d'una figura e forma."* And so once more Tartaglia's ideas seem to be derivative in character.

33. M. Clagett, "A Medieval Archimedean-type Proof of the Law of the Lever," *Miscellanea André Combes*, Vol. II (Rome, 19€7), 409–21.

34. See my *Archimedes*, Vol. 1, 9.

35. We have mentioned influences on Giovanni Fontana, Nicholas of Cusa, Leon Battista Alberti, Luca Pacioli and Leonardo da Vinci. We can also mention the influence of the Arabo-Latin tradition of Archimedes on the *Artis metrice practice compilatio* of Leonardo de Antoniis, a Franciscan who composed some geometrical notes on Campanus about 1404/5 in Bologna.

The Italian translation of the *Compilatio* was edited by M. Curtze in the *Abhandlungen zur Gesch. d. Math.*, 13. Heft (1902), 339–434. Curtze misidentified the author with Leonardo Mainardi (fl. 1488). The correct identification was given by A. Favaro in *Bibliotheca Mathematica*, 3. Folge, Vol. 5 (1904), 326–41; cf. his article in *Atti R. Intituto Veneto*, Vol. 63 (1904), 377–95. I am giving an edition of the Archimedean parts of this tract based on several Latin manuscripts in my *Archimedes*, Vol. 3.

36. J. L. Heiberg, "Neue Studien zu Archimedes," *Abhandlungen zur Geschichte der Mathematik* 5. Heft (1890), 83–84, concluded that Jacobus Cremonensis had made use of Moerbeke's translation.

My own study of the Cremonensis translation leads me to conclude that occasionally Cremonensis did indeed consult the Moerbeke translation. While Cremonensis did give better translations than Moerbeke in a number of cases, in one spectacular case Moerbeke's version was better; see S. Heller, "Ein Fehler in einer Archimedes-Ausgabe, seine Entstehung und seine Folgen," *Abhandlungen der Bayerischen Akademie der Wissenschaften. Mathematisch-naturwissenschaftliche Klasse, Neue Folge*, 63. Heft (1954), 21.

Incidentally, Heller believed that Jacobus had not known the work of Moerbeke (*Ibid.*, 19). But all his evidence shows is that Jacobus did not pay much attention to the earlier work in certain given passages, which is certainly true.

37. The first important (but quite incomplete) study of the relation between Leonardo and Archimedes was made by A. Favaro, "Archimede e Leonardo da Vinci," *Atti del Reale Instituto Veneto di Scienze, Lettere ed Arti*, Vol. 71 (1911–12), 953–75. Also incomplete but more interesting are the various sections on Archimedes and Leonardo in R. Marcolongo, *Studi Vinciani: Memorie sulla geometria e la meccanica di Leonardo da Vinci* (Naples, 1937).

I have prepared for Volume 3 of my *Archimedes* a new and detailed study of all the relevant passages in Leonardo's notebooks that may reflect some direct knowledge of Archimedean texts. I have shown some traces in Leonardo's notebooks of the following Archimedean works: (1) *On the Measurement of the Circle*, (2) *On Spirals*, (3) *On the Sphere and the Cylinder*, (4) *On the Equilibrium of Planes*, (5) *On Floating Bodies*, and perhaps (6) Eutocius' *Commentary on the Sphere and the Cylinder*.

Although Leonardo mentions two manuscripts of Archimedes—one apparently of the Moerbeke translation and the other of the Cremonensis translation—some of his knowledge (particularly of works 1, 2 and 3) seems to be second-hand and based on medieval works. Most important was the influence of *On the Equilibrium of Planes*, which Leonardo uses and cites specifically and which he appears to have read in both translations. Also of interest is a long fragment from Book II, Proposition 10, of Moerbeke's translation of *On Floating Bodies* that seems to be in Leonardo's hand (*Codice Atlantico*, 153rb, rc, ve).

38. See Leonardo da Vinci, MS L, 2r.

39. In 1522, at Nürnberg, Johann Werner published a *Commentarius seu Paraphrastica ennaratio in undecim modos conficiendi eius problematis quod cubi duplicatio dicitur* which was stimulated by Valla's translation of the pertinent sections from Eutocius' *Commentary*.

XIII

THE WORKS OF FRANCESCO MAUROLICO

I. A CHRONOLOGY OF THE MATHEMATICAL WORKS OF FRANCESCO MAUROLICO

In the course of preparing a detailed examination of Francesco Maurolico's knowledge of Archimedes for my *Archimedes in the Middle Ages*, Volume 3, I found it necessary to work out the chronology of Maurolico's works. It seemed possible that this exercise might have some interest for scholars interested in the activity of the great mathematician of Messina and hence my decision to publish it separately. In the first section I have attempted to record every instance of a dated work. In the second my aim has been to republish Maurolico's own lists of his works and so far as possible to relate the works mentioned in these lists to the established chronology of the first section.

The chronology of Section I has been compiled with special attention to the sundry dated notes and works in Maurolico's extant notebooks. These include the following MSS.: Bibliothèque Nationale, lat. 6177, 7249, 7251, 7459, 7462-68, 7471, 7472, 7472A, 7473, 17859; Real Bibl. del Escorial, cod. lat. &.IV.22, J.III.31; Madrid, Acad. de la Historia, Bibl., Cortes 2787 (formerly 675); Molfetta, Bibl. Seminarii Minoris Servati, Cod. Melph. 5-7, H.15; Rome, Vittor. Eman., S. Pant. 115/32, 116/33, 117/34; Vatican, Vat. Barb. lat. 2158; Vat. lat. 3131; Parma, Bibl. Palat., Pal. 1023; Naples, Bibl. Naz., I. E. 56; Lucca, Bibl. Govern., Cod. 2080 and Florence, Bibl. Naz. Centr., Magliabech. XIV, 39. (In the chronology below the Parisian MSS. will be cited simply « BN lat. .... » and the Roman manuscripts as « Cod. S. Pant. .... ».) Unfortunately I have not yet been able to locate one more important manuscript that belonged to the Villadicani family of Messina but at least I have been able to cite Macrì's notes on that manuscript[1]. In ad-

---

[1] G. Macrì, *Francesco Maurolico nella vita e negli scritti*, 2nd ed. (Messina, 1901), pp. 101-103: « Diremo in ultimo d'un volumetto conservato in Messina, ed appartenente alla nobile famiglia Villadicani. Porta sulla prima pagina in diversa scrittura: *Opere originale e di proprio carattere del Rev. Abb. D. Fra. Maroli* e seguono siffatta leggenda le parole: *Di N 426 fogli o sia pagine costa;* in fatto poi le pagine si vedono al *recto* ed al *verso* numerate con tanta cura, da comprendervi anche i foglietti bianchi. È cartaceo, di centimetri 16 × 11 con legatura in pergamena, e con margine di otto millimetri costantemente mantenuto; ha scrittura minuta, uguale, con poche abbreviature, ed i titoli segnati in rosso od in nero, imitano sempre le maiuscole a stampa. Nel rimanente non può dubitarsi che sia autografo, poichè al termine dei *Dati aritmetici* di Giordano si legge: *Manu ejusdem Francisci transcripta, Messanae*

dition to the manuscripts, I have used the various editions of Maurolico's works. Though the chronology is primarily one of Maurolico's
mathematical and astronomical works, it includes other works. Excluded
are the dates of important events in his life which I have outlined in

20° decembris 1532. L'autografia del volume riesce inoltre confermata per altre considerazioni: le cancellature apprestano ritocchi di forma o di sostanza possibili solo all'autore, il quale taglia spesso il
superfluo, o riduce il pensiero a migliore espressione. Il codice offre quasi in ogni opuscolo nuove de-
duzioni o posteriori notizie, aggiunte sul margine di molte pagine, quando non al termine del capitolo;
e vi sono assai volte frammezzati foglietti bianchi, per più copiose dimostrazioni o per isvolgimenti più
estesi. La mano del Maurolico si mostra sempre ferma, quantunque molti quaderni siano scritti o copiati
nell'anno medesimo della sua morte; soltanto si vedono a quando a quando cancellature, addizioni mar-
ginali, o correzioni. Alle pagine 278 e 279 si leggono ricette e consigli igienici, e quindi sembra sicuro
che gli autografi differenti si fossero legati in un solo volume dopo la morte di lui. Molti ricordi o
sunti sembrano materiali da adoperare nello scriver di cose letterarie o scientifiche, come fa chiaro il
difetto della data, aggiunta sempre alle opere destinate alla stampa ». The contents of the manuscript are
described in the Appendix, pp. xxiii-xxvi:

« IV.
AUTOGRAFI CONTENUTI NEL CODICE VILLACANENSE.

1. Tavola sinottica della matematica senza titolo (pag. 1). Divide in astratto la scienza in *aritmetica*
e *geometria (matematiche pure)*; la divide poi per applicazione ai corpi, in *astronomia, geografia, pro-
spettiva, musica, ponderatoria.*

2. *In breviaturam elenchi Conradiani paralipomena* (pag. 3 a 6). Notizie intorno a Giorgio Valla pia-
centino, ed enumerazione delle opere originali e delle traduzioni di costui, stampate da Aldo Manuzio.

3. *Viri quidam mathematici viennensis gymnasii* (pag. 7 ad 11). Note sulla vita e sulle opere di En-
rico d'Assia, Giovanni Gmunden, Giorgio Peurbach, Giovanni Regiomontano. Son tratte da ciò che Giorgio Tanstetter scrisse de' tedeschi e delle loro opere.

4. *De quibusdam aliis authoribus, quaedam notatu digna sequentur* (pag. 12 a 15). Note intorno a
Giorgio da Trebisonda, Giorgio Tanstetter, Pietro Apiano, Giovanni Bianchini, Abramo Zaculo ed altri
minori.

5. *De cosmographia Claudii Ptolemaei* (pag. 14). Elenco delle traduzioni di questo libro, sino alla
versione italiana di Pier Andrea Mattioli, ed alle addizioni di Jacopo Gastaldo e di Geronimo Ruscelli.

6. *Strabonis libri quindecim de situ orbis, a quibus tralati* (pag. 15). Poche linee sulla traduzione
del veronese Guarino, fatta per ordine di Nicolò V.

7. *Instrumenta et observationes Jo. de Monte Regio, in libro quodam impresso Norimbergae apud
Jo. Montanum et Ulricum Neuber, anno Domini 1544* (pag. 16). Enumerazione delle opere del Regiomontano, del Peurbach, del Valter, contenute nel volume. Vi si rilevano alcune erronee censure di Ber-
nardo Valter alle tavole alfonsine.

8. *De quibusdam aliis authoribus sphaerae et computi* (pag. 17). Cenni su quanto scrissero intorno
all'argomento Jacopo da Cremona, Giovanni Petsan, Giovanni Sacrobosco, Giovanni Campano.

9. *De quibusdam aliis recentioribus* (pag. 19 a 29). Notizie sulle opere di Erasmo Osvaldo, di Nicolò Copernico, di Daniele Barbaro. Vi si leggono giudizi severi intorno a diversi scrittori, e due distici
contro Daniele Sambech.

10. *De horis canonicis* (pag. 29). Poche linee sull'ordine delle ore canoniche.

11. *De Matthiolo Dioscoridis expositore* (pag. 30 a 37). Sunto de' libri di Pier Andrea Mattioli, con
poche addizioni del Maurolico.

12. *De Abrahamo Ortelio cosmographo* (pag. 38 a 46). Sunto del *Theatrum orbis terrarum*, con os-
servazioni sulla difficoltà e sul valore dell'opera.

13. *De insulis totius orbis per Bordonem et Porcaccium collectis* (pag. 47 a 50). Descrizioni somma-
rie delle isole, tratte dai libri di Benedetto Bordone e di Tommaso Porcacchi.

14. *Petri Criniti de poetis latinis, ad Cosmum Paccium pontificem aretinum, breviarum* (pag. 55 a
72). Compendio destinato alla stampa ed enumerato sempre negli *Indices lucubrationum.* Seguono due
libri del Maurolico (sesto e settimo) che stampiamo in quest'appendice. Il settimo fu pubblicato nella
Messana Illustrata del Samperi.

15. *De grammaticis per Svetonium* (pag. 100 a 106). Notizie tolte dal libro di Svetonio, alle quali
l'autore a pag. 103 del manoscritto, soggiunge altre con l'avvertenza: *Quae sequuntur ex commentariis
Raphaeli sumpta sunt.* Si parla in quest'ultima parte d'Ambrogio Calepino, di Lorenzo Valla, di Cristo-
foro Scobar. V'è pure un breve capitolo: *De rethoribus ex Svetonio.*

16. *Jordani arithmetica data* (pag. 124 a 200). Opera composta col metodo dal Maurolico usato per
gli antichi, siccome avverte nella chiusa (pag. 197): *Arithmeticorum Jordani ex traditione Francisci Maurolycii messanensis, liber quartus et ultimus hic completus est. Fuerunt autem haec Jordani data e quodam*

Part III, Chapter 5, Section 1 of my *Archimedes in the Middle Ages,* Volume 3. I must finally observe that not every mathematical piece or note is mentioned since there are numerous fragments in the manuscripts that are not dated. Thus my chronology can only be considered preliminary to a detailed description of all of the manuscripts [2].

---

vetusto, verum infinitis mendis et defectibus maculoso codice, magno labore multisque vigiliis extracta, et manu eiusdem Francisci transcripta Messanae 20 sept. 1532.

17. *Boetii arithmetica* (pag. 208 a 212). Le più utili teoriche di Boezio ridotte in un compendio, che termina con le parole: *Catanae die 28 januarii 1554. Lector vale; caetera in quibus Boetius speculatur, plus habent fastidii quam jucunditatis; ideoque negligenda duximus.*

18. *Musica Boetii* (pag. 213 a 218). Opuscolo del tutto simile al precedente.

19. *Perspectivae Rogerii Bacchonis breviarium* (pag. 219 a 269). Vi si abbreviano le cose utili, dando in questa forma commiato al lettore: *Sic lector, habes hic libri frumentum, et paleis mundatum et locupletius.*

20. Dalla pagina 270 alla pagina 288, si vedono ricordi staccati di geometria, d'aritmetica, d'algebra, con figure e numeri. Profittando dello spazio, il Maurolico scrisse alle pagine 278 *retro* e 279, due inni sacri, la ricetta delle pillole arabiche, il regime per evitare le vertigini, e quello per allontanare il tintinnio alle orecchie.

21. Dalla pagina 289 alla pagina 338, si leggono inni sacri e cose liturgiche; solamente la pagina 322 contiene note brevissime d'aritmetica.

22. Viaggio (1548) senza titolo, da Napoli a Messina (pagine 348-49), stampato dall'Arenaprimo, il quale omise le avvertenze finali: *Haec in diurno, dum in itinere essem cum D. Jo. Vigintimillia, fuerunt inter eundum adnotata: hic autem rescripta 14 junii 1575.* A queste parole in rosso, seguono in nero le altre: *Caetera cum his, in narratione quadam,* le quali danno a credere che il matematico avesse di quel viaggio fatto più estesa narrazione in altro lavoro.

23. *In Alphonsi tabulas canones* (pag. 351 a 368). Regole generali astronomiche con annotazioni (*scholia*), e con una appendice sulle ecclissi.

24. *In tabulas Joannis de Monte Regio brevissimi canones* (pag. 369 a 387). Sunto delle tavole delle direzioni del Regiomontano, con lunga nota finale, e con un capitolo sulla tavola del primo mobile: *In tabulam magnam primi mobilis brevissimi canones.*

25. *In tabulas eclipsium Georgii Peurbachii canones* (pag. 392 a 399). Sommario con molte note.

26. *In tabulas Joannis Blanchini canones brevissimi* (pag. 400 a 424). Dopo alcune notizie sul Bianchini, vi si espone l'uso delle tavole di costui, con diverse annotazioni ».

Presumably Macrì has not reported in his description all dates mentioned in the codex, for elsewhere (p. 227), he notes that item 12 was completed 18 October, 1574.

[2] Brief and inadequate descriptions of the Parisian manuscripts have been given by F. Napoli, « Nota intorno ad alcuni manoscritti di Maurolico della Biblioteca Parigina », *Rivista sicula di scienze, letteratura ed arti,* Vol. 8 (1872), pp. 185-192 and « Intorno alla vita ed ai lavori di Francesco Maurolico », *Bullettino di bibliografia e di storia delle scienze matematiche e fisiche,* Vol. 9 (1876), pp. 1-22. For the reader to get some idea of how inadequate and erroneous the descriptions of Napoli are, he need only compare Napoli's brief desciption of BN lat. 7472A in his « Nota », p. 188 (« Radices motuum, latera diametrorum, distantiae, tabellae tum radicales ad calculum pertinentibus et demonstrationes 28 settembre 1571)») with the actual table of contents I have published below under the entry 1571, 28 September. Macrì, *Francesco Maurolico,* 2nd. ed., pp. 97-98 pointed out the important fact that the manuscripts were part of the Colbert collection and he reasoned properly that they were sent to Paris by the French occupying force before it left Messina in 1678. The conclusive piece of evidence was the publication in Paris in 1679 of the cancelled passages from the autograph manuscript of Maurolico's *Sicanicarum rerum compendium.* The editor was Stephanus Baluzius (Étienne Baluze) in his *Miscellaneorum liber secundus, hoc est collectio veterum monumentorum, quae hactenus latuerunt in variis codicibus ac bibliothecis,* p. 323. The fragments of Maurolico are entitled: « Collectio locorum quorumdam insignium consilio omissorum, in libro sexto Rerum Sicanicarum Maurolyci abbatis, edito Messanae anno MDLXII ». In the table of contents Baluze designates the fragments as follows: « Ex authographo Maurolyci quod extat in Codice 1823 bibliothecae Colbertinae ». The codex mentioned is of course that now numbered BN lat. 6177. Baluze's additions were added by Giacomo Longo to the second edition of Maurolico's *Compendium* (Messina, 1716). Turning from the Parisian manuscripts, we should note that the three manuscripts of the Biblioteca Vittorio Emanuele have been briefly (and incompletely) described by L. de Marchi, « Di tre manoscritti del Maurolicio (!) che si trovano nella Biblioteca Vittorio Emanuele di Roma », *Bibliotheca Mathematica,* No. 3 (1885), cc. 141-44, No. 4 (1885), cc. 193-95. Cf. Macrì, *Francesco Maurolico,* 2nd ed., Appendix, pp. xxvii -xxxii.

1508, January. This date has been inscribed on the last folio of one of Maurolico's notebooks (Paris, BN lat. 7472, 39r) in a hand that may be that of Maurolico: « data Messanae anno incar. domini (?) M.C. VIII mense Ianuario. » On the same page, in a different ink, Maurolico abortively wrote: « Archimedes de isoper. » This could have been written on any date.

1518, 23 January. This date appears on Book IV of Maurolico's *De momentis aequalibus*. It is a typographical error for 1548, 23 January, which see.

1521, 19 October. Maurolico completed the *Photismi de lumine et umbra* at Messina (see MS Bibl. Govern. di Lucca, Cod. 2080, 25r; cf. ed., Naples, 1611, p. 29).

1522, 3 January. Maurolico completed at Messina the first two parts of his *Diaphaneon sive transparentium libellus* (see BN lat. 7249, 11v; this date is missing from the ed. of Naples, 1611).

1528, 14 August.. Maurolico dedicated to Ettore Pignatelli his *Grammaticorum rudimentorum libelli sex* (Messina, 1528), IV. Mentions there (7v) that he has proved Archimedes' conclusions concerning the measurement of the circle, the sphere and the cylinder, and equal moments before seeing Archimedes' works.

1529, November. Maurolico has dated his explanation of some astronomical tables in Paris, BN lat. 7472A, 63v: « Mense novembri 1529. Infrascripta tabula constructa est ad erigendas duodecim domos ad latitudinem loci ubi polus elevatur graduum 38, quanta est fere latitudo Messanae in freto siculo. » The table is given on 64r-69v. Then follows another table apparently written at the same time (7or: « Sequitur tabula positionum ad latitudinem graduum 38, quanta est latitudo Messane in freto siculo, » with the table on 70v-76v).

1529, 9 December. Completion of Maurolico's *Libellus de impletione loci quinque solidorum regularium* at Messina (see Cod. S. Pant. 117/34, 1r-20v; date on 20v; cf. De Marchi, « Di tre manoscritti, » c. 195). Cf. 1535, 21 September.

1531 - 1532. Probable period of the composition of Book I of *De lineis horariis*, dedicated to the *stradigò* Francesco Santapace (see *Opuscula mathematica* [Venice, 1575], p. 161; and see below, the entry for 1553, 9 July).

1532, 21 January. Completed a terse treatment of Book II of Euclid's *Elements* (Cod. S. Pant. 115/32, 23r-24v [21r-22v]; date on 24v [22v]).

1532, 9 July. Completed at Messina dedication to Gerolamo Barresi, *stradigò* of Messina, of his (Maurolico's) compendium of Books XIII-XV of Euclid's *Elements*. See *Opuscula mathematica* (Venice, 1575), pp. 105-106, noting that they had studied the first twelve books books together. Cf. Cod. S. Pant. 115/32, 2r[1r]: « 13us, 14us et 15us de solidis regularibus in alio libello ad Petrae dominum olim dedicato. » Gerolamo Barresi was the «signore di Pietra Perzia.» (Cf. De Marchi, « Di tre manoscritti, » c. 142.)

1532, 20 September. Completed at Messina a paraphrase of the *Arithmetica* of Jordanus (for the manuscript, see above, note 1, item 16).

1533, 30 December. Notes on the problem of finding two mean proportionals between two given lines (Cod. S. Pant. 115/32, 44r-v).

1534, 23 July. Completed at Messina his version of Moerbeke's translation of Archimedes, *De quadratura parabolae* (ed., 1685, p. 195).

1534, 16 August. Completed his version at Messina of Serenus, *Cylindricorum libelli duo* (BN lat. 7465, 1r-20r, date on 20r: « Messanae in freto siculo manu industria francisci Maurolycii (!) 16 Aug. 1534. »). Napoli, « Nota, » p. 187 misreads this as 16 July.

1534, 19 August. Completed his version of Archimedes, *De circuli dimensione* (BN lat. 7465, 21v-28v, date on 28v; and ed., 1685, p. 36). Same date, completed his own *Tetragonismus* (BN lat. 7465, 31r; ed., 1685, p. 39). No doubt his *Hippocrates tetragonismus* in the same manuscript (29r-v) and edition (pp. 36-37) was done at the same time.

1534, 10 September. Completed at Messina his version of Archimedes, *De sphaera et cylindro* (in reality a reworking of the medieval *Liber de curvis superficiebus*). See ed., 1685, p. 85.

1534, 3 October. Completed at Messina his version of Autolycus, *De sphaera quae movetur liber* (BN lat. 7472, 1r-7v, date on 7v). Published with truncated comments instead of proofs and without date at Messina in *Theodosii sphaericorum... libri III etc.* (Messina, 1558), 61r-62r. As Mogenet notes, the published text is a pale resumé of the manuscript edition, which Mogenet shows is partially dependent on Valla's version but which has considerable original material (see J. Mogenet, « Pierre Forcadel traducteur d'Autolycus, » *Archives internationales d'histoire des sciences*, No. 10 [1950], pp. 13-15).

1534, 8 October. Completed his version of Theodosius, *De habitationibus liber* (BN lat. 7472, 9r-16r, date on 16r). Published with brief comments instead of the proofs in the MS and with some changes in the enunciations (and without date) in *Theodosii sphaericorum... libri III etc.* (Messina, 1558), 62r-62v.

1534, 20 October. Completed at Messina his version of Autolycus, *De ortu et occasu syderium.* See BN lat. 7472, 17r-36v, date on 36v. Though dependent on Valla, Maurolico's version is highly original (cf. Mogenet, « Pierre Forcadel, » pp. 15-17).

1534, 25 October. A note entitled *Situla horaria* completed at Messina on this date (Cod. S. Pant. 115/32, 48v). It would appear from the handwriting and ink that his *Ex Heronis et aliorum spiritalibus* (*Ibid.*, 45r-46v), his *Turris eiaculatoria* (*Ibid.*, 48r-v) and notes on similar instruments (*Ibid.*, 49r-v) were written at the same time. Cf. Baron della Foresta, *Vita dell'Abbate del Parto D. Francesco Maurolyco* (Messina, 1613), p. 5.

1534, November. Completed treatments of the following books of the Campanus version of Euclid's *Elements*: Book V on 5 November (Cod. S. Pant. 116/33, 1r-8v, date on 8v), Book VII on 9 November (*Ibid.*, 9r-19v, date on 19v), Book VIII on 14 November (*Ibid.*, 20r-29v, date 29v), Book IX on 19 November (*Ibid.*, 30r-39v, date on 39v). And see 1541, 2 August for Book X.

1535, 21, 23, 24 September. Additions to Maurolico's *De impletione loci.* (Cod. S. Pant. 117/34, 21r-v). Cf. 1529, 9 December.

1535, 1 October. Further notes on the proportional means problem (Cod. S. Pant. 115/32, 44r, 47r).

1535, 21 October. Completed at Messina his *Cosmographia.* See ed., Venice, 1543, 103r.

1536, 4 May. Completed at Messina his letter to Pietro Bembo on the eruption of Mt. Etna, etc. Cf. MS Vat. Barb. lat. 2158, 143r-46v, date on 146v. See also

XIII

F. Guardione, « Francesco Maurolico nel secolo XVI, » *Archivio storico siciliano*, Anno 20 (1895), pp. 36-39.

1536, 19 May. An observation on the conjunction of Saturn and Mars on this date (BN lat. 7472A, 80r). On the same folio an observation of Mars for 29 May of that year.

1536, 3 July. Some notes for his *De compaginatione solidorum regularium*, i.e. pattern forms for the construction of regular solids by folding together the patterns (Cod. S. Pant. 115/32, 26v).

1537, Month and day not clear. Apparently completed notes entitled *Inventio solaris diametri per hipparchum (del.), Heronem et Proclum* and *Inventio solaris diametri per Ptolemaeum (del.) per Hipparchum quem sequitur Ptolemaeus* (Cod. S. Pant. 115/32, 50r).

1537, 17 March. Some notes on planetary conjunctions (BN lat. 7472A, 32v).

1537, 25, 26, 28 and 29 December. More fragments of his *De compaginatione solidorum regularium* (Cod. S. Pant. 115/32, 27r-v). Cf. 1536, 3 July.

1538(?), 15 January. Completed at Messina some gnomonic tables. (BN lat. 7472A, 57r-62v, date on 62v). Notice also individual notes on the observation of Mars on 13 February and of Saturn and Mars on 9 June (*Ibid.*, 80r).

1538, 12 August. Completed his *Modus fabricandi astrolabium cum demonstrationibus geometricis* (BN lat. 7464, 9r-16v, date on 16v; other fragments are dated 11 August on 3v and 13 August on 4r). This work is to be distinguished from his section on the astrolabe in his *De instrumentis astronomicis* published in his *Opuscula mathematica* (Venice, 1575), pp. 61-77.

1539, 18 April. Planetary observations for that date (BN lat. 7472A, 11r) and a solar eclipse of that day (*Ibid.*, 80r, at Messina). He also notes on the latter folio a conjunction of Jupiter and Mars on 14 August, 1539.

1539, 22 September (i.e., x. Kalends of October). Composed at Messina a letter to Francesco Cardinal Quignonio (MS Parma, Bibl. Palatina, Pal. 1023 – a part of the Fondo Beccadelli). Religious content rather than mathematical, but so far as I know has not been published : « Reverendissime Pater. XV. Kal. octobris recepi literas tuas una cum enchiridio diurnarum precum. Utrumque fuit mihi munus quovis thesauro preciosius. Quantum enim fuit a tanto praesule salutari ac donari? Sedasti mihi titubantem animum : atque adeo argute obiectionibus respondes; ut tecum omnino sentiam, ac nihil non optime factum existimem. Vellem tamen ut id, quod de Dionysii Areopagitae historia scripsi, maturius, sicubi vacarit, discuteres : ne quid ambigui relinquatur. Offers mihi praeterea patrocinium curamque tuam. Utinam tantam gratitudinem aliquo servitio promeruissem. Audebo tamen aliquid petere, quando cum clementissimo antistite loquor. Sed nihil gratius praestare poteris, quam si facultatem mihi breviarii tui, cuius incredibili amore sum affectus, recitandi a pontifice maximo ' Impetres, atque impetrata ' mittas. Vale et vive felix. Messanae in freto siculo. Xº kal. Octobres. M.D. XXXVIIII. Ad omne tibi servitium paratus Franciscus Maurolycus. » Note that Maurolico had sent to the Cardinal something he wrote on Dionysius the Areopagite. I have not discovered any copy of this treatment. Perhaps it was only a letter.

1539, 28 September. The following deleted passage appears on folio 2r (old pag. 1r) of Cod. S. Pant. 115/32 : « Arithmeticae Speculativae Supplementa, completa ad primam noctis horam, dici dominicae 28ᵉ Sept. XIIIᵉ Ind. 1539. » The same date (indeed the same hour) is added on folio 11v: « Caetera curiosiores inquirere. Sed obscura minusque necessaria minus curanda sunt: quod et Cicero in officiis praecipere videtur. Die ☉ [i.e. solis]: 28 Sept. 1539, ad primam noctis horam. » This seems to terminate a section (8r-11v) entitled on 8r: « Caput 16ᵘᵐ et arithmeticae postremum ». The passage beginning on 11r: « Denique tam super... » and ending with the Ciceronian reference and date was added by Maurolico to the end of Book II of his *Arithmetica* (Venice, 1575), pp. 174-75.

1540. Composed at Messina his *De gestis apostolorum* (published by Maurolico with the *De vita Christi* of Matteo Caldo in 1555, q. v.). Dated on 50v:
« Ad mille, quadraginta quinquies centum
dum Paulus Papa tertius habebat
Carolus quintus Cesar regimentum,
Franciscus haec maurolycus scribebat
Messanae dum sedebat secus rivi
fontem, specumque sui Nicandri divi. »

1540. *Radices superationum et motuum pro tabulis Blanchini* (BN lat. 7472A, 33r-36v) and *Radices motuum ad meridiem Messanae* (*Ibid.*, 40v-42v) were calculated for this year.

1540, 8 January. An extract « Ex magna Ptolemaei constructione » (BN lat. 7472A, 30r) is so dated. It is essentially the piece added to the *Cosmographia* (Venice, 1543), fol. ✠r, entitled « ut Ptolemaeus » and extending to the statement about Mercury's revolutions (« ...In spatio annorum solarium 46.d.1. 1/30. »).

1540, 24 January. Proemium of the *Cosmographia* (Venice, 1543), addressed to Pietro Bembo and executed at Messina (sign. a iv verso, « Messanae in freto siculo. Nono Cal. Febr. M.D.XL. »).

1540, 11, 19-21 June. Brief lunar and planetary observations (BN lat. 7472A, 80v).

1540, 5, 9, and 12 August. Some arithmetical notes on proportional terms (Rome, Cod. S. Pant. 115/32, 12v, 13v).

1541. The probable year of Maurolico's preparation of a map of Sicily for Giacomo Gastaldo (see Macrì, *Francesco Maurolico*, 2nd. ed., p. 220; cf. Baron della Foresta, *Vita*, p. 7). See also 1545.

1541, 6 April. Completed the main section of his *Quadrati fabrica* (Venice, 1546), 8r: « Completum in palatio D. Ioannis baptistae prioratus Hierosolymitanae religionis, equitum rhodiorum, die mercurii circa meridiem. 6. Aprilis. 14. Ind. MDXLI. » On 18 April, a scholium was completed (*Ibid.*, 9r). See below 1541, 7 and 9 December and 1542, 11 January.

1541, 2 August. Completed his summary of Book X of the *Elements*, occupying Cod. S. Pant. 116/33, 40r-105r, dated on 105r: « Euclidis Elementorum 10ᵘˢ finis. Cuius omnes propositiones a 5ᵃ ad 70ᵃᵐ inclusive scriptae fuerunt in aedibus prioratus S. Joannis hierosolymitani Messanae. Caeterae in aedibus nostris. Die ♂ [i.e. Martis] 2° Aug. 14ᵃᵉ Ind. 1541... ». Some *Emendenda* (105v-107v) were completed at Messina on 7 August (107v).

1541, 12 November. Composed an *Additio in 26ᵃᵐ 11ⁱ elementorum* [*Euclidis*] (Cod. S. Pant. 115/32, 6v). A postscript (7r-v) and another proposition (7v) were completed on 14 November (dates on 7v).

1541, 7 December. The first 19 of the *quaestiones* of the *Paralipomena* to his *Qua drati fabrica* (Venice, 1546), 9r-11r were completed. This date is not mentioned in the edition but rather in BN lat. 7464, 51v. This manuscript contains only the *Paralipomena* and not the first part of the tract. It begins on 48r at the point in the text « arcus noti et statim... ») (= ed., 1546, 9r, line 8 from bottom) and runs to the end of the text. However the table of stellar positions given at the end of the edition (12r) is inserted on folio 50v of the manuscript. It is noted there (but not in the edition) that the table is « ad annum dominum 1540 vel circa. » The date after Quest. 19 is given as follows (51v): « Messanae per franciscum Maurolycum paulo ante meridiem diei ☿ [i.e. Mercurii] 7º Decembris 15ᵃᵉ Ind. M.D.XLI. »

1541, 9 December. A 20th question is added to the above noted *Paralipomena* of the *Quadrati fabrica* completed (according to the same manuscript, 52v on this date The same date is repeated at the bottom of the page below the main part of an added corollary. This date is missing in the edition, which gives after the complete corollary as its final date: 11 Jan., 1542, q.v. The latter date is not given in this manuscript on folio 52r where the corollary (and tract) ends without date but with a brief notation of the instrument's utility.

1542, 11 January. *Quadrati fabrica* (Venice, 1546) completed at Messina on this date according to the edition (11v), but see the preceding date.

1542, 27-30 July. Some astronomical tables. First note a table for planetary latitudes (BN lat. 7472A, 77v-78r). On 77v he says: « Sequitur tabula latitudinum quinque errantium, 28 Julii 1542, » while on the bottom of 78r he gives the date « 30 Julii. » At about the same time he completed a similar table for the moon (78v-79r). On 78v he says: « Sequitur tabula latitudinis lunae... 28 Julii 1542. » Then on folio 83v he notes: « Sequitur tabula stationum primarum quinque planetarum... 3ᵃ hora noctis 30 Julii 1542 ». The table is on folio 84r. A succeeding table (84v-85r) on the *diversitas aspectus solis et lunae* appears to have been done at the same time. In fact, another table (*Tabula minutorum proportionalium triplicium: pro triplici videlicet lunaris epicycli*) linked with it (85v-86r) is dated on 86r, 27 Julii 1542. Cf. also the tables on 86v-89r.

1542, 31 July, 1 August. More tables (*Ibid.*, 54v: *declinatio solis*, dated « 1542, ultimo Julii »; 55r: *ascensio recta*; 55v: *Differentia ascensionalis*, dated « 1º Aug. 1542 »; 56r: *quantitas diei*; 56v: *Tabella equationis dierum*).

1542, 5 December. Completion at Messina of the various tables of astronomical parameters printed at the end of the *Cosmographia* (Venice, 1543), verso of last sheet. Much of this material is found in BN lat. 7472A, 25r-27r, 29v-30r, 49r-53r, but the order of presentation has been considerably changed.

1543, The *Cosmographia* was published at Venice.

1543, 1 March. Completed at Messina his letter on Sicilian fish, sent to Petrus Gillius (MS Florence, Bibl. Naz. Centr., Magliab. XIV, 39, 2r-10r; date on 10r: « Messanae in freto Siculo Kalendis Martiis MDXLIII. »). Not published until 1893: *Tractatus per epistolam Francisci Maurolici ad Petrum Gillium de piscibus siculis codice manu auctoris exarato Aloisius Facciola messanensis nunc primum edidit* (Palermo, 1893). Cf. p. 13 of the latter edition for the date. Note that Maurolico disclaimed any detailed study of this branch of philosophy (MS, 9v; *ed. cit.*, p. 13): « Postremo scies me non huic philosophiae parti sed mathematicis apprime disciplinis operam dedisse... »).

1543, 17 and 26 December. Notes following a table of declinations and ascensions. The table appears on BN lat. 7472A, 90v-91r (it is identical with the table published in the omnibus edition of 1558, 66v). The notes start on folio 91r. On the bottom of 92r we find the date 17 December, 1543 and on folio 92v the date of 26 December, 1543. We also find on the latter folio the note: « Sequitur tabella differentiarum ascensionalium ad latitudinem graduum 38¹/₆. » No table follows, but see folio 55v and the entry for 31 July and 1 August, 1542.

1544, 14 January. Notes following a table « ad sciendum cum quo puncto zodiaci oriatur aut occidat astrum quodpiam cuius latitudo non excedit 12 grad. ad latitudinem graduum 38¹/₆. » The introductory note is on BN lat. 7472A, 82r. The table is on folios 82v-83r; the concluding notes are on folio 83v and bear the date of 14 January, 1544.

1545. The publication of Maurolico's map of Sicily with the following legend: « Descrittione della Sicilia con le sue Isole, della qual li nomi Antichi et Moderni et altre cose notabili per un libretto sono brevemente dicchiarate per Giacomo Gastaldo Piemontese Cosmographò. In Venetia 1545. » See Macrì, *Francesco Maurolico*, 2nd. ed., p. 221.

1545, 4 January. Completed *Arithemeticae praxeos demonstrationes. Cap. XIIII⁰* and *Geometricae praxeos demonstratio, Cap. XV* on the same date. The first occupies folios 28r-35r, the second folio 35r-v of Cod. S. Pant. 115/32. The former is dated « 4º Jan. 1545 » on 35r, the latter « eodem die 1545 » on folio 35v.

1546. The publication at Venice of Maurolico's *Quadrati fabrica et eius usus.* Also the publication at Venice of an anonymous *La descrittione della isola di Sicilia,* which may have been written by Maurolico (see Macrì, *Francesco Maurolico,* 2nd. ed., pp. 221-22).

1546, 16 October. Completed at Palermo a table of squares, cubes and the differences of successive cubes for numbers 1 to 100 (BN lat. 7464, 27r-28v, date on 27r).

1547, 2 June. Completed at Palermo Book III of his *Emendatio et restitutio conicorum Apollonii Pergaei* (MS Real Biblioteca del Escorial, J.III.31, 98r; cf. ed., Messina, 1654, p. 128, where the date appears erroneously as 29 June).

1547, 24 June. Completed Book IV of his version of the above-noted work of Apollonius, again at Palermo (*MS cit.,* 115r; *ed. cit.,* p.. 150): « Panhormi. VIII Kalendas Iulias MDXXXXVII. Franciscus Maurolycus, multa diligentia emendatum ac restitutum scribebat. »

1547, 25 October. Completed at Castelbuono his reconstruction of Books V and VI of the lost books of the *Conics* of Apollonius; and so was completed his whole *Emendatio et restitutio conicorum Apollonii Pergaei.* See *ed. cit.,* p. 192. Books V and VI are not in the Escorial codex cited in the preceding entries.

1547, 6 December. Completed at Castelbuono Book I of his version of Archimedes, *De momentis aequalibus* (ed., Palermo, 1685, p. 111).

1547, 19 December. Completed at Castelbuono Book II of Archimedes, *De momentis aequalibus* (*ed. cit.,* p. 132).

1547, 30 December. Completed at Castelbuono Book III of Archimedes, *De momentis aequalibus* (*ed. cit.,* p. 155).

1548, 23 January. Completed at Palermo, Book IV of Archimedes, *De momentis aequalibus* (*ed. cit.,* p. 180; the edition gives MDXVIII, but surely this is an error for MDXLVIII).

1548, 24 October. Completed at Palermo a table of sines with entries « by the hours and minutes, » i.e. a sexagesimally divided table of sines with 360 entries (BN lat. 7472A, 47v-48r). The table is described on folio 47r. Cf. a further note on astronomical tables on folio 48v, dated at Palermo on 26(?) October.

1549, 7 June. A note on spherical triangles, written at Castelbuono (BN lat. 7472A, 90r).

1549, 2 July. Deleted this date and reference to Castelbuono in the canons for astronomical tables (*Ibid.*, 54r).

1549, 18 October. Completed at Castelbuono his version of Archimedes, *De lineis spiralibus* (ed., 1685, p. 225).

1549, 17 December. Completed at Castelbuono his version of Archimedes, *De conoidibus et sphaeroidibus* (*ed. cit.*, p. 275).

1550, 13 February. Completed at Thermae (i.e. Termini-Imarese) his *Praeparatio ad Archimedis opera* (*ed. cit.*, p. 25).

1550, 13 August. Completion of his *Tabula foecunda* (i.e. a table of tangents) at Pollina; the table is included in BN lat. 7472A, 45v-46r. On the bottom of folio 45v is the date 12 Aug., 1550 and on folio 46r, we read « Hanc tabulam foecundam a Joanne de monte regio calculatam sive authoris incuria sive impressorum neglegentia multis in locis mendosam correximus. Pollinae 13° Aug. die ☿ [i.e., Mercurii] 1550. » The table was published in the omnibus volume of 1558, 65v.

1550, 28 August. Completion at Pollina of a table of sines (BN lat. 7472A, 45r). It was included in the omnibus volume of 1558, 65r. It was at about this time that he prepared his *Tabella benefica*, i.e. a secant table, also included in the work of 1558, 66r. In the latter work at the end of his *Demonstratio Tabulae beneficae*, 61r, he says: « Ex quibus quidem manifestum est quod omnia quae Ioannes de Monte regio per tabellam foecundam vitato divisionis fastidio per multiplicationem elaborabat, hic per Beneficam nostram haud difficilius supputari possunt. Sed practica huiusmodi regularum exempla inferius una cum tabellis ipsis exponemus. Haec in arce Apollinari [i.e. Pollina], dum cum D. Ioanne Vigintimillio Hieraciensium Marchione degeremus, olim mense Augusto, 1550, speculabamur. » The *Tabella benefica* itself is not in BN lat. 7472A, but a preliminary version of the *Demonstratio* is found on folio 90r, preceded by some notes on the *Tabella declinationum generalis* on folio 89v, dated « Pollinae, 6... 1550. » Observe that on 90r at the bottom there is a separate note on spherical triangles written at a different time and dated at Castelbuono, 7 June, 1549. The *Tabella declinationum et ascensionum* appears on folio 90v-91r and was later published in the omnibus edition of 1558, 66v (see above, 1543, 17 and 26 December).

1551, 23 October. An addition to *Ad inveniendos dictarum linearum terminos in numeris* (Cod. S. Pant. 115/32, 20r). A similar addition on folio 5r, dated 24 October, 1551.

1552, The *Rime del Maurolico* was published at Messina. The only copy that I know of is in Messina, but the leaves which would include the date are missing. However, this is the date assigned to it by Maurolico in his *Index lucubrationum* (see the next section, note 73) where he gives the title in Latin: *Rhythmi quoque vernaculo sermone.*

1553, 12 February. An *Ad lectorem* that is part of the additions to Parts I and II of his *Diaphana* (BN lat. 7249, 14v; cf. edition of Naples, 1611, p. 68). For the pristine text, see the entry for 1522. Thus all of the additions to Parts I and II through 14v appear to have been done in 1553. See also 1554, 19 May.

1553, 9 July. Completed at Castelbuono Book II of his *De lineis horariis* (see *Opuscula mathematica*, Venice, 1575, p. 262). Book I is not dated, but because of its dedication to Francesco Santapace as *stradigò*, Macrì, *Francesco Maurolico*, 2nd. ed., pp. 201-202, believes it to have been composed in the period 1531-1532. This work is to be distinguished from his *De lineis horariis brevis tractatus* appearing in the same volume, pp. 80-100.

1553, 19 July. Completed at the monastery of S. Maria a Parte (! Partu) Book III of his *De lineis horariis (ed. cit.,* p. 285). Both Books II and III were dedicated to Juan de Vega.

1553, 1 August. A lunar table completed and dated (BN lat. 7472A, 16r).

1553, 6 November. This is the earliest date attached by Maurolico to an *Arithmetica* or a tract on figured numbers that appears in BN lat. 7473, 41r-85r. This work, while having some material that resembles the *Arithmetica* completed in 1557, existing in Vat. lat. 3131 and published at Venice in 1575, is a distinct work, or better « works » since it consists of chaotically arranged individual tracts and sections. The date 6 November appears in the colophon of a tract in which the tract is distinguished from his other arithmetical treatise (61v): « Sic in hoc et in aliis duobus arithmeticae nostrae libellis quos antea scripseramus supplevimus ea quae Boetius et Iordanus circa formas numericas omiserant. Catanae. 6 No. 1553. » There is a difficulty with this date of 6 November, which appears to embrace all of the material on numerical figures and 15 propositions that appear between 50r and 61v. For on the bottom of folio 58r after Proposition 7 we find « Catanae. kal. decemb. 1553. » It could be that he blocked out the tract on 6 November, composing the terminal note on that date, and then at a later date filled in the blank pages preceding it, thus completing at least through Proposition 7 by 1 December, 1553.

1553, 7 November. Maurolico was given an annual salary of 100 gold pieces by the Senate of Messina for two years to complete his mathematical works and his compendium of Sicilian history (see Maurolico's *Sicanicarum rerum compendium,* 2nd ed. [Messina, 1716], p. 248).

1553, 1 December. Completed at Catania Proposition 7 of a brief arithmetical essay (see above, 6 November, 1553).

1553, 8 December. Maurolico began a tour of Sicilian towns at some of which he dated the notes and works mentioned in the succeeding items (through 5 January, 1554). Maurolico jotted down his itinerary in BN lat. 7473, 85v-86r. It is quite difficult to read but I shall give it as accurately as I can. Macrì, *Francesco Maurolico,* 2nd. ed., Appendix, pp. XXXIII-XXXIV, transcribes the part on folio 85v with some errors; e.g. he misassigns the diary to folio 41v (on page 43) and believes the journey to have begun on 1 December, when the manuscript clearly has « 8 December. » Furthermore, he mistranscribes « ad abbatiam nostram » (= S. Maria a partu) as « abbatiam maiorem. » The phrases I cannot read on folio 86r are replaced by dots. « 8° decembris 1553 discedimus Catana post prandium et ad horam primam noctis appulimus Leontinos. 10° decembris post prandium discedimus Leontinis et ad horam primam noctis appulimus Xortinum. 11° decembris post missam discedimus Xortino. Pransi sumus in abbatia S. Mariae de Arcu fundata per Sinibaldum de Norvegia comitem Noti anno domini 1212, et ad horam 23½ ingressi sumus Notum. Apud Notum multo me officio comitatus est per oppidum d. Joannes Datus baro Fregentini. Dantes (?) bartullius ex Marturano Kalabriae puerorum praeceptor, eius ex filio nepotem docebat. Antonius Scala cupediarius, cuius filius Josephus Catanae studebat, tentacula mihi obtulit et palmarum foliata cephalea. 14° decembris post missam profecti Noto. Pransi sumus Rendae et ad horam 23^{am} appulimus Motycam, recepti in aedibus Alexandri Marabelli. Franciscus Januarius epistolio me laudavit me (!). Jo. Salvus panormius iam can...(?) agnovit(?). 17° discessimus Motuca post missam in aede S. Spiritus et pransi in medio itineris in aedibus baroniae S. Jacopi. Ad occ[asum] solis appulimus Vizinium, suscepti in aedibus Marii Cannizzarii medici. Dies discessus nubilus et sequens cum nocte tota pluviosissimus. 20° decembris die ♀ [i.e. Veneris] 4^{or} temporum pluvioso post missam in aede Magdalenae Vizinio profecti. Calagurae pransi sumus. Postridie missam audivimus in aede S. Juliani. Recepti in aedibus Jo. Jac. Adami. 22° decembris digressi Calagura. Pransi sumus Plocii in aedibus Francisci Catalani Iurati, et postridie post missam et prandium et epigramma recitatum pertransito lacu Proserpinae, quo processerat magistratus cum equitatu ad horam 22^{am} appulimus Ennam. Suscepti in aedibus Francisci Bormaldi baronis Buzzettae. Hic supervenit filius comitis Asari Hernandum officio excipiens. In festo natali missam audivimus in aede S. Francisci proxima et una nobiscum filius comitis Asari et baro buccettae. 26° decembris discessit Hernandus cum dictis barone et filio comitis versus Asarum, nos autem ad cenobium Engii liberaliter excepti. Et audita abbatis d. Laurentii Alegrii florentini missa venimus cum pluvia Hieracium et inde ad abbatiam nostram [i.e., S. Mariae a partu] (86r).... Inde in Cast. b. [i.e. Castello bono] descendentes vidimus Franciscum (?).... Ultimo mensis ascendimus et inde (?) 3° Ianuarii. 4° profecti. Pransi Hieracii (?) et (?) ad (?)... horam (?) cum pluvia appulimus ad cenobium Engii. 5° cum pluvia venimus Nicosiam.... 6° Argyram tempore iam serenato. 8° profecti (?) per scapham [i.e. scaphulam].... Suscepti in aedibus Josephi Stellae. 9° ad horam mer[idiei] Catanam. » Most of the stops on his trip are readily identifiable: Leontini (= Lentini), Xortinum (= Sortino), Noto, Motuca (= Modica), Vizzini, Calagura (= Caltagirone), Plocium (= Piazza Armerina), Enna, Engium (= Gangi), Hieracium (= Geraci Siculo), S. Maria a Partu, Castelbuono, Nicosia, and Argyra (= Agira). Note that the trip was completed on 9 January. Note also that Fernando (= Hernandus) and Maurolico went their separate ways on 26 December.

1553, 15 December. Completed at Modica a table of figure numbers (BN lat. 7473, 61v-62r). The word « Motucae » (= Modicae) appears on the margin of 61v, and after the table on 62r we read: « Haec scribentes speculabamur dum Motucae essemus. 15 de. 1553. » This table is identical with that prefaced to the *Arithmetica* appearing in Vat. lat. 3131, 8r-v and published in Venice, 1575, sign. T 2, except that the first section « numeri lineares » and the last two sections « Solida Regularia in numeris » and « Quadrati Quadratorum » of the published tables are missing in BN 7473. Furthermore, the table in BN lat. 7473 is free of the printing errors found in the published table (the table in Vat. lat. 3131 is also free of these errors).

1553, 16 December. The arithmetical sections following the above-noted table in BN lat. 7473 and occupying folios 62r-64r are dated on 64r. They were also composed at Modica: « Hactenus Motucae 16 dec. »

1553, 18 December. Further sections of the same work noted in the two preceding entries (*Ibid.*, 64r-65v, date on 65v): « Haec speculabamur ad horam noctis 3ᵃᵐ diei ☽ [i.e., lunae] 18. decem. 1553 dum Vizinii cum Fernando Vega diversaremur. »

1553, 19 December. More of the same work (*Ibid.*, 66r): « Haec ibidem postridie. »

1553, 21 December. More of the same work (*Ibid.*, 66r-67r, date on 67r): « Calathagyri sive Calagurae. 21° de. 1553. » This is the town now called Caltagirone. Incidentally, marginal references to « Calag. » (= Calagurae) appear on folios 62v, 65v and 66r though the accompanying texts appear to have been composed elsewhere according to their specific designations by Maurolico.

1553, 24 December. Completed at Enna another arithmetical section (*Ibid.*, 41r-44v), which contains 10 propositions from Jordanus' *Arithmetica* with Maurolico's proofs. The date appears twice. The first time at the end of Proposition 10 on 44r: « Ennae 24° decem. 1553. » The second time in an immediately following colophon on 44v: « Ecce igitur quanto facilius ac brevius ostendimus id quod Iordanus scabra et fastidiosa demonstratione ostenderat. Lector vale. Trinacriae centrum, Cereri gratissima sedes. Hic complet numeros fertilis Enna meos. Ennae ad horam noctis quartam, die ☉ [i.e., solis], 24 decem. M.D. L.III. » See also 24 (?) December for preface.

1553, 24 (?) December. The preface to the brief section described in the entry of 24 December appears on folio 41r of BN lat. 7473: « Maurolycus Abbas lectori. S. Cum olim Catanae degerem, dum Ferrandus (*!*) Vega Ioannis proregis filius nostris lectionibus operam daret: atque inde per Nectinae Vallis oppida cum eo milites equitesque recensente peregrinarer quidquid ocii supererat, solitis speculationibus impendebam; cumque in primis multa Catanae completa Motucae, Vizinii atque Calagurae scripsissem, veluti per adscriptas locorum et tempore notas in istoc quaternione legentibus constare potest; hic ea quae Plocii speculantibus occurrerant, Ennae Servatoris natalem celebrantes, adnotabimus rursus a principio exorsi, quoniam scilicet haec quamvis posterius coanimadversa praecedere debuerant. Nec te lector perspicacissime perturbet si quid ex his a Iordano fuerit antea demonstratum. Non enim ob id labori parsimus: semperque hic et alibi conati sumus sylvosas ac difficiles Iordani demonstrationes in brevitatem pariter ac facilitatem redigere atque nonnulla aliter demonstrare ab Euclide interim demonstrata, ut supervacua et inanem laborem adiicientia, omittentes et multa magis fastidiosa quam aut trivialibus utilia aut speculativis lucunda resecantes. Hic igitur nonnulla circa cubos demonstratura praemittam aliquot elementares propositiones quamquam alibi demonstratas, ut sequentia facilius et in promptu magis lectoribus exponantur. Verte igitur paginam et feliciter lege. Nam Trinacriae centrum, Cereri gratissima sedes, hic complet numeros fertilis Enna meos. 24° (?) dec. 1553. » Note his mention here of visits to Modica, Vizzini, Calagura (i.e. Caltagirone), Plocium (i.e. Piazza Armerina) and of course Enna. The date looks more like « 29 » than « 24 » but we know from his itinerary that Maurolico was not in Enna on the 29th (see above, entry for 1553, 8 December). It could perhaps be « 23. » Macrì, *Francesco Maurolico,* 2nd. ed., p. 42 quoted the first few lines of this preface and assigned it to the impossible date of 29 December.

1553, 25 December. A diagram relating various pyramidal and columnar numbers is so dated at Enna on folio 68r of BN lat. 7473.

1554, 5 and 12 January. More on figure numbers (BN lat. 7473, 68v): « Excogitatum Herbitae 5° Jan. et scriptum Catanae. 12°. » « Herbita » is the town of Sperlinga in Enna province. This is the last of the dated works on Maurolico's tour that began on December 8, 1553. See that date.

1554, 16 January. Calculations at Catania (BN lat. 7473, 49v): « Catanae 16° Jan. 1554. »

1554, 28 January. Completed a tract on Boethius' *Arithmetica* (for the manuscript, see above, n. 1, item 17). Perhaps he completed the tract on Boethius' *Musica* in the same manuscript about this same time (*Ibid.*, item 18). However, the fact that he listed this work in his Catalogue of 1540 (see below Sect. 2, Intro., n. 9, item 29) implies that some form of this work was executed much earlier.

1554, 7 April. Completed at Catania Book I of his version of *Theonis ex traditione Pappi data* (BN lat. 7467, 5v).

1554, 13 April. Completed at Catania Book II *Theonis... data* (*Ibid.*, 16r). The complete work thus occupies 1r-16r.

1554, 16 April. Completed at Catania his version of Euclid's *Phaenomena* (*Ibid.*, 18r-23v, date on 23v). This work was published in the omnibus *Theodosii Sphaericorum elementorum libri III etc.* (Messina, 1558), 63r-64v, but in a considerably different form. For example, an interesting passage (present only in truncated form in the published version) appears at the end of the work in the manuscript (23v): « Illud vero notandum est Euclidem in his demonstrandis multum laborasse, cum ipsius tempore non extarent adhuc, sphaericorum traditiones quibus τῶν φαινομένων tota demonstratio innitetur. Coactus est igitur Euclides totum demonstrandi Maurolycus exercitii gratia in arce Catanensi scribebat 16 Aprilis 1554. » Maurolico made the same kind of observation concerning the *Mechanica* attributed to Aristotle in relationship to the later work of Archimedes' *On the Equilibrium of Planes* (see Maurolico's *Problemata mechanica* [Messina, 1613], pp. 8-9).

1554, 17 April. An *Appendix quaedam* to the *Theonis... data* was added on this date (BN lat. 7467, 16v). It includes two propositions, the square root of a given magnitude will be given and the cube root of a given magnitude will be given.

1554, 3 May. Completed at Catania a table of numbers for formation of regular solids (BN lat. 7473, 85r).

1554, 17 May. Completed at Catania a *Dialectica* occupying folios 20r-40v of BN lat. 7473. Dated on folio 40v.

1554, 19 May. Completed at Catania an epilogue to the additions made to Parts I and II of the *Diaphana* (BN lat. 7249, 15r). This date is not in the edition of 1611 where the epilogue appears on pages 66-67.

1554, 20 May. Completion of the first section of Part III of the *Diaphana*, entitled *De organi visuali* (BN lat. 7249, 17r; Lucca, Bibl. Govern., cod. 2080, 56r; ed., 1611, p. 73).

1554, 27 May. Completed at Catania a chart entitled *Praedicamentum et Substantiae* (BN lat. 7473, 19v).

1554, 29 May. Completed at Catania the remainder of Part III of the *Diaphana* (Lucca, Bibl. Govern., cod. 2080, 63r; ed., 1611, p. 80 – the *ed.* has « die 8 Maii 29.... 1551 », thus misprinting « die ♂ » as « die 8 »). This second section of Part III entitled *De conspiciliis* is incomplete in BN lat. 7249. It stops at the bottom of 19v with « susceptam lucem » of the edition, p. 80, lines 13-14. Since the piece on perspective added to the edition (see below, the entry for 12 October, 1567) is also missing in the manuscript, it looks like the last folio of the manuscript has become separated from the codex.

1554, 14 June. Completed at Catania a *Prologus de divisione artium* (MS. Molfetta, Bibl. Sem. Min., Cod. Melph. 5-7, H. 15, 9v; ed., 1968, p. 26).

1554, 18 June. Completed at Catania a *Prologus de quantitate* (Cod. Melph. 5-7, H. 15, 20v; ed., 1968, p. 47).

1554, 22 June. Completed at Catania a *Prologus de proportione* (Cod. Melph. 5-7, H. 15, 26r; ed., 1968, p. 61).

1555. Published in a single volume Matteo Caldo's *De vita Christi* (which Maurolico had reworked somewhat) and his own *De gestis apostolorum*, Venice, 1555. See above, 1540. This edition is dated 1556 in the *Index lucubrationum* (see the text in the next section, footnote 97). The divergent dates can be reconciled if we assume that the work was published after 1 January and before 1 March, 1556 and if we further assume that the Venetian style of dating was used where the new year began on 1 March.

1555, 1 February. In an arithmetical section generally dated 1557 he has a proposition numbered as « prima » on the multiplication of a root by the collateral odd number (BN 7473, 48v): « Quod Motucae [= Modicae] nos demonstraturum promisimus. Catanae p° feb. 1555. » Hence he is fulfilling at Catania the promise made at Modica in December of 1553.

1555, 12 March. Completed at Catania Book I of his *Geometricae quaestiones* (BN lat. 7468, 21v; ed. Napoli, « Scritti inediti di Francesco Maurolico, » (*Bullettino di bibliografia e di storia delle scienze matematiche e fisiche*, Vol. 9 (1876), p. 84).

1555, 29 March. Completed at Catania Book II of his *Geometricae quaestiones* (MS. cit., 42v; ed. cit., p. 113).

1555, 13 June. Completed an appendix to the *Photismi de lumine et umbra* entitled *De erroribus speculorum* (ed., 1611, p. 30). The date is missing in Lucca, Bibl. Govern., Cod. 2080.

1556. Maurolico says that the *De vita Christi* was published this year, but see 1555 above.

1556, 30 January and 15 February. Grammatical-logical extracts from Peter of Spain (BN lat. 7473, 19r).

1556, 2 March. A letter written to Simeone Ventimiglia (R. Bibl. del Escorial, &. IV.22, 185r-v).

1556, 27 May. A brief note from Hippocrates (BN lat. 7473, 18v).

1556, July. A letter to the Emperor Charles V (*Theodosii sphaericorum... libri III*, Messina, 1558, 2*v-3*r).

1556, 8 August. Completed at Messina his long letter on mathematics addressed to Juan de Vega (BN lat. 7473, 1r-16v, date on 16v; see ed., Napoli, « Scritti inediti, » pp. 23-40, date on p. 40).

1556, 22 September. Note on the difference of transit of a star with diagram (BN lat. 7472A, 92v).

1556, 1 October. Notes some entries from the Alphonsine tables to be corrected (BN lat. 7472A, 15r). On the same day Maurolico completed a *Tabella diversitatis aspectus ad latitudinem graduum 38¹/₆* (*Ibid.*, 15v).

1556, 7 and 8 October. A diagram for an eclipse of the moon in 1556, drawn at Messina on 7 October (BN lat. 7472A, 13v), followed by some notes entitled *Semi-diametri visuales* (14r) and dated 8 October.

1556, 3 November. Some brief *Regulae affirmationum et negationum tam universalium quam particularium* to be added to his *De dialectica* (BN lat. 7473, 18v).

1556, 22 December. Completed at Messina the epilogue to the tract called *De solidis regularibus numeris* that is a part of the long collection of arithmetical tracts found in BN lat. 7473. This tract occupies folios 69r-80v, with a supplement on folios 81r-84v. It ends (80v): « Sic nihil omisisse videmur ex iis quae ad numeralium formarum consyderationem faciunt. Nam et ipsas superficialium formas et solidorum tam columnas quam pyramides omnimodas et nunc demum ipsa regularia solida complexi sumus. Sed instat iam Decembris pluviosissimi dies 22ᵘˢ et approximat Servatoris (*!*) dies natalis. Et alia nobis cura incumbit. Lector vale. Messanae die ♂ [i.e., Martis] 22⁰ Decembris Ind. XVᵉ 1556 et deo gratias. »

1556, 23 December. Table of numbers for the figured numbers dated « postridie » (*Ibid.*, 81r).

1556, 25 December. Supplementary note to the preceding work (*Ibid.*, 81r-v, date on 81v).

1557, 5 January. Further supplementary notes (*Ibid.*, 81v-82r, date on 82r).

1557, 12 January. Still further supplementary notes (*Ibid.*, 82v-84v), dated at Messina on folio 84v.

1557, 16 January. Some arithmetical theorems (BN lat. 7473, 49r).

1557, 7 February. A note on Proposition IX.3 of the *Elements* of Euclid (Cod. S. Pant. 116/33, 30v).

1557, 9 and 10 February. More arithmetical notes (BN lat. 7473, 48r).

1557, 3 March. *Diffinitiones numerorum solidorum regularium centralium* (*Ibid.*, 45r).
1557, 6 March. A continuation of the notes of 9 and 10 February (*Ibid.*, 47v-48r).

1557, 24 March. A further continuation of the notes of 6 March (*Ibid.*, 47v). The works of 10 February, 6 and 24 March are continuously numbered as rules or propositions 1-6 in the margins.

1557, 18 April. Completed at Messina Book I of his *Arithmetica* (Vat. lat. 3131, 83r; ed., 1575, p. 82).

1557, 1 May. Brief grammatical-logical extracts from Paul of Venice and Peter of Spain (BN lat. 7473, 19r).

1557, 24 July. Completed at Messina Book II of his *Arithmetica* (ed., 1575, p. 175; not dated in Vat. lat. 3131). See also 1 December, 1568.

1557, 8 October. A letter of Maurolico concerning a problem posed by Commandino, namely, to divide a given hyperboloid of revolution by a plane parallel to the base so that the segments have a given ratio (see MS Vat. Barb. lat. 304, 284r).

1558, August. Dated his dedication at Messina to Juan de la Cerda, now viceroy of Sicily, of his *Theodosii sphaericorum elementorum libri III etc.,* published that month and year at Messina (1*v, 72v). The works included in that edition were: 1. the beginning of Maurolico's *Index lucubrationum* (2*r), 2. Maurolico, *De sphaera sermo* (3*v-6*v), 3. Theodosius, *Sphaericorum libri* (1r-17r), 4. *Praefatio Maurolyci in sphaerica Menelai* (17r), 5. Menelaus, *Sphaerica,* 17v-45r, 6. Maurolico, *Sphaerica* (45v-60r), 7. Maurolico, *Demonstratio tabulae beneficae* (60r-61r), 8. Autolycus, *De sphaera quae movetur* (61r-62r), 9. Theodosius, *De habitationibus* (62r-62v), 10. Euclid, *Phaenomena* (63r-64v), 11. Maurolico, *Habitationum collatio* and *De astrorum fulsionibus* (64v), 12. Maurolico, *Tabella sinus recti* (65r), 13. Maurolico (from Regiomontanus), *Tabella foecunda* (i.e. a table of tangents) (65v), 14. Maurolico, *Tabella benefica* (i.e., a table of secants) (66r), 15. Maurolico, *Tabella declinationum et ascensionum* (66v), 16. Maurolico, *Tabellarum canones* (67r-68v), and 17. Maurolico, *Compendium mathematicae* (68v-72r). See MS Madrid, Academia de la Historia, Cortes 2787 (formerly 675), 1r-212v, which is perhaps the manuscript sent by Maurolico to the printer, though it is not in Maurolico's hand. Items 1. and 2. of the printed edition are not in the manuscript.

1561, 14 September and 8 October. Logical notes on « contingence » (BN lat. 7473, 18r).

1562, 10 June. Definitions of « unum, » « res, » « aliquid, » « ens, » « verum, » and « bonum » (*Ibid.*).

1562, After 1 November. Publication of *Sicanicarum rerum compendium* at Messina. The autograph manuscript is BN lat. 6177. The deleted passages from the manuscript were collected and printed first in 1679 by Baluzius and then added to the second edition of the work (Messina, 1716). See footnote 2 above. The scarcity of datable works between August 1558 and 1562 is perhaps a reflection of Maurolico's concentrated effort on the *Compendium.* I have added « After 1 November » because the dedication is so dated. At the end of the first edition there is included a letter from Maurolico to the prelates and legates of the Council of Trent dated at Messina, 1 October, 1562.

1563. Various astronomical observations for sundry dates in 1563 and 1564 (BN lat. 7472A, 80v-81r).

1563, 18 May, 11 June, 13 and 28 October. Notes *De subiecto et praedicato* (BN lat. 7473, 16v-17r).

1563, 5 July. Dates a brief discussion of a lunar eclipse seen at Messina (BN lat. 7472A, 14r).

1563, 3 December. Completed an epitome of Books XI and XII of Euclid's *Elements* (Cod. S. Pant. 115/32, 36v-41v, date on 41v).

1564, 14 July or 18 November. Completed the dedication of his *Martyrologium* to Cardinal Amulio. Some editions have 14 July and some 18 November (see E. Rosen, « Maurolico's Attitude Toward Copernicus, » *Proceedings of the American Philosophical Society,* Vol. 101, p. 189, n. 71).

1564, 1 December. Partly defaced title perhaps not in Maurolyco's hand: « Quadrati horarii Astrolabi, Speculatio, fabrica et usus », occurs on a binding page before 1r of BN 7464. No treatise. See entry of 1538, 12 Aug.

1565, 5 May. Completed his *Brevis demonstratio centri in parabola* (BN lat. 7466, 7v-13r, date on 13r; cf. ed., Napoli, « Scritti inediti, » p. 121).

1565, 13 October. Additional notes on the same subject (*Ibid.,* 13v-17r, date on 16v).

1565, 14 November. Conclusion of the added notes (*Ibid.,* 17r).

1566, 30 December. Earliest of a series of passages on music, musical ratios and proportions (BN lat. 7462, 6r-36v, this date on 33r). This series carries the following additional dates: Ultimo Januarii, 1567 (12r); 6 February, 1567 (12v); 21 June, 1567 (7v); 20 November, 1567 (7r). The material is largely Boethian, with attention also to the views of Guido of Arezzo and Jacques Le Fèvre. See also 1569, 17 March.

1567, 28 January. The first dated piece in a series of dated parts of a compendium of the first ten books of the *Elements* of Euclid, Book I (BN lat. 7463, 5r). Other parts are Book II, dated 30 January (8v); Book III, 2 February (10r); Book IV, 4 February (11v); Book V, 6 February (15r); Book VI, 26 October, 1567 (18v); Book VII, 13 February [1568?] (22r); Book VIII, 17 February (26v); Book IX, 22 February (32v); Book X, 23 and 27 December, 1569 (49r). Then follows *Regulae circa figurarum isopleurarum et solidorum regularium latera ex XIII⁰ Elementorum,* 24 December (50r); an *Additio in finem 10ⁱ* (no date, 50v-51r); and a slightly different treatment of Book II, 29 January, 1570 (54r).

1567, 6 May. Completion of a discursive index of the sciences (BN lat. 7471, 14r-23r, date on 22v).

1567, 9-10 May. Completed *In sphaeram communem adnotationes* (*Ibid.,* 7r-11v, dates on 11r-v).

1567, 10 May. Completed *In magnae Ptolemaicae constructionis libros argumenta per Maurolycum* (*Ibid.,* 2v-6v, date on 6v).

1567, 11 June. A marginal note to the Ptolemaic work given in the next entry (*Ibid.,* 36r).

1567, 16 June. Completed a *Breviarium* of Ptolemy's *Almagest* (BN lat. 7471, 25v-43v with added notes on 44r-v). On 25v we read: « Sequitur Breviarium sive Epitome brevissima in Almagestum Ptolemaei, sive idea Ptolemaica per Maurolycum diligentissime ordinata. » The rest of the page is blank. Then on 26r the tract begins with the following title: « Ptolemaicae traditiones ex singulis magnae constructionis libris. » The date appears on 43v: « die ☽ [i.e. lunae] 16⁰ Junii dum Sol medium limen fenestrae occiduus precutit. M.D.LX.VII. » See also 1569, 6 June.

1567, 12 October. Completed *Problemata ad perspectivam et iridem spectantia* (in the ed. of the *Photismi de lumine et umbra,* Naples, 1611, pp. 81-84).

1567, December. *Ecclesiasticus computus* was written. It has no colophonic date but in the body of the text the 18th of December, 1567, is mentioned as « today » (*Opuscula mathematica,* Venice, 1575, p. 34; cf. p. 43, « in hoc anno 1567 »).

1567. See the various dates of musical pieces given under the entry 1566, 30 December, and also the various dates regarding Euclid's *Elements* under 1567, 28 January.

1567 or 1568. Maurolico's *Martyrologium.* I have not been able to discover any edition earlier than the two editions of Venice, 1568. But Maurolico seems to say in his *Index lucubrationum* (see Section 2[1] of this article line 166) that it was printed in September, 1567, with the small format edition [16⁰] appearing in July, 1568. Perhaps the entry in the *Index lucubrationum* could be read to mean that the work was completed in September, 1567, and then published twice in 1568, which would agree with the known facts about the editions of 1568. Rosen, « Maurolico's attitude, » p. 189, n. 71, mentions no edition earlier than 1568.

1568. See the various pieces of Euclid's *Elements* given in the entry of 1567, 28 January.

1568, 20 April. Most of the *Index lucubrationum* completed by this date (BN lat. 7466, 1r-3v, dates on 3r-v; cf. my text in section 2[1] below; note that the date of 22 April also appears on 3v). Part of the *Index* was published in the omnibus edition of 1558, virtually all of it following the *Arithmetica* (Venice, 1575). A copy also appears in Vat. lat. 3131, 168r-171v, where it also follows the *Arithmetica*, and in the Codice Villacanense (see below, Sect. 2[1]).

1568, 1 December. Dedicated his *Arithmetica* to Cardinal Amulio (Vat. lat. 3131, 6r-7r, date on 7r; the dedication was edited by Boncompagni, *Bullettino*, Vol. 8, pp. 55-56). This dedication was not included in the edition of 1575. Maurolico notes in the dedication that the work was prepared eleven years ago and was transcribed by his nephew.

1569, 17 February. Completed *Circa magnetem problemata* (ed. in *Opuscula mathematica*, Venice, 1575, pp. 100-102, date on p. 102). This is attached to his *De lineis horariis brevis tractatus* and thus the latter work must have been composed about this time. Its prologue mentions the *Liber de horologiorum descriptione* of Commandino published in 1562 and so the work ought to be later than that date.

1569, 17 March. Compiled a brief musical note: « Notularum proportio » (BN lat. 7462, 20v). This is the latest date in his musical notebook. The numbering skips from 20v to 27r.

1569, 16 April. A letter to Francesco Borgia (see M. Scaduto, « Il matematico Francesco Maurolico e i Gesuiti, » *Archivum historicum Societatis Jesu*, Vol. 18 (1949), pp. 134-37, where the letter is misdated MDLXXX).

1569, 20 April. An outline of four volumes of profane and sacred poetry (BN lat. 7466, 6r-v, date on 6v; see Section 2[4] below).

1569, 4 May. Dedicated his *Problemata mechanica* to Cardinal Amulio (see ed., Messina, 1613, p. 5). The same date appears at the end of his appendix on the magnet (*Ibid.*, p. 55).

1569, 6 June. An addition to his epitome of Ptolemy's *Almagest* (BN lat. 7471, 44r-v, date on 44v).

1569, 7 October. Completed his *Demonstratio algebrae* (BN lat. 7466, 19r-23v, dated on 25v, with an added note of 18 January, 1570; cf. the edition of Napoli, « Scritti inediti, » pp. 41-49).

1569, 17 November. Additions to his short work on the centers of gravity of a parabola and a paraboloid (BN lat. 7466, 13v-17r, date on 16v). See above, 1565, 5 May.

1569, December. Dates of parts of his compendium of Euclid's *Elements*. See above, 1567, 28 January.

1570, 18 January. See above, 1569, 7 October.

1570, 29 January. The last date in his compendium of Euclid's *Elements*. See above, 1567, 28 January.

XIII

1570, 2 September. *Species quantitatum rationalium et irrationalium* (BN lat. 7462, 3r).

1570, 17 September. Rearranged the works listed in his *Index lucubrationum* in the order they should be read (BN lat. 7466, 5v; see my text in Section 2[3] below).

1571, 24 August. Maurolico's weather prediction for Don Juan of Austria before the battle of Lepanto (see F. Guardione, « Francesco Maurolico nel secolo XVI, » *Archivio storico siciliano,* Anno 20 (1895), pp. 13-14 and Macrì, *Francesco Maurolico,* 2nd. ed., pp. 79-85).

1571, 11 September. Maurolico's letter to Girolamo Barresi (Cod. S. Pant. 115/32, 42v-43r, date on 43r; cf. Guardione, « Francesco Maurolico, » pp. 35-36).

1571, 27 September. Completed his *Sphaericorum epitome.* The *Epitome* occupies folios 93r-95v of BN lat. 7472A. On folio 93r he mentions his earlier *Sphaerica* with its four tables (i.e. the work published in the omnibus edition of 1558): « Nos quoque duos sphaericorum libellos fecimus. In quorum primo de sinuum arcuum in [sphaeralibus] triangulis proportione. In secundo trianguli sphaeralis orthogonii discussimus proprietates. Et in calce subiunximus practicam calculi, eius fundamenta demonstrantes, trium tabularum usu ad negotium applicato. Harum prima fuit Tabula sinus recti. Altera foecunda, quam Joannes Regiomontanus invenit. Tertia Benefica, quam nos adiunximus. Per has absolvimus quaestiones quae circa sphaeralia triangula proponi solent. Et quoniam harum praecipuae versantur circa declinationes, ascensiones et caeteros primi mobilis arcus, idcirco ad alleviandum calculi laborem construximus tabulam generalem declinationum et ascensionum, quae per calculum praedictum facile construitur. Ut autem intelligatur modus et structura dictarum trium tabularum, esto triangulum *abc...* ». The tract is dated on 95v: « Haec aggregata sunt ut in promptu sit speculatio et praxis. 27 Sep. 1571. » Compare the somewhat different treatment of the theory of the tables in his *Geometricae quaestiones* (ed., Napoli, « Scritti inediti, » pp. 77-84).

1571, 28 September, 1 and 2 October. Additions to the *Sphaericorum epitome* (BN lat. 7472A, 95v-97r, with « postridie » on 95v; 1 October, 1571 on 96v; and « postridie » on 97r).

1571, 28 September. A long list of the tables and other material included in BN lat. 7472A, beginning (1r) « Calculi nonnullarum eclipsium solarium [11r-14v, 1v-10v blank]. Tabula diversitatis aspectus ad latitudinem graduum $38^1/_6$ [14v-15v]. Tabula motus lunae horarii [16r, 16v blank]. Quantitatum irrationalium species [17r]. Latera figurarum equilaterarum et corporum regularium [17v-20v]. Radices motuum pro tabulis Alfonsi [20v-24v]. Motus medii, eccentricitates, semidiametri, distantiae, termini, equationes, diversitates, latitudines, arcus visionum, ordo magnarum revolutionum [etc., 25r-32r, a series of tables of parameters, referring to the Alphonsine tables, Ptolemy, Albategnius, Thabit, Peurbach, some of which were appended to the *Cosmographia* of 1543]. [De conjunctionibus, 32v.] Radices superationum et motuum pro tabulis Blanchini [33r-35v, 36r-40r blank except for a brief table with parallel columns entitled « Argumentum trepidationis » and « equatio 8^{ae} sphaerae » on 39v]. Radices motuum ad meridiem Messanae [40v-42v, 43r-44v blank]. Tabula sinus recti per singulos gradus supponens sinum totum per 100000 [45r]. Tabula foecunda correcta [45v-46r]. Tabula [sinus recti] pro computo horario (?) [47r-48v]. Semidiametri, eccentricitates, distantiae, magnitudines corporum celestium et terrae [49r-53v, mostly published at end of *Cosmographia*]. Tabellae declinationum, ascensionum, differentiae ascensionalis, quantitatis dierum, aequationis dierum [54r-56v]. Anguli, altitudines, distantiae pro lineis horariis ad latitudinem graduum $38^{1}/_{6}$ [57r-62v]. Tabula aequationis dierum [63r]. Tabula domorum [63r-69v]. Tabula positionum pro latitudine graduum $36^{1}/_{6}$ (! $38^{1}/_{6}$) [70v-76v]. Tabula semidiametrorum visualium solis et lunae [77v]. Tabula latitudinum planetarum et lunae [77v-79r]. Tabula stellarum primae magnitudinis [79v]. Observationes quaedam [80r-81r]. [Nomina stellarum, 81v-82r]. Cum quo puncto zodiaci oriatur aut occidat stella [82v-83v]. Tabula stationum quinque planetarum [84r]. Tabula diversitatis aspectus solis et lunae [84v-85r], cum tabulis duobus minutorum proportionalium [85v-87r]. Tabula arcuum et angulorum pro ecclipsibus ad latitudinem graduum $38^{1}/_{6}$ [87v-89r]. Tabula declinationum et ascensionum generalis cum regulis et demonstrationibus [89v-92v]. Tabula differentiarum ascensionalium et latitudinis ortus pro latitudine graduum $38^{1}/_{6}$ [see note on 89v, but table is missing here]. Tabula Benefica [missing here]. Fructus Almagesti [missing here]. Demonstrationes quaedam tabularum faecundae et beneficae [90r]. Tabellae motuum diurnorum planetarum, solis, lunae, in hora cum regulis [missing here]. Sphaericorum Epitome, cum regulis ad tabulam sinuum, faecundam et beneficam pertinentibus demonstratis [93r-97r]. 28° Sep. 1571.» (I have added the folio numbers to this table of contents and a few missing titles). The dates of many of the pieces have been included above in the Chronology: 1529, November; 1536, 19 May; 1537, 17 March; 1540, 8 January, 11, 19-21 June; 1542, 27-30 July, 31 July and 1 August; 1543, 17 and 26 December; 1544, 14 January; 1548, 24 October; 1549, 7 June; 1550, 13 August, 28 August; 1553, 1 August; 1556, 22 September, 1 October, 7 and 8 October; 1563; 1563, 5 July; 1571, 28 September, 1 and 2 October. By adding this table of contents Maurolico had in effect closed out this notebook.

1572, 6 November. Completed his tract *Super nova stella quae hoc anno iuxta cassiepes apparere cepit considerationes* (Naples, Bibl. Naz., cod. I.E. 56, 10r; cf. ed., C. D. Hellman, « Maurolyco's 'Lost' Essay on the New Star of 1572, » *Isis*, Vol. 51 (1960), pp. 322-336).

1574, 18 October. Completed a summary of Abrahamus Ortelius' *Theatrum orbis terrarum*. (See above, n. 1, at the end.) For some of Maurolico's notes to this work, see Macrì, *Francesco Maurolico,* 2nd. ed., pp 227-28. Maurolico composed in October of this year a geographical work entitled *De insulis totius orbis per Bordonem et Porcaccium (Ibid.,* pp. 228-29).

1575, 21 or 22 July. Maurolico died (see E. Rosen, « The Date of Maurolico's death, » *Scripta mathematica,* Vol. 22 [1956], pp. 285-86).

This essentially completes my chronology of the mathematical works of Maurolico. However, I can briefly add the following posthumous editions for the sake of completeness:

(1) *Opuscula mathematica* (Venice, 1575), including the following works: *De sphaera liber unus, Computus ecclesiasticus in summam collectus, Brevis tractatus instrumentorum, [Brevis] tractatus de lineis horariis, Euclidis propositiones Elemen-*

*torum libri tredecimi solidorum tertii regularium corporum primi, Musicae tra-
ditiones, De lineis horariis libri tres.* The *De sphaera liber unus* and *Computus
ecclesiaticus* were republished from this volume at Mexico in 1578. The latter
work was again published in Cologne in 1581 in an edition of Georg Peurbach's
Theoricae novae planetarum, which was reprinted at Cologne in 1591.

(2) *Arithmetica* (Venice, 1575). Has its own title page but published with the
   Opuscula.

(3) *Index lucubrationum.* Appended to the *Arithmetica* in the same volume.

(4) The proofs, notes and scholia of Maurolico's versions of Autolycus' *De sphaera
   quae movetur* and Theodosius' *De habitationibus* as given in the omnibus edition of 1558 were included in *Autolyci De sphaera quae movetur liber et Theodosii De habitationibus liber... de Vaticana Bibliotheca deprompta et nunc primum in lucem edita. Iosepho Auria... interprete. His additae sunt Maurolyci
   annotationes* (Rome 1587). Maurolico's version of Euclid's *Phaenomena* as presented in the 1558 edition was similarly used in Auria's *Euclidis Phaenomena
   post Zamberti et Maurolyci editionem... de Graeca lingua in Latinam conversa
   a J. Auria... His additae sunt Maurolyci breves aliquot annotationes* (Rome,
   1591).

(5) *Photismi de lumine et umbra.... Diaphanorum partes seu libri tres* (Naples,
   1611; 2nd. ed., Lyon, 1613).

(6) *Problemata mechanica cum appendice et ad magnetcm et ad pixidem nauticam
   pertinentia* (Messina, 1613).

(7) M. Mersenne, *Synopsis mathematica* (Paris 1626; 2nd ed., Paris, 1644). Some
   enunciations, lemmata and scholia without proofs from various works in Maurolico's omnibus edition of 1558. An exception is Maurolico's version of the
   Phaenomena of Euclid, which is completely given from the edition of 1558.

(8) Michael Psellus, *Compendium mathematicum* (Leiden, 1647). This edition (pp.
   142-94) includes Maurolico's *Compendium mathematicae* and his *De sphaera sermo* drawn from the edition of 1558.

(9) *Emendatio et restitutio conicorum Apollonii Pergaei* (Messina, 1654).

(10) *Admirandi Archimedis Syracusani monumenta omnia mathematica quae extant*
   (Palermo, 1685).

(11) *Sicanicarum rerum compendium* (2nd. ed., Messina, 1716). Includes the passages deleted from BN lat. 6177 and missing in the edition of 1562 (see above,
   n. 2).

(12) F. Napoli, « Scritti inediti di Francesco Maurolico, » *Bullettino di bibliografia e
   di storia delle scienze matematiche e fisiche,* Vol. 9 (1876), pp. 23-121. Includes
   his long letter on mathematics to Juan de Vega (pp. 23-40), his *Demonstratio
   Algebrae* (pp. 41-49), his *Geometricarum quaestionum libri duo* (pp. 50-113),
   and his *Brevis demonstratio centri in parabola* (pp. 114-121).

(13) Two books (Bks. VI and VII) added by Maurolico to Petrus Crinitus, *De poetis
   latinis,* published by G. Macrì, *Francesco Maurolico nella vita e negli scritti,* 2nd
   ed. (Messina, 1901), Appendix, pp. xxxv-xlviii. Macrì also published three versions
   of the *Index lucubrationum*: that of the 1575 edition (*Ibid.,* pp. iii-viii), that
   given by the Baron della Foresta (pp. ix-xiv) and that in the Codice Villacanense (pp. xv-xxii); a part of Maurolico's diary of his trip with Fernando de

Vega in December, 1553 (xxxiii-xxxiv); and Maurolico's letter to Juan de Vega of 1556 (pp. xlix-lxxvi).

(14) *Super nova stella.* See full title and bibliographical reference to the edition of 1960 in the entry for 1572, 6 November.

(15) *Prologi sive sermones quidam De divisione artium. De quantitate. De proportione,* ed. of G. Bellifemine (Melphicti, 1968).

I have omitted here any reference to the second edition of Maurolico's *Cosmographia* (Paris, 1558), his work on Sicilian fish (see entry for 1543, March 1), and the later editions of his *Martyrologicum.* My main purpose was to establish Maurolico's year-by-year efforts to compose mathematical and astronomical works. Details of the printed editions have been admirably given by E. Rosen, « The Editions of Maurolico's Mathematical Works, » *Scripta mathematica,* Vol. 24 (1959), pp. 59-76.

2. THE « INDEX LUCUBRATIONUM » OF FRANCESCO MAUROLICO

INTRODUCTION

Readers of my extended treatment of Francesco Maurolico in Volume 3 of ' my *Archimedes in the Middle Ages* and the first section of this article will perhaps conclude that Maurolico had a compulsion for listing and relisting the crucial mathematical works from Antiquity and the Middle Ages and for indicating his own role in correcting and extending these works. I refer to such passages or lists in his *Grammaticorum rudimentorum libelli sex* (Messina, 1528)[1], his letter to Pietro

---

[1] (7v-8r): « Omnes Menelai de Sphaericis conclusiones ostendi: nec dum Menelai Sphaerica vidi. Verum nullam ex hoc quaero laudem: quid enim feci nisi quod antea factum erat? Decretum est itaque nobis nonnulla de huismodi egregiis disciplinis emittere: ne frustra vigilasse et cum magna valetudinis nostrae iactura laborasse videamur. Sed operae precium est, quando in eum sermonem incidimus, opuscula quaedam a nobis in lucem danda enumerare. Emittenda est in primis ad bonarum artium amatores [1] Epistola: in qua singularum artium mathematicarum, et earundem authorum laudes explicabuntur: et quorundam audaculorum errores patefient. [2] Deinde libellus de Sphaera mobili: in qua circulorum et arcuum diffinitiones, et circa eosdem omnimoda Theoremata Theodosii Sphaericis in haerentia disponentur. [3] Tertio dehinc loco, quaedam in Peurbachii Theorias oportunae et necessariae additiones. [4] Post haec astronomicorum problematum libelli quatuor: in quibus omnis Astronomiae calculus geometricis innixus fundamentis explicabitur. [5] Quibus annectetur sinuum siue chordarum Tabella circuli semidiametrum millies mille particularum supponens. His adiicientur duo libelli, [6] unus de praxis arithmeticae theoria: alter [7] de arithmeticis Datis. His et alii duo, [8] de Photismis unus, [9] alter de Diaphanis. In illo, praeter caetera, patescet cur solaris radius per qualecunque foramen transmissus in circularem redigatur formam: in hoc ratio rotunditatis et colorum Iridis aperietur: Quorum utrunque fuit Ioanni vulgatae Perspectivae authori incognitum: [10] Ad haec quosdam locos annotabimus circa linearum Symmetriam: circa Solidorum structuram: circa maximas planetarum aequationes: ubi ostendemus quod circulo, cuius dimetiens rationalis supponatur, Octogonum atque Dodecagonum aequilaterum circumscribente: ipsius octogoni latus trt irrationalis linea que Minor: Dodecagoni vero latus ea quae Apotome a praestantissimis mathematicis appellatur. [11] Determinabimus etiam locus lunaris deferentis: in quo maxima centri contingat Aequatio: ubi turpiter erravit is, qui planetarum Theorias exposuit. [12] Postremo dabitur totius mathematicae disciplinae Compendium quoddam ex Euclide, Theodosio, Archimede, Menelao, Iordano, Boetho (!), Ptolemaeo, caeterisque acutissimis mathematicis excerptum ». I have added the numbers in brackets. Title [1] may be a preliminary form of the letter to Juan de Vega of 1556 (see below, note 4). For title [2], see notes 36 and 37 to the text of the *Index lucubrationum* below. For titles [3], [4] and [11], see below, note 2, the passage before title [1] and note 65 of the

Bembo of 1536 [2], his dedicatory letter of 1540 to Pietro Bembo [3], his
letter to Juan de Vega of 1556 [4], his preliminary *Index lucubrationum*

Index. Fort title [5], see notes 59 and 82 of the *Index*. For titles [6]-[9], [12] see notes 33, 34, 47, 48, 56, of the *Index*. For title [10], see below, note 2, titles [7] and [11], note 4, title [4] : note 9, titles [26] and [66].

[2] MS Vat. Barb. lat. 2158, 143v-144v; cf. F. Guardione, « Francesco Maurolico nel secolo XVI », *Archivio storico siciliano*, Anno 20 (1895), pp. 37-38: « Ex Euclide vix sex elementorum libri leguntur: ad caeteros nemo progreditur, seu ad ignotas oras. Nam Theodosii et Menelai Spherica, Appollonii (!) Conica. Archimedis opera de circuli dimensione, de Sphaera et Cylindro, de Isoperimetris, de momentis aequalibus, de Quadratura parabolae, de Speculis ignificis nusquam apparent, non secus ac si admisso inexpiabili perpetuum meruerint exilium. Et horum si quid circumfertur, tot tantisque scatet mendis, ut vix etiam ab authore ipso emendari possit. Astronomica quoque studia adeo exoleverunt (!) ut Ptolomaeo, caeterisque optimis authoribus neglectis, nil nisi Sphaeram Ioannis de Sacro Bosco lega[mus] [et] celebremus, quasi opus egregium et notatu dignum. Sed quid mirum? Nemo potest hominis illius errores deprehendere nisi Ptolemaeum praelegerit. Quod si Ioannes a regio monte aut eius praeceptor Georgius [Peurbachius] uterque mathematicus consummatissimus, sicut planetarum theorias, sic Sphaerae rudimenta nobis edidisse (!) iam pridem cum Gerardo cremonensi una Ioannes hic de Sacro Bosco rudis astronomus exibilatus et explosus fuisset. Eorum autem quae ad tabularem pertinent calculum, nihil illustrius Alphonsinis tabulis; et tamen illarum Canones multis magnisque mendis foedantur: quae a nemine, quem sciam, animadversae sunt: et quoties opus illud a celebratis doctoribus recognitum impressioni traditum est, toties canones ipsi suis inquinatae maculis in lucem exierunt. Libet mihi, Bembe vir doctissime, de tanta doctrinae huius calamitate tecum conqueri: qui omnis generis scientiarum et amator et defensor es eximius. Ego, quantum mihi licuit, assiduis studiis, mentisque agitationibus, nisus sum collatis priscis exemplaribus et dictorum authorum et aliorum opera complura emaculare et in suum restituere nitorem; quae tibi gratissima fore spero, si quando curis necessariis, laboribusque vacuus his vacare poteris. Scripsi quoque per me nonnulla videlicet [1] de figuris locum implentibus, ubi Averrois error patebit, qui putavit sicut cubos inter regularia quinque solida, et ita pyramides per se locum implere. [2] Item Sphaericorum libellos V. [3] Arithmeticam. [4] Arithmetica data: in quibus multa a Boethio, Iordanoque pretermissa demonstrantur. [5] Photismos, in quibus solaris radii per qualevis foramen transmissi rotunditas demonstratur. [6] Diaphana, in quibus multa de iride, quae necubi leguntur. [7] De motuum symmetria. [8] Arithmeticas, [9] geometricasque quaestiones. [10] De Sphaera mobili. [11]Speculationes multas, ubi inter caetera demonstramus latus octogonii lineam esse minorem, dodecagoni vero latus Apotomen, circulo inquam, rationalem habente diametrum inscriptorum, et alia complura. [12] Dialogos quoque tres, primum de forma, numero et ordine tam elementorum, quam coelorum. Alterum de circulis Sphaerae et planetarum motibus. Tertio de anno et reliquis temporum spaciis. Quos dialogos, quoniam mundi formam et ordinem continent, placuit appellare Cosmographiam: quam ego tibi, si lubet, dedicare decrevi... [13] Nunc operam de historiis antiquis recentibusque colligendis, quaecumque de Sicanicis agunt: ut hinc demonstratibus rebus (que multae sunt et a multis celebratae utriusque linguae authoribus) una texatur historia ». For titles [1]-[6], [7]-[10], [12]-[13], see in succession the following notes of the *Index lucubrationum*: 41, 36, 33, 34, 47, 48; 49, 50, 37; 38, 68. For titles [7] and [11], see above, note 1, title [10]; and below note 4, title [4]; note 9, titles [26] and [66].

[3] See below, note 9 for a text of the catalogue of the works listed there.

[4] F. Napoli, « Scritti inediti di Francesco Maurolico », pp. 31-39: « [1] Scripsimus in primis Arithmeticam quamdam, in qua primum practicae operationes per numerarios terminos cuius adhibitae quantitati demonstrantur. Complura de progressionibus, de quadratis, hexagonis, cubis, pyramidibus aliisque numerorum speciebus, ab aliis omissa dicutiuntur. Item irrationales magnitudines ad numerarios terminos rediguntur. [2] Post haec fecimus arithmetica data, unde questionum multarum calculus derivatur. Hic multo plura nec minus necessaria, quae in Iordani datis demonstrantur. [3] Adiecimus his positionum regulas, quas arabico verbo, Algebra, vulgus appellat, et alias quasdam cum demonstrationibus. Redegimus autem totum algebraticum negocium ad praecepta quatuor: Adeo ut pro traditionibus Diophanti, qui XIII libris ea de re graece pertractavit, his meis interea contentus sim. [4] Speculationes item varias subtexui circa linearum et motuum symmetriam, circa octogoni et dodecagoni latera: circa triangulorum rectilineorum calculum: circa locos maximarum aequationum in planetis et alia circa triangulòrum latera et areas. In sphaericis multum laboravi: [5] Nam cum Theodosius horum elementa tribus ordinatissime tradat libellis: in quorum primo de polis et circulis se invicem orthogonaliter secantibus: in secundo de circulis se tangentibus ac de parallelis: in ultimo de arcuum quorundam collatione agit; [6] cum totidem Menelaus voluminibus caetera Sphaericarum exequatur, in primo de triangulis: in secundo de arcuum collatione: meliori via quam Theodosius: in postremo de sinuum proportione disserens: [7] ego multa a Menelao praetermissa conspiciens, ea supplevi; atque in duos libellos redacta disposui. In primum enim quaecunque ad rationem sinuum facilius et aliter demonstrata contuli. In secundum multa copiose de triangulo sphaerali orthogonio profundissima et non contemnenda tradidi.

of 1558 [5], his *Compendium mathematicae* printed in the same volume
of 1558 [6], his dedicatory letter to Cardinal Amulio of 1568 [7] and his
completed *Index lucubrationum* of 1568 [8].

---

et [8] in calce operis praxim tabularis calculi, quo Ioannes de Monteregio utitur. In supputandis (p. 32) universaliter stellarum declinationibus et rectis ascensionibus, quoque utitur. [9] In tabella faecunda apertissime demonstravi rem a nemine hactenus perspectam. [10] Sed et tabellam, quam Beneficam appellare placuit, exaravimus; cuius videlicet beneficio nec pauciora nec minori facilitate supputanda tradidimus; et huius cum secunda collationem... [11] Neque me paenitebit unquam de momentis aequalibus libellos quatuor scripsisse: cuius materiae invento et laus Archimedi nostro debetur. Ego tamen multa copiosius super ea re demonstravi. In primo quidem de momentorum proportione: in secundo de centris triangulorum et planarum figurarum: in tertio de portionibus paraboles. In his Archimedes succincte nimium se praestitit: si modo quod extat, opus integrum est, pondus enim et momentum, cum sint, magnitudinum notandae species, erant multo latius tractandae. In quarto denum libello totam mihi laudem vendicare non erubescam, nam de centris solidorum quod ab Archimede praetermissum magnopere admiror, disserui. Et in pyramide centrum gravitatis id puntum esse ostendi, quod utcumque positi solidi quartam celsitudinis partem versus basim relinquit. [12] Fecimus praeterea et Epitomen quamdam in problemata Aristotelis mechanica, brevissime scilicet ad quaesita singula respondentes: sed his addidimus complures ad similia spectantes, scituque iucundas quaestiones. [13] Demonstravimus etiam quae planae solidaeque figurae locum impleant: hoc est, quarum anguli sic compaginari possunt, ut adamussim congruentes nihil vacui relinquant: ubi Averroes pueriliter errasse convincetur, dum asserit duodecim pyramides, quemadmodum octo cubos, locum implere posse, suam in mathematicis inscitiam manifeste declarans. [14] Scripsimus et de compaginatione talium solidorum: hoc est quemadmodum eorum unumquodque ex prosecta, ut decet, pagina conglutinari et effingi possit. [15] De circuli quadratura movit me ad scribendum Phineus, nam dum eius errores demonstratione atque calculo arguere sum exorsus, multa super ea re disserui. Et tandem certis coniecturis ac rationibus inductus hoc mihi persuasi, omnis rectilineae figurae circulo cuipiam inscriptae sive extrinsecus adscriptae perimetrum, talem cum circuli diametro rationis colligantiam habere, ut si diameter ipsa rationalis supponatur, perimeter quoque figurae aut rationalis sit, aut sub aliquam speciem irrationalis lineae per numeros determinatae cadat: circuli vero peripheriam sicut omnis figurae inscriptae perimetro maior est, circumscriptae vero minor, ita cum diametro nullum tale sortiri commercium: ac sicut tam inscriptorum quam circumscriptorum laterum numerositatem singuli ambitu ac dexteritate praeterlapsa circumflectitur, ita nullius cum eis participare cognitae rationis terminos: et perinde circumferentiae proportionem ad diametrum, unde te- tragonismus dependent, inexplicabilem esse, quanquam aliquo modo sine mechanico artificio assignari potest: sicut ibi latius tradidimus. Deinde Archimedis praece(p. 33)ptum secuti multoque maioribus numeris usi (quod cuique licet) intra limites angustiores memoratam clausimus proportionem. Semper enim inquirentes hanc magis ac magis vero appropinquare possumus, nunquam tamen ipsum praecise consequi quoniam scilicet inter numerarias proportiones non existit, sed inter incognitas et nullo pacto nominabiles. Tanta est inter rectam flexamque lineam, fortasse propter dissimilitudinem, inimicitia. Veniam nunc ad perspectivae negocium ubi neque operam mihi videor lusisse: quandoquidem et hic compluscula tam ab aliis omissa quam notatu digna demonstravi. Geminos enim composui libellos, [16] de Photismis, hoc est radiationibus, unum: [17] alterum de diaphanis. In illo radiosas foraminum incidentias ad formam luminosi corporis redigi, certis ostendimus argumentis; ubi Ioannes Petsan laborando vix emergit. Item inversas lucis aut illuminatarum rerum imagines repraesentari radiorum intersectione aut concursu. In hoc autem figuras similiter rerum per convexa diaphana transparentium ad terminum quemdam inversas imprimi: propter radiorum concursum. Adiecimus demonstrationes super Iridem minime contemnendas: et quibus nullae sint potiores, ad causas formae, situs atque colorum utriusquae Iridis concludendas. Postremo visuum qualitates, conspiciliorumque eis adcommodatorum formas discussimus, rem quanquam experientia notam, a nullo tamen literis mandatam. [18] Nec me continui ab astronomicis, sphaeraeque mobilis opusculum a diffinitionibus et elementis exorsus, et quaecumquae ad primi mobilis circulos spectant per propositiones executus, et in theorias annotamenta quaedam... (p. 34)... Cogor itaque egomet meis mihi satisfacere conclusionibus, quando aliud, quo me consoler, non superesset. [19] Nam dialogi de forma, numero situque eleméntorum et caelorum, quos appellare libuit Cosmographiam, et quos Petro Bembo dedicavi, fuere potius exercitium quoddam et repastinatio, quam ordinatum ac propositionibus distintum opus;... (p. 35)... [20] Excudimus et nos aestate proxima libellos tres de lineis horariis: in quorum primo de circulis horas e meridie distinguentibus: deque circulis horas ab ortu vel occasu numerantibus, per quos horarii officii eiusdem lineae in horologiis plano gignuntur, locuti sumus. In secundo autem de flexis lineis sive' periferiis quas lineae a meridiano horas exorsae secant, et in sectionum punctis tangunt lineae horas ab ortu vel occasu inceptas dimensae: quae periferiae aut sunt circuli aut aliae conicarum sectionum species, ut sunt parabolae, hyperbolae aut ellipses. In tertio de huiusmodi sectionum proprietatibus, unde omnis theoria et situs linearum per

Of these various lists two are of great importance as detailed indications of his own efforts, the letter of 1540 printed in his *Cosmographia*

gnomonis sive indicis umbram horas indicantium, intelligetur. Quod a nemine hactenus, qui de hoc negocio scripserunt, tentatum est, ut qui nihil ultra praxim descriptionis sint progressi. Quo ad caelestium motuum supputationem, astronomum vulgus omnibus horis habet in manibus, veluti pinguia succina, tritas temporarias ephemerides: atque ii praecipue qui nimium constellationibus tribuentes, omnia caelesti influxu non minus superstitiose, quam curiose credunt fieri: nec, nisi consulto prius diario, passum quidem obambulant, aut nares sibi emungunt: Hi quidem contenti sunto, arithmeticas figuras, et planetarum, signorum atque aspectuum caracteres tantum intellexisse: Hec enim illis satis superquae sunt, a limine caetera salutantibus. Absit a nobis, ut adeo perditi simus, ut professionis nobilissime finem in iudiciaria statuamus: atque scientiam in prima certitudinis arce constitutam, incerto exitu terminemus. Ambiguae divinationis studium non avertet me a iucundissima speculatione. Itaque dum a rivulis ad indeficientem aquarum ortum converto gressum, non destiti quidquid ad calculi rationem pertinet ab initio rimari. Et post ptolemaicas traditiones, post albategnios, tebitios ac toletanos numeros, ad Alfonsinum cum me abacum contulissem animadverti magnum ex tali opere authoris ingenium (p. 36). Qui dum Octavi orbis trepidationem quas Tebitius fuerat commentus, motu altero longitudinis adornat: censetur omnium optime super eo motu sensisse, quanquam pauci admodum animadvertunt, quae ratio, seu coniectura moverit Alfonsum. Illud autem mihi minime dubium est, hunc caeteris tabularum authoribus, vel ob id esse praeponendum, quod circulo ac tempore similiter divisis, per eosdem numeros diversis satisfiat dimensionibus. [21] Harum ego tabellarum canones dum relego, animadverti erratum fuisse alicubi, praesertim in supputanda duorum inferiorum latitudine. [22] Item motuum diurnorum calculum elimavi. Et in eclipsium supputatione multa notatu digna supertexui. [23] Idem in tabulas Blanchini et [24] eclipsium Georgii Peurbachii nec non in tabulam magnam primi mobilis: et [25] in tabulas directionum Ioannis Regi montani diligenter effeci: adiectis quam plurimis oportunis ad omne propositum tabellis. [26] Et ne ulli parcerem labori, omnia praeterea, quae ad anni, mensium atque kalendarii rationem pertinent, scrutatus regulas collegi: nec non his quoque qui circa aequinoctii aut lunationum retrocessionem, circa Paschatis solemnitatem, aut intercalationem bissextilis diei quaestiunculas movent, plene satisfacerem. Quod alioqui et officio nostro sacerdotum incumbebat, quippe qui tam stata quam mobilia festa indicimus, et quotidianam lunae variabilis aetatem in recitanda sacri martyrologii lectionem, pronunciamus... (p. 37)... [27] Itaque conflavi super astrorum iudiciis quoddam isagogicum compendium ex Ptolemaeo, Hermete, Dorotheo, Alkindo, Saphare, Messealla, Aomare, utroque Hali, Albumasare, Zaele, Jerge, Abraamo, Alfragano, Albupatro, Alcabitio, Bonato, Leopoldo, Campano, Gazulo, Firmico, Manilio, Abbano, Joanne Saxone, Regimontano, Bellantio, Gaurico, Cardano, Schonero, aliisque optimis quibusque authoribus congregatum. [28] Adiecimus huic epitomen de temporum signis Augustini Sessae, ex Aristotele, Theofrasto, Ptolemaeo, Theone, Arato, Hesiodo, Callisthene, Anaximandro, Alexandro, Virgilio, Plynio, Alberto, aliisque philosophis excerptam... Utque totam studiorum meorum rationem aperiam, consultoque prius literatorum iudicio, aliquid deinde quod hodiernis academiis acceptum, studiosis viris utile, populoque gratum sit, excudam; recitabo tandem quorundam breviariorum argumenta. Ita enim me praestiti ut nunquam authorem quempiam legerim, quin inde sive epitomen, sive notatu digna excerpserim, aut super obscuriora necessarias expositiones, aut scholia conscripserim. Omitto [29] primum grammaticarum institutionum libellos sex, quorum primus elementa et declinabilium partium flexiones: secundus generum ac declinationum nominis ac verborum: tertius regulas: quartus partium accidentia et orthographiam: quintus constructionum: sextus pedum, metrorum et accentuum precepta complectitur. Qui quamvis festinanter impressi et immature editi (p. 38) usui tamen nobis quandoque fuerunt, dum ingenuis adolescentibus instituendis olim Messanae operam daremus et ad reliqua properabo. [30] Genealogias Boccaccii verborum superfluitate amputata, eo redegi, ut sub multo minori volumine, multo plura carptim hinc inde collecta, et ab illo pretermissa comprehenderem. Quod opus quanquam semper necessarium fuit poeticis lectionibus: tam hactenus nescio qua incuria neglectum iacuit: utinam hoc praestitisset Erasmus et annosas ineptias tacuisset. Adiecimus his Asterismorum fabulas utpote assimilem materiam [31] Diodori nostri siculi epitomen, [32] Phornuti ac Palephati allegorias: [33] deinde alias quasdam regum et primatum genealogias, et [34] chronicum breviarium cum ordine veterum patriarcharum, pontificum, ac principum, per annos ab initio mundi distinctos. [35] Post breviaturam Polydori de rerum inventoribus: [36] breviarium item pontificalium Synodorum, [37] Itinerarium Syriacum, [38] historiam rerum Sicanicarum ex diversis ac vetustis exemplaribus collectam. [39] Historiarum quoque, navigationum ac diversarum rerum epitomen... [40] Ad haec decreveram quandoque coaptare quatuor liberalium quaestionum libellos, ita ut in primo scientiarum distinctionem, ordinem, subiecta, principia, diffinitiones, au(p.39)thorum discrepantias, interpretum ac expositorum errata, et plurima quaesita colligerem non dico conderem: nam et si tantus non sum ut de omnibus disseram, non tamen adeo rudis sum ut ab aliis tradita nesciam digerere. In secundo autem arithmeticas, in tertio geometricas, in postremo astronomicas quaestiones per exempla proponerem ac propositas solverem. Sic enim vel universae praxi satisfacturus videbar. Haec itaque sunt

(Venice, 1543) and several times since[9], and his *Index lucubrationum*. The *Index lucubrationum* is obviously the more important since it re-

illustrissime Prorex, in quibus versatus sum et in quibus ingenium meum exercui ». For titles [1]-[3], [5]-[7], [9]-[21], [23]-[27], [29]-[38], see in succession the following notes of the *Index lucubrationum*: 33, 34, 35; 2, 3, 36; 82, 82, 42 and 43, 16, 41, 40, 6, 47, 48, 65, 38, 46, 53; 54, 55, 55, 63, 83; 80, 72, 78, 87, 72, 74, 78, 78, 75, 68. For title [4], see above note 1, title [10]; note 2, titles [7] and [11]; and below, note 9, titles [26] and [66]. For title [8], see Sect. 1: 1558, August, item 15. For title [22], see perhaps Sect. 1: 1571, 28 September, items on folios 54r-56v, 63r. Title [28] is perhaps a part of the same work as title [27]. Title [39] is not mentioned elsewhere. Title [40] may reflect the work noted in Sect. 1: 1567, 6 May.

[5] The *Index lucubrationum* through line 71 below was published as part of the front matter of the *Theodosii Sphaericorum elementorum libri III* (Messina, 1558).

[6] *Ibid.*, 68v-72r. This properly speaking does not contain a list of his works but merely summarizes the main points of many mathematical works. Similar general accounts of various geometrical works were given in Maurolico's *Geometricae quaestiones* (see Napoli, « Scritti inediti », pp. 23-40) and his *Prologi vel sermones* (ed. Bellifemine).

[7] It appears in MS Vat. 3131, 6r-7r, before the *Arithmetica*. It has been edited by B. Boncompagni, « Intorno ad una proprietà de' numeri dispari », *Bullettino di bibliografia e di storia delle scienze matematiche e fisiche*, Vol. 8 (1875), pp. 55-56n. There is no real list of his own works here but he makes the usual comments about the erroneous translations and versions of earlier works.

[8] It is this text I have edited below.

[9] The list from the *Cosmographia* was of course also printed in the second edition of that work (Paris, 1558), and in the various editions [with some variation] of Conrad Gesner's *Bibliotheca universalis*, first published in Zurich in 1545, 252r-253v. The list was printed in D. Scina', *Elogio di Francesco Maurolico* (Palermo, 1808), pp. 114-121, and in G. Libri, *Histoire des sciences mathèmatiques en Italie*, Vol. 3 (Paris, 1840), pp. 241-247. I give the catalogue here from the *Cosmographia* (Venice, 1543), adding bracketed numbers and altering the punctuation somewhat:

« In prima sectione.

[1] Euclidis elementa in libellos xv ita distincta, ut primi quatuor Planorum, quintus ac sextus pro portionum, septimus, octavus, nonus Arithmeticorum vocentur, decimus Symmetria, quinque reliqui solidorum; ex traditione Theonis, ut transtulit Zambertus: nec exclusis Campani additionibus quibusdam. Adiectis praeterea circa regularia solida speculationibus complurimis: quae ad plenam ipsorum solidorum, quo ad perpendiculares, bases, superficies et corpulentias, collationem, erant necessariae, ubi plane quivis animadvertet Zambertum quamvis graece peritum, exemplaris tamem vitio deceptum peccasse. Campanum vero authoris alicubi terminos temere pervertisse. [2] Theodosii sphaerica: quae hactenus incorrecta ac neglecta iacuerunt: quasi non sint astronomiae totius et praesertim sphaerae fundamenta. [3] Apolloni Pergaei Conica emendatissima: ubi manifestum erit, Io. Baptistam Memmium in eorum tralatione pueriles errores admississe Mathematicae praesertim ignoratione deceptum. [4] Sereni Cylindrica. [5] Archimedis Syracusani de circuli dimensione libellus cum calculo nostro ad mensuram peripheriae propius accedente. [6] Eiusdem de sphaera et Cylindro ex traditione Eutotii Ascalonitae. [7] Eiusdem de isoperimetris figuris tam planis, quam solidis: ubi planarum circulus, solidarum vero figurarum isoperimetrarum sphaera concluditur esse maxima. [8] Menelai sphaerica cum Tebitii, nostrisque additionibus; unde tota sphaeralium triangulorum scientia scaturiit. [9] De figuris p'anis, solidisque regularibus locum implentibus libellus noster: quamquam de hoc negocio Ioannem a Regio monte accuratissime scripsisse certum sit: verum opus nondum, quod sciam, editum; demonstramus autem in libello e solidis regularibus cubos per se: pyramides vero cum octahedris compactas duntaxat implere locum, qua in re Averroem pueriliter errasse, manifestum erit. [10] Euclidis data ex traditione Pappi; tralatio est Zamberti. [11] Inventio duarum mediarum proportionalium ex traditione praestantissimorum authorum Platonis, Architae, Menaechmi, Heronis, Philonis Byzantii et Pappi. [12] Modus secandi datam sphaeram ad datam rationem ex Dionysodoro, quae quamvis a Georgio Valla tralata sint: tamen vix intelligi poterant: tum quod fuerant obscure, ne dicam male tradita: tum quod ad ea perpendenda opus erat in Menechmo (!) et Dionysodoro quibusdam Apollonii et Archimedis locis.

In secunda sectione.

[13] Boetianae Arithmeticae compendium. [14] Iordani Arithmeticorum libelli decem ad miram tum facilitatem, tum brevitatem redacta, [15] Eiusden Data arithmetica. [16] Arithmetica nostra speculativa: in qua multa circa triangulos, quadratos, hexagonos, cubosque numeros et alias eorum species, ab olim praetermissa acutissime demonstrantur; tum circa praxim arithmeticam tam rationalium, quam irrationalium magnitudinum, quae in decimo elementorum, praecepta cum minime negligenda, tum ad practicas quaestiones necessaria. [17] Data arithmetica nostra, in quibus multa sunt a Iordano praetermissa. [18] Euclidis Optica, in quibus agitur de his quae ad visum et visibilia pertinent. [19] Eiusdem Catoptrica, hoc est specularia: in quibus de iis quae in Speculis apparent. [20] Ptolemaei specula:

ferred to almost all of his works over his long and active career. It too has been printed on several occasions. The first part was printed, as I have said, in the front matter of his *Theodosii Sphaericorum elemen-*

---

ubi optimis ipsc argumentis refractiones ad angulos aequales omnino fieri demonstrat. [21] Archimedis libellus de speculis comburentibus: in quo docet ac ostendit, speculo, ut sit ad comburendum efficacissimum, formam dandam esse a parabola: quae est una ex conicis sectionibus; quare negocium huiusmodi intelligere volenti opus esse notitia conicorum elementorum. [22] Photismi nostri, sive radiationes: in quibus de lumine et umbra, quo ad perspectivam spectat, satis agitur; tum lucem per qualecunque foramen admissam adipisci forman ad certum intervallum radianti corpori similem: et perinde Solis radium in circularem formam, aut si deficiat in lunulam similemve deficienti proiici, demonstravimus, locum scilicet non satis intellectum a Ioanne vulgatae perspectivae authore. [23] Diaphana nostra: in quibus ostendimus ea quae per corpus aliquod perspicuum transparent, magnitudine, numero, situ, formaque diversis spectari, iuxta formam perspicui corporis, tum etiam multa super Iride discussimus. [24] Ioannis Petsan Perspectiva emendata. [25] Rogerii Bacchonis Perspectiva utilissima. [26] De motibus et motuum symmetria demonstrationes nostrae scitu iucundae. [27] Archimedis de momentis aequalibus, sive de aequiponderantibus libellus ex traditione Eutotii Ascalonitae. [28] Eiusdem libellus de quadratura parabolae acutissimus: quem intelligere volenti opus est conicorum et momentorum aequalium notitia. [29] Boetianae musicae compendium. [30] Musicae speculativae ac practicae compendium ex Guidone, aliisque authoribus: in quo vocum consonantium ac dissonantium ratio plene discutitur. [31] Arithmeticae quaestiones nostrae. [32] Geometricae quaestiones nostrae. [33] Tetragonismus, sive quadratura circuli, Hippocratis, Archimedis et aliorum. [34] Positionum regulae: quae vulgo Algebra barbaro nomine appellantur, cum demonstrationibus et exemplis ad quatuor praecepta redactae.
In tertia sectione.
[35] Magnae ptolemaicae constructionis compendium, cum demonstrationibus Tebitii circa ea in quibus Ptolemaei demonstratio deficit: item cum quibusdam Albategnii, Georgii Peurbachii et Ioannis de Regio monte, aliorumque additionibus, ubi quivis totam astrorum theoriam facile adipisci potest. [36] Sphaera nostra mobilis in octo capita, multasque conclusiones distincta. [37] Georgii Peurbachii theoriae cum scholiis nostris. [38] Procli sphaera. [39] Campani sphaera. [40] Theodosii de habitationibus. [41] Eiusdem de noctibus et diebus libellus. [42] Autolyci de sphaera quae movetur. [43] Autolyci de ortu et occasu Syderum, sive Phaenomena. [44] Euclidis Phaenomena ad miram facilitatem redacta. [45] Alphagrani compendium. [46] Tebit[ii] rudimenta. [47] Eiusdem de motu octavae sphaerae. [48] Albategnii et aliorum quorundam traditiones. [49] Georgraphiae (!) ptolemaicae compendium. [50] Astronomica problemata nostra: in quibus totus astronomiae calculus, modusque ad tabulas emendandas sive restituendas exponitur. [51] Tabella nostra sinus recti distincta per singulos quadrantis gradus, graduumque minutias, supponens sinum maximum, hoc est circuli semidiametrum in millies mille pluresve particulas sectam: ac geometricae astronomicaeque praxi per quam necessaria. [52] In Alfonsi tabellas problemata, nam canones, qui circunferuntur, non carent omnino mendis. [53] In directionum tabulas Ioannis de Monte regio, problemata: in quibus nonnulla ab authore praetermissa ingeniose discutiuntur. [54] In tabulam magnam primi mobilis eiusdem authoris, brevissimi et ad omnia generales canones. [55] In tabulas eclipsium Georgii Peurbachii canones. [56] In diarium perpetuum canones: in quibus calculi ad eas tabulas pertinentis summa brevibus exponitur.
In quarta sectione.
[57] Quadrati geometrici fabrica et usus cum demonstrationibus. [58] Quadrantis fabrica et usus. [59] Astrolabi fabrica et usus. [60] Quadrati horarii fabrica et usus. [61] Solariorum fabrica ad omnem horizontem. [62] Vitruvianae Architecturae compendium: in quo complures loci enodantur. [63] Aristotelis problemata mechanica. [64] Trochilia nostra in quibus rotarum contextus in horologiorum machinis exponitur. [65] Heronis inventa spiritalia: ac nonnullae machinae hydraulicae a recentioribus inventae. [66] Speculationes mathematicae nostrae: in quibus circa linearum symmetriam: circa optica et catoptrica, circa determinationes maximarum aequationum in deferentibus planetarum, et alias quaestiones, multa discutiuntur». For titles [1]-[9], [14]-[17], [27]-[32], [34]-[37], [40], [42], [44], [50]-[55], [57]-[61], [63], see in succession the following notes of the *Index lucubrationum*: 1, 2, 4, 5, 6, 6, 6, 3, 41; 7, 8, 33, 34; 11, 12, 47, 48, 10, 10; 6 and 42, 6, 31 and 61, 31 and 61, 49, 50; 35, 60, 37, 65; 14; 13; 15; 51, 59, 53, 55, 55, 55; 45, 45, 45 and 66, 45 and 81, 45; 16. Titles [10] and [13] do not appear in the other lists. For titles [11] and [12], see my *Archimedes*, Vol. 3, Part III, Chap. 5, Sect. II, n. 8. Titles [18]-[19] do not appear in his other lists. For title [26], see above, note 1, title [10]; note 2, title [7]; note 4, title [4]. For title [33], see Sect. 1: 1534, 19 August. Titles [38]-[39], [41] are not given in the other lists. Title [43] is not in the *Index* but is extant (see Sect. 1: 1534, 20 October). Titles [45]-[49] are not in the other lists. For title [56], see perhaps Sect. 1: 1571, 28 September, items on folios 54r-56v, 63r, and see the letter of 1556 (note 4 above, title [22]). Titles [62], [64]-[66] are not in the other lists, but for [64] and [65] see Sect. 1: 1534, 25 October.

torum libri III (Messina, 1558), 2*r (*siglum*: *Ed1*) and the whole text
was printed at the end of his *Opuscula mathematica* and *Arithmetica*
(Venice, 1575) (*siglum*: *Ed2*). The latter was republished by his nephew
the Baron della Foresta [10], by Libri [11], by Francesco Guardione in 1895 [12],
and by G. Macrì in 1896 and again in 1901 [13]. In addition to these printed
editions of the *Index,* three manuscripts contain it: Vat. lat. 3131, 168r-
171v (*siglum*: *Mv*), where it is appended to the *Arithmetica* and is apparently not in Maurolico's hand, BN lat. 7466, 1r-3v (*siglum*: *Mp*),
which is in his hand, and Codice Villacanense published by Macrì
(*siglum*: *Vi*) [14], also in Maurolico's hand. The Parisian manuscript was
his personal copy to which he continued to add notes after he had sent
the work to be published. In view of this fact I have decided to publish
this version, adding notes that constitute a first step toward integrating
it with the extended chronology of Section 1 of this article. Complete
integration can be accomplished only when the contents of the manuscripts have been more thoroughly studied.

I have also added in the footnotes references to parallel titles in
the earlier lists (whose texts are given here in the footnotes to this introduction). One caution must be observed in judging the significance
of the titles listed in the earlier lists. We are not always sure whether
a title listed is that of a completed work, or of a work partially completed, or of a work intended but not yet begun and existing only in the
form of notes. Apparently the *Index lucubrationum* itself refers only to
works that have been completed, though on occasion Maurolico neglec-
ted to list some of the works also mentioned in the earlier lists that
were completed and are extant. A case in point is his version of Autolycus' *De ortu et occasu syderum,* which appears in the list of 1540 (see
above, note 9, title [43]) and is extant (see Sect. 1: 1534, 20 October)
but is not mentioned in the *Index.* Another case is that of Hero's *Spiritalia* (title [65] in the catalogue of 1540). It is not mentioned in the
Index but at least some material from it is extant (see Sect. 1: 1534,
25 October). Part of the difficulty in determining a true canon of Maurolico's works is that he customarily prepared epitomes or excerpts of,
made notes on, or reconstructed almost every work that he read in his
long lifetime, as he himself confesses (see above, note 4, the passage
following title [28]). Hence he must have felt some hesitation in includ-

---

10 *Vita dell'Abbate del Parto D. Francesco Maurolyco* (Messina, 1613), pp. 36-41.
11 Libri, *Histoire*, Vol. 3, pp. 247-55.
12 « Francesco Maurolico nel secolo XVI », *Archivio storico siciliano,* Anno 20 (1895), pp. 40-44.
13 G. Macrì, *Francesco Maurolico nella vita e negli scritti* (Messina, 1896), pp. 157-62. He also republished (pp. 163-68) the version of the *Index* published by the Baron della Foresta. Both of these
texts were repeated in Macrì's second edition of 1901 (Appendix, pp. iii-viii, ix-xiv).
14 Macrì, *op. cit.,* 2nd. ed., pp. xv-xx.

ing works in the *Index* which he had not significantly changed or improved. This may be the case with various titles in the Catalogue of 1540 (see titles [18], [19], [38], [39], [41], [45], [49], [62], all of which are absent from the *Index*). Some works are absent from both the Catalogue of 1540 and the *Index*. For example, note the titles added by the Baron della Foresta to the *Index* (see below, text 2 [1], lines 126-143). Of these items at least two are extant: *De piscibus siculis* (see Sect. I: 1543, I March) and the *Epitome de grammaticis Suetonii* (see Sect. I, n. I, item 15). In addition, there are other titles missing in the *Index* and the *Catalogue* that appear in the Codice Villacanense (see Sect. I, n. I, items 1-13, 20-21?, 22).

In my text of the *Index lucubrationum* below I have added within brackets references that appear in no previously published editions except that of the Codice Villacanense published by Macrì where the readings are often different. Most of the bracketed readings appear to have been added by Maurolico at some time later than the original text. I have also added within double brackets a group of added titles that are missing from the 1575 edition but were included by the Baron della Foresta in his *Vita* (and from him by Libri). These latter entries are not present in the three manuscripts. The *sigla* of the five versions of the *Index* used in my text have been noted above; the marginal folio numbers are those of BN lat. 7466.

In addition to giving the text of the *Index lucubrationum* that appears in BN lat. 7466, I have also published in subsections [2], [3] and [4] further lists that follow the *Index* in that manuscript. Lists [2] and [4] were also included in the Codice Villacanense and have been published by Macrì, *Francesco Maurolico,* 2nd. ed., Appendix, pp. xx-xxii. However, the lists as published by Macrì are somewhat truncated in comparison with the lists in the Paris manuscript. Thus in list [4] as published by Macrì only three volumes of poetry are described rather than the four volumes of the Paris list. At any rate, I have given a complete set of variant readings for the Codice Villacanense. The material in brackets consists of additions made by Maurolico to his pristine lists. I have not added any notes to these added lists since the titles given there have been identified in the notes accompanying my text of the *Index*.

[1]

1r / INDEX LUCUBRATIONUM MAUROLYCI

Euclidis Elementa, discussis interpretum erroribus, tam Campani nimium sibi confidentis quam Zamberti professionem ignorantis. Cum ad-

5 ditionibus quarundam propositionum, praesertim ad regularia solida spectantium [1].

Theodosii Sphaerica elementa, lib. 3. astronomicis principiis necessaria [2].

Menelai Sphaerica. lib. 3. multis demonstrationibus adaucta ad scientiam sphaeralium triangulorum [3].

10 Apollonii Conica elementa lib. 4. et demonstrationibus et lineamentis oportunis instaurata [4].

Sereni Cylindrica lib. 2 [5].

Archimedis opera: De dimensione circuli. De sphaera et cylindro. De isoperimetris. De momentis aequalibus. De quadratura parabolae. De sphaeroidibus et conoidibus figuris. De spiralibus. Cum additione demon-
15 strationum et artificio facilitatis [6].

Jordani Arithmetica [7]. Et Data [8]. Theonis Data [9].

Rogerii Bacchonis et Joannis Petsan Perspectivae [10]. Cum adnotationibus errorum.

Ptolemaei Specula [11]. Et De speculo comburente libellus [12].

---

[1] See Section 1, entries for 1532, 21 January and 9 July; 1534, November; 1541, 2 August and 12 November; 1557, 7 February. Since this first title appeared also in the partial text of the *Index* of 1558, the above-noted entries from Section 1 appear to be the ones intended by this item. But see also Sect. 1: 1563, 3 December; 1567, 28 January. For the only published piece of the *Elements,* see the *Opuscula mathematica* (Venice, 1575), pp. 103-144. Cf. the reference to the *Elements* in the catalogue of 1540 (see above, Introduction, note 9, title [1]).

[2] Sect. 1: 1558, August, item 3. Cf. the earlier references of 1528 (Intro., note 1, title [2]), 1540 (*Ibid.*, note 9, title [2]), and 1556 *Ibid,*. note 4, title [5]).

[3] Sect. 1: 1558, August, item 5. Note the entry in the catalogue of 1540 (Intro., note 9, title [8]) and in the letter of 1556 (*Ibid.*, note 4, title [6]).

[4] Sect. 1: 1547, 29 June, 25 October, where the edition of 1654 is cited. See the catalogue of 1540. (Intro., note 9, title [3]).

[5] Sect. 1: 1534, 16 August. See the catalogue of 1540 (Intro., note 9, title [4]).

[6] Sect. 1: 1534, 23 July, 19 August, 10 September; 1547, 6, 19, 30 December; 1548, 23 January; 1549, 18 October, 17 December; 1550, 13 February, in all of which entries the edition of 1685 is mentioned. For references in the catalogue of 1540, see Intro., note 9, titles [5]-[7], [27]-[28], [33], that catalogue omitting references to *On Conoids and Spheroids* and *On Spiral Lines.*

[7] Sect. 1: 1532, 20 September; 1539, 28 September; 1545, 4 January; 1553, 6 November; 1, 15, 16, 18, 19, 21, 24, 25 December; 1554, 5, 12, 16 January; 1556, 22, 23, 25 December; 1557, 5, 12, 16 January; 1557, 9, 10 February, 3, 6, 24 March; 1557, 18 April, 24 July. I assume that it is the first of these references which is intended, namely the text in the Codice Villacanense. Cf. the catalogue of 1540 (Intro., note 9, title [14]).

[8] Again the reference is no doubt to the work of Sect. 1: 1532, 20 September. It is listed in the catalogue of 1540 (Intro., note 9, title [15]). He also gives as title [17] a *Data arithmetica nostra.* Cf. perhaps Sect. 1: 1553, 24 December.

[9] Sect. 1: 1554, 7, 13 April. The *Theonis data* does not appear in the catalogue of 1540, nor in the 1558 edition of the *Index.*

[10] At least the first of these redactions is that contained in the Codice Villacanense, item 19 (see Sect. 1, n. 1). In fact the work there described by Macrì may be a work combining both redactions but misdescribed by Macrì. Both redactions are mentioned in the catalogue of 1540 (Intro., note 9, titles [24] and [25]).

[11] I have not found Maurolico's redaction of this work, also mentioned in the catalogue of 1540 (Intro., note 9, title [20]).

[12] Not located. Appears in the catalogue of 1540 under the name of Archimedes (Intro., note 9, title [21]).

20   Autolyci De sphaera quae movetur [13].

Theodosii de habitationibus [14].

Euclidis Phaenomena brevissime demonstrata [15].

Aristotelis problemata mechanica. Cum additionibus complurimis, et iis quae ad pyxidem nauticam, et quae ad iridem spectant [16].

25   PROPRIA

Prologi, sive Sermones quidam: De divisione artium [17]. De quantitate [18]. De proportione [19]. De mathematicae authoribus [20]. De sphaera [21]. De
cosmographia [22]. De conicis [23]. De solidis regularibus [24]. De operibus Archi-
IV   medis [25]. De quadratura circuli [26]. De centris [27]. De instrumentis [28]. / De
30   calculo [29]. De perspectiva [30]. De musica [31]. De divinatione [32].

---

[13] Sect. 1: 1534, 3 October; 1558, August, item 8. It is briefly noted in the catalogue of 1540 (Intro., note 9, title [42]).

[14] Sect. 1: 1534, 8 October; 1558, August, item 9. Noted briefly in the catalogue of 1540 (Intro., note 9, title [40]), which also adds a reference to *De noctibus et diebus libellus* (*Ibid.*, title [41]) not mentioned in the *Index lucubrationum*.

[15] Sect. 1: 1554, 16 April; 1558, August, item 10. Mentioned in the catalogue of 1540 (Intro., note 9, title [44]). I have found no early redaction.

[16] Sect. 1: 1569, 4 May. It is mentioned in the catalogue of 1540 (Intro., note 9, title [63]) but I have found no early redaction. The title is given in a somewhat different form in the 1558 edition of the Index: « Problemata mechanica Aristotelis cum oportunis et notatu dignis additio-nibus ». Cf. also the reference in the letter of 1556 (Intro., note 4, title [12]).

[17] Sect. 1: 1554, 14 June. Not mentioned in the 1540 Catalogue.

[18] Sect. 1: 1554, 18 June. Not mentioned in the 1540 Catalogue.

[19] Sect. 1: 1554, 22 June. Not mentioned in the 1540 Catalogue.

[20] Perhaps the letter to Juan de Vega noted in Sect. 1: 1556, 8 August.

[21] Probably the *De sphaera sermo* in the edition of 1558; see Sect. 1: 1558, August, item 2.

[22] Probably the proem to his *Cosmographia*, dated 1540 and published in Venice, 1543. See Sect. 1: 1540, 24 January.

[23] Probably the short proems to his version of Apollonius' *Conica*. See Sect. 1: 1547, 29 June and 25 October. See ed. of 1654, pp. 3, 151.

[24] Sect. 1: 1556, 22 December. But perhaps this refers rather to the *Libellus de impletione loci quinque solidorum regularium* (see Sect. 1: 1529, 9 December). The chief argument against such an identification is that there seems to be no separate *prologus* or *sermo* preceding that work.

[25] Sect. 1: 1550, 13 February. Thus I take this to be a reference to the short essay on the works of Archimedes that is a proem to the *Praeparatio ad Archimedis Opera*.

[26] This could be a reference to the *Maurolyci Tetragonismus* (see Sect 1: 1534, 19 August) or the *Modus alius quadrandi circulum* which must have been done about the same time. See the cata-logue of 1540 (Intro., note 9, title [33]).

[27] This may be a reference to the preface of Book IV of his *De momentis aequalibus* (see Sect. 1: 1548, 23 Jaunary), which begins: « Superest nunc agere de centri gravitatis inventione in solidis ».

[28] See the *In tractatum instrumentorum ad lectorem prologus* (in *Opuscula mathematica*, Venice, 1575, p. 48). The fact that this item is in the 1558 edition of the *Index* indicates that this work was composed before that date.

[29] I do not know to which prologue this refers.

[30] This may refer to some lost preface to an earlier version of the *Problemata ad perspectivam* (see Sect. 1: 1567, 12 October). At least, the inclusion of this item in the 1558 edition of the *Index lucubrationum* points to some earlier tract. Perhaps the prologue introduced his lost redactions of Peckham and Bacon mentioned above in note 10.

[31] This possibly refers to the musical pieces in the *Opuscula mathematica* (Venice, 1575), pp. 145-160. If so, perhaps only to the various remarks addressed « Ad lectorem » (see pp. 145, 150, 155, 158). Or perhaps it refers to the introductory part of the work on Boethius' *Musica* referred to in Sect. 1, n. 1, item 18. I do not think it refers to the somewhat similar material found in BN lat.

Arithmetica speculativa libris duobus [33]. In quorum primo multa de
formis tam planis, quam solidis numerorum a nemine hactenus animad-
versa. In secundo autem theoria et praxis rationalium et irrationalium
magnitudinum per numerarios terminos cum multis novis, quae ad deci-
35   mum Euclidis faciunt, conclusionibus abunde tractatur.

Arithmetica Data libellis quatuor demonstrata [34].

Positionum et rei demonstrationes ad quatuor praecepta sive capita
redactae [35].

Sphaericorum libelli duo [36]. In quibus multa a Menelao neglecta sup-
40   plentur pro sphaeralium triangulorum scientia.

Sphaera mobilis, in octo capita, pro circulis primi motus [37].

Cosmographia de forma, situ, numeroque caelorum et elementorum olim
Petro Bembo dicata [38].

Conicorum elementorum quintus et sextus, post quatuor Apollonii li-
45   bros locandi [39].

De compaginatione solidorum regularium [40].

7462 (see Sect. 1: 1566, 30 December), for the latter appears to have been composed later than 1558
and this entry appears in the 1558 edition of the *Index lucubrationum*. Note, however, that the
catalogue of 1540 has two musical titles (Intro., note 9, titles [29] and [30]). The first is to the
work on Boethius' *Musica* in the Codice Villacanense mentioned above (Sect. 1, n. 1, item 18). The
second could refer either to the material published in 1575 or to some of the material in BN lat.
7462, or to both, granting however that the titles could not refer to the items specifically dated
in 1566 or 1567.

[32] I have seen no trace of this prologue.

[33] Sect. 1: 1557, 18 April, 24 July. These entries refer to the work published in Venice, 1575.
But an *Arithmetica* had been prepared in some form quite early in his career (see Sect. 1: 1539, 28
September; 1553, 6 November). He refers to such a work in 1528 (Intro., note 1, title [6]), in
1536 (*Ibid.*, note 2, title [3]), and in the catalogue of 1540 (*Ibid.*, note 9, title [16]). Cf. also his
letter of 1556 (*Ibid.*, note 4, title [1]).

[34] This is the work referred to in 1528 (*Ibid.*, note 1, title [7]), in 1536 (*Ibid.*, note 2, title
[4]) and in the catalogue of 1540 (*Ibid.*, note 9, title [7]). It may have some relation to the tract
mentioned in Sect. 1: 1532, 20 September, or to the collection of tracts mentioned in Sect. 1: 1553,
6 November and appearing in BN lat. 7473. See also Sect. 1: 1554, 28 January for his tract on
Boethius' *Arithmetica*.

[35] Sect. 1: 1569, 7 October. While this is the date of completion given in the manuscript, he
seems to have made an earlier version or at least contemplated one (see the catalogue of 1540, Intro.,
note 9, title [34]) and the letter of 1556 (*Ibid.*, note 4, title [3]). It also appeared in the 1558
edition of the *Index*.

[36] Sect. 1: 1558, August, item 6. A separate treatise of his own is not singled out in the catalogue of 1540, but see Intro., note 9, title [8] for a reference to his additions to Menelaus' *Sphaerica*. See also the reference in the letter of 1536 (*Ibid.*, note 2, title [2]) and the one in the letter
of 1556 (*Ibid.*, note 4, title [7]).

[37] I do not believe that this work is extant. It may be an earlier version of the *De sphaera
liber unus* published in the *Opuscula mathematica* (Venice, 1575), pp. 1-26. The latter has more
than eight chapters and is not limited to « circulis primi motus ». This work is mentioned in 1528
(Intro., note 1, title [2]) and in the catalogue of 1540 (*Ibid.*, note 9, title [36]).

[38] Sect. 1: 1535, 21 October; 1540, 24 January; 1543. This is the work published in Venice
in 1543.

[39] Sect. 1: 1547, 25 October.

[40] Sect. 1: 1536, 3 July; 1537, 25, 26, 28 and 29 December. It is not mentioned in the catalogue of 1540 but is mentioned in the letter of 1556 (Intro., note 4, title [14]).

Quae figurae tam planae, quam solidae locum impleant, ubi Averroes geometriam ignorasse indicatur [41].

50 De momentis aequalibus libri quatuor [42]. In quorum postremo de centris solidorum ab Archimede omissis agitur [43]. Et de centro solidi paraboles [44].

De quadrati geometrici, Quadrantis et Astrolabi speculatione, fabrica usuque [45].

55 De lineis horariis libri tres [46]. In quibus tota huiusmodi linearum theoria, quo ad situm, colligantiam et descriptionem ipsarum plene tractatur. Nam lineae horariae a meridie ceptae secant periferiam quandam in iis

2r punctis in quibus eandem tangunt lineae ho/rariae ab occasu et ortu exorsae. Talis autem periferia vel circulus est, vel ex conicis sectionibus aliqua, scilicet parabole, ellipsis, vel hyperbole.

60 Photismi de lumine et umbra ad perspectivam et radiorum incidentiam facientes [47] [ad illuminationem et calorem].

Diaphana in tres libros divisa [48]. In quorum primo de perspicuis corporibus. In secundo de iride. In tertio autem de organi visualis structura et conspiciliorum formis agitur.

65 Quaestionum arithmeticarum libri tres [49]. Geometricarum libri duo [50]. Astronomicorum problematum tres [51]. In quibus regulae et exempla traduntur.

---

[41] Sect. 1: 1529, 9 December; 1535, 21, 23, 24 September. Also noted in the letter of 1536 (Intro., note 2, title [1]), in the catalogue of 1540 (*Ibid.*, note 9, title [9]), and in the letter of 1556 (*Ibid.*, note 4, title [13]).

[42] Sect. 1: 1547, 6, 19, 30 December; 1548, 23 January.

[43] As I have indicated in my *Archimedes in the Middle Ages*, Vol. III, Part III, Chap. 4, Sect. III, the fact that this reference to his treatment of the centers of solids was included in the 1558 edition of his *Index lucubrationum* alerted Commandino to Maurolico's efforts in this subject.

[44] Sect. 1: 1565, 5 May. This reference to his tract of 1565 is of course missing in the 1558 edition of the *Index lucubrationum*.

[45] For the tract on the astrolabe, see Sect. 1: 1538, 12 August; for that on the *quadratum*, see 1541, 6 April, 7 and 9 December; 1542, 11 January; and 1546. For the third instrument, the quadrant, see *De instrumentis astronomicis*, and, as I have noted above (Sect. 1: 1538, 12 August), the section on the astrolabe in the *De instrumentis* differs from the tract in BN lat. 7464, 9r-16v. Works on the three instruments are noted in the catalogue of 1540 (Intro., note 9, titles [57]-[59]). To these are added in that catalogue: « [60] Quadrati horarii fabrica et usus. [61] Solariorum fabrica ad omnem horizontem ». These last entries may refer to tracts that eventuated in the *De lineis horariis*.

[46] Sect. 1: 1531-1532; 1553, 9 July, 19 July. It was published in the *Opuscula mathematica* (Venice, 1575), pp. 161-285.

[47] Sect. 1: 1521, 19 October; and see the editions of 1611 and 1613. This and the next item were mentioned in the lists of 1528 (Intro., note 1, titles [8] and [9]), 1536 (*Ibid.*, note 2, titles [5] and [6]), 1540 (*Ibid.*, note 9, titles [22] and [23]), and 1556 (Ibid., note 4, titles [16] and [17]).

[48] Sect. 1: 1522, 3 January; 1554, 19, 20, 29 May; and see editions of 1611 and 1613.

[49] I know of no tract bearing this title. I do not believe it refers to any of the material noted in Sect. 1: 1545, 4 January, or 1553, 6 November, since the title *Arithmeticae quaestiones nostrae* also appears in the catalogue of 1540 (Intro., note 9, title [31]).

[50] Sect. 1: 1555, 12, 29 March. This work may have developed from an earlier tract since he refers to a work with this title in the catalogue of 1540 (Intro., note 9, title [32]).

[51] I have not found this title among the various astronomical pieces published in the omnibus editions of 1558 or 1575. It could perhaps refer to something in BN lat. 7472A (see Sect. 1: 1571, 28 September). It is perhaps the work referred to in the catalogue of 1540 as *Astronomica problema nostra* (Intro., note 9, title [50]). At any rate, the work intended by the entry seems to have been completed as early as 1558 since it appears in the 1558 edition of the *Index lucubrationum*.

Adnotationes omnimodae in diversos Mathematicae locos [52].

70   Canones tabularum Alfonsi [53], Blanchini [54], Eclipsium, Directionum, primi mobilis [55].

Compendium Mathematicae brevissimum [56].

Elementorum Euclidis Epitome [57]. Cum novis et artificiosissimis praesertim circa proportiones demonstrationibus [et diffinitionibus].

75   Conicorum Apollonii breviarium libris tribus, facilius et directe demonstratum [quo ad res precipuas et necessarias] [58].

Tabula sinus recti supponens sinum maximum, sive semidiametrum, plurium quam millies mille particularum, quod est totius geometrici astronomicique calculi necessarium instrumentum [59].

80   Compendium Magnae Constructionis Ptolemaicae omnium observationum astronomicarum seriem paucis comprehendens ex Breviario Jo. Regimontii [60].

Compendium Musicae Boetii, cum quibusdam scholiis ad intervallorum proportionem facientibus [61].

Sphaera in compendium, breviter omnia complectens [62].

85   Computus ecclesiaticus, brevis et exactus [63].

---

[52] I have found no work with this title. A somewhat different title appears in the 1558 edition of the *Index lucubrationum*: « Adnotationes omnimodae, et in diversas tabulas canones », which perhaps arose by the conflation of this and the next title in the text of the full *Index lucubrationum*. It may, however, merely refer to the *Tabellarum canones* published in the omnibus edition of 1558 (see Sect. 1: 1558, August, item 16).

[53] See Sect. I, n. 1, item 23 for the copy in the Codice Villacanense. The work is listed in the catalogue of 1540 (Intro., note 9, title [52]). It appears to be referred to in the letter of 1556 (Intro., note 4, title [21]).

[54] See Sect. 1, n. 1, item 26 for the copy in the Codice Villacanense. It appears to be the item referred to in the letter of 1556 (*Ibid.*, title [23]).

[55] The reference to canons on eclipses is to Maurolico's *In tabulas eclipsium Georgii Peurbachii canones* found in the Codice Villacanense (see Sect. 1, n. 1, item 25). The reference to « canones tabularum... directionum, primi mobilis » is the work on Regiomontanus found in the same codex (*Ibid.*, item 24). See the references in the catalogue of 1540 (Intro., note 9, titles [53]-[55]) and in the letter of 1556 (*Ibid.*, note 4, titles [24]-[25]). Note that the entries to which I have added notes 53, 54 and 55 were not given in the 1558 edition of the *Index*.

[56] This is the work published at the end of the omnibus edition of 1558 (see Sect. 1: 1558, August, item 17). We know that Maurolico intended to write such a work as early as 1528 (see Intro., note 1, title [12]). Also note that the 1558 edition of the *Index* terminates with this entry.

[57] See above, footnote 1. Particularly note Sect. 1: 1567, 28 January.

[58] This epitome in three books does not appear to be extant. But notice that Book III of Maurolico's *De lineis horariis* is itself an epitome of the subject of conic sections (see *Opuscula mathematica* [Venice, 1575], pp. 263-285). For the date of that book, see Sect. 1: 1553, 19 July.

[59] Sect. 1: 1550, 28 August. He mentions such a title in 1528 (Intro., note 1, title [5]) and in 1540 (*Ibid.*, note 9, title [51]). See also below, note 82.

[60] Sect. 1: 1567, 10 May, 16 June. He certainly intended if not completed an earlier compendium for he gives a very full title of such a work in his catalogue of 1540 (see Intro., note 9, title [35]).

[61] For the copy in the Codice Villacanense, see Sect. 1, n. 1, item 18. Cf. above, note 31, where the reference to this work in the catalogue of 1540 is mentioned.

[62] See note 37 above. I believe this item to be the *De sphaera liber unus* published in the *Opuscula mathematica* (Venice, 1575), pp. 1-26.

[63] Sect. 1: 1567, December. This is the work published in the *Opuscula mathematica* (Venice, 1575), pp. 26-47. Compare the title in the letter of 1556 (Intro., note 4, title [26]).

Adnotationes in Jo. Sacroboscum [64]. In Theorias planetarum [65].

2v

/ Quadrati, Quadrantis, Astrolabi, Instrumenti armillaris et Sphaerae solidae demonstratio, fabrica, et usus, per novam et artificiosam brevemque speculationem [66].

90

De lineis horariis regulae brevissimae et theoria pro quocunque horizonte [et quocunque horologio] [67].

Historiae rerum sicanicarum [68].

Martyrologium Sanctorum correctum et instauratum. Cum Topographia urbium et aliis appendicibus [69].

95

Hymnorum ecclesiasticorum liber unus.... Poemata Phocylidis et Pythagoriae latina [70].

Carminum et epigrammatum libelli duo [71].

Genealogia deorum Jo. Boccaccii adaucta. Cum multis illustrium virorum et principium carptim collectis genealogiis ad poesim et historiam

100

necessariis [72].

Rhytmi vulgari seu vernaculo sermone in laudem crucis [73].

Chronologia ab orbe condito principum, praesulum et notabilium rerum brevissima [74].

Itinerarium Syriacum, cum historiis ad loca sacra pertinentibus [75].

105

Ad Petrum Bembum de Aetnaeo incendio [76].

Ad Synodi Tridentinae patres epistola [77].

---

[64] Sect. 1: 1567, 9-10 May.

[65] Perhaps this is the work or works referred to in 1528 (Intro., note 1, titles [3], [4], or [11]), in 1540 (*Ibid.*, note 9, title [37]), and in 1556 (*Ibid.*, note 4, title [18]). See also note 60 above.

[66] This is the *De instrumentis astronomicis* published in the *Opuscula mathematica* (Venice, 1575), pp. 48-79.

[67] This is the *De lineis horariis brevis tractatus* published in the *Opuscula mathematica* (Venice, 1575), pp. 80-102. See Sect. 1: 1569, 17 February.

[68] Sect. 1: 1562, after 1 October. The most important editions are those of Messina, 1562 and 1716. See also the autograph copy of BN lat. 6177. Maurolico was working on the compendium as early as 1536 (Intro., note 2, title [13]).

[69] Sect. 1: 1567 or 1568.

[70] Both of these collections seem to be missing. But see Sect. 2 [4] where Maurolico outlines the contents of four volumes of poetry.

[71] The manuscript of this collection also seems to have disappeared. Again see Sect. 2 [4] below. Some examples of his poetry have been given by his nephew the Baron della Foresta in the *Vita*, pp. 49-55.

[72] This work is lost. It is mentioned in the letter of 1556 (Intro., note 4, titles [30] and [33]).

[73] This was published in Messina, 1552, under the title of *Rime del Maurolico* by Giovan Pietro Villadicani, a rich and noble collector. For the latter's friendship with Maurolico, see Macrì, *Francesco Maurolico*, 2nd. ed., pp. 44-46. See also pages 44-46 describing the poetry. The date of 1552 is that given below in the text over notes 93 and 96.

[74] This work is lost. It was mentioned in the letter of 1556 (Intro., note 4, title [34]).

[75] Also missing. See the reference to it in 1556 (*Ibid.*, title [37]).

[76] Sect. 1: 1536, 4 May. This is the letter to Bembo of 1536.

[77] Sect. 1: 1562, 1 October.

Breviarium Platinae de vitis pontificum. [78].

Breviarium sex librorum de vitis patrum.

Breviarium Laertii de vitis philosophorum.

110   Breviarium Petri Criniti de vitis poetarum.

Breviarium Polydori de inventoribus rerum.

Breviarium conciliorum [de here].

[L. flori. Iustini. Herodoti. Leonardi Aretini. Belli neopolitani] [79].

Breviarium fabularum Diodori Siculi.

115   Grammaticarum institutionum libri sex [80].

3r   / Quadrati horarii fabrica et usus [cum demonstratione] [81].

Demonstratio et praxis trium tabellarum, scilicet sinus recti, faecundae et beneficae ad scientiam et calculum sphaeralium triangulorum utilis [quae cum sphaericis posita est] [82].

120   Compendium iudiciariae ex optimis quibusque authoribus decerptum [83]. In quo de naturis signorum, domiciliorum, planetarum, constellationum; de regulis aspectuum, directionum, profectionum, revolutionum, nativitatum, electionum, et quaestionum praesertim ad agricolas, nautas, medicos et milites spectantibus summatim agitur [84]. [Item de divinationibus
125   concessis] [85].

[[De piscibus siculis brevis tractatus [86].

Palephati de non credendis historiis Epitome [87].

Fulgentii Mythialogiorum (*ex cor.*) Epitome.

Ciceronis de natura deorum, et de divinatione Epitome.

130   Scholia in Asinum aureum Lucii Apulii.

Epitome de grammaticis Suetonii.

Tractatus de placitis philosophorum.

---

[78] This and at least five of the six following *Breviaria* are lost so far as I know. Maurolico refers to the epitomes of Diodorus and Polydorus, and to the breviary on councils or synods, in his letter of 1556 (Intro., note 4, titles [31], [35], and [36]). For the Breviarium of Petrus Crinitus' work with Maurolico's additions, see Sect. 1, n. 1, item 14. Macrì, *Francesco Maurolico*, 2nd. ed., Appendix, pp. xxxv-xlviii has published the two books (VI and VII) added by Maurolico. The latter books were composed at Messina and sometime later than the year 1550 mentioned therein (*Ibid.*, p. xlv).
[79] These extracts are also lost.
[80] Published in Messina, 1528. See Sect. 1: 1528, 14 August.
[81] Published in Venice, 1546. See Sect. 1: 1541, 6 April, 7 and 9 December; 1542, 11 January; and 1546.
[82] Sect. 1: 1550, 13 and 28 August; 1558, August, items 7, 12-14, 16.
[83] I have not located this work. It is referred to in the letter of 1556 (Intro., note 4, title [27], and perhaps [28]). Some of the tabular material useful for such a work was given in BN lat. 7472A (see Sect. 1: 1571, 28 September).
[84] Note that the edition of 1575 substitutes for « spectantibus summatim agitur » the phrase « et exclusis superstitionibus directae ».
[85] See above, note 32. This added entry appears exclusively in BN lat. 7466.
[86] Sect. 1: 1543, 1 March. This and the remaining works within double brackets were added by the Baron della Foresta, *Vita*, pp. 40-41. Presumably he had his uncle's manuscripts of these works.
[87] See the reference in the letter of 1556 (Intro., note 4, title [32]).

Opus epistolarum ad diversos viros Illustres.

Quam plures epistolae ad multos.

135   Plurimorum sanctorum vitae, videlicet: Sancti Pancratii Tauromini-
tanorum Pont., Historia Sanctorum Alphii Philadelphii et Cirini, Vita
Agatonis Liparitani, Vita Sancti Angeli Carmelitae, Vita Sancti Alberti
Carmelitae, Vita Cononis naxii viri Sanctissimi mon. ord. Sancti Basilii,
Vita Sancti Calogeri, Vita B. Gullielmi, Vita Sancti Philippi Praesb. ar-
140   gyritae, Vita Sancti Corradi Placentini, Vita Laurentii Presb. qui floruit
in villa Frazano, Vita Sanctae Vennerae Siculae, Vita Sancti Nicandri
Hermitae, et sociorum ex quibusdam graecis historiis decerpta, Vita B.
Eustochii Virginis Franciscanae Mess.]].

Notandum quod ex suprascriptis operibus

145   Theodosii, Menelai et Maurolyci Sphaerica, item Autolyci Sphaera,
Theodosii de habitationibus, Euclidis Phaenomena, Demonstratio et praxis
trium tabellarum sinus recti, faecundae ac beneficae, Compendium Mathematicae brevissimum, simul in uno volumine, Messanae impressa fuerunt
a Petro Spira filio Georgii Spirae germani, anno salutis 1558 [88].

150   ## Item

Cosmographia olim Petro Bembo dicata. Impressa fuit Venetiis apud
Iunctas, anno sal. 1543 [89]. Et rursum Basileae apud Jo. Oporinum. [Item
Parisii apud Gulielmum Cavellat, 1558] [90].

Item

155   Quadrati horarii fabrica et usus Jo. Vigintimillio dicata. Impressa fuit
Venetiis apud Nicolaum Bassarinum, anno sal. 1546 [91].

---

[88] Sect. 1: 1558, August, for a detailed table of contents of this edition. Incidentally, the Baron
della Foresta in listing this and the succeeding published works does not give the dates and places
of publication. Further, he leaves out the last two items, i.e., the Rhythmi quidam vulgares and
the De vita Christi but adds the following: « Insulae Siciliae Topographia cum eius inscriptione.
De Sphaera liber unus. De Lineis Horariis lib. tres acutissimi. Computus Ecclesiasticus strictim
collectus. Tractatus Instrumentorum Astronomicorum, Musicae traditiones, Euclidis Propositiones elementorum tredecimi, solidorum tertii. Regularium corporum primi. Arithmeticorum lib. duo subti-
lissimi. Photismi de umbra ». All of these additional items but the first and the last were published
in the combined Opuscula mathematica and Arithmetica of Venice, 1575. His list of the contents
of this volume is quite inaccurate. The last item is of course a reference to the 1611 edition of the
Photismi and Diaphana. The first item refers to a desciption and map of Sicily. The Baron della Foresta made further reference to the latter (Vita, p. 7): « come altresi a Giacomo Castaldo Piemontese Cosmografo il desegno di tutta l'Isola di Sicilia che stampossi poscia in Roma più d'una volta ».
For this map, see Sect. 1: 1545. For the description of Sicily, see Ibid.: 1546.
[89] Sect. 1: 1535, 21 October; 1540, 24 January; 1542, 5 December; and 1543.
[90] This note, missing in the 1575 edition of the Index lucubrationum, decisively answers the
question raised by Rosen as to whether Maurolico knew of the second edition of the Cosmographia
(see Rosen, « The Editions », p. 72, n. 34).
[91] See note 81.

Item

Grammatica quaedam rudimenta, Messanae per eundem Georgii Spirae filium et nepotes, anno sal. 1528 [92].

160   Rhythmi quoque vernaculo sermone, ibidem per eosdem anno sal. 1552 [93].

[*Inf. mg.*: 20. apr. 1568.]

3v   / Item

Martyrologium correctum et instauratum Reverendissimo domino M.
165   Antonio Amulio Card. dedicatum. Cum Topographia et multis appendicibus, anno praeterito 1567 in mense Septembri, Venetiis apud Iunctas impressum fuit, et iterum in forma parva mense Julio 1568 [94]. [Et tertio ibidem 1570 et Neapoli 1572.] [*mg.*: 20. apr. 1568.]

Item

170   Historiae Sicanicae compendium. Cum epistola simul ad patres Tridentinae synodi, impressum fuit Messanae per eundem Georgii Spirae filium et nepotes, anno sal. 1562 [95]. [*mg.*: 22. apr.]

Item

Rhythmi quidam vulgares de passione et resurrectione domini, Mes-
175   sanae impressi per eundem Petrum Spiram, anno salutis 1552[96].

Item

De vita Christi, eiusque matris, et gestis apostolorum senariis rhythmis vulgaribus libelli 8, Venetiis per Augustinum Bindonem 1556 impressi [97].

[2]

4r   / Ordo congruus compendiorum

Elementorum Euclidis cum demonstrationibus novis et convenientibus ac brevibus in 5um et sequentes libros, exclusis superfluis. In 13um, 14um,

---

[92] See note 80.
[93] See note 73.
[94] See note 69.
[95] See note 68.
[96] See note 93.
[97] Guardione, « Francesco Maurolico », pp. 31-33, gives verses from the *De gestis discipulorum Dei*, using the copy at the Biblioteca Comunale in Palermo, which I have also read. Note that this copy was published in Venice in 1555. Maurolico here reports the date of publication as 1556. Cf. Sect. 1: 1540 and 1555.

5 15um de solidis regularibus additis multis necessariis conclusionibus. [De compaginatione ipsorum; de impletione loci contra Averroem.]

Sphaericorum Theodosii, Menelai, Maurolyci. Autolyci de sphaera quae movetur. Phaenomnon (!) Euclidis. Theodosii de habitationibus. Regulae in tabulas sinus recti, faecundam, beneficam.

10 Conicorum Apollonii in 3. lib. cum directis et magis necessariis conclusionibus. Item by (del. Maurol) maxime spectantibus ad demonstrationem speculi comburentis et ad quadraturam parabolae [et areas ellipsium, quae maxime intersunt Conicorum ad lineas horarias. Item cylindricorum Serenii, superfluitatibus omissis].

15 Operum Archimedis: de dimensione circuli, de sphaera et cylindro, de spiralibus, de sphaeroidibus et conoidibus, de momentis aequalibus [cum inventione centri gravitatis in solidis ab Archimede omissa], de quadratura parabolae, de isoperimetris, [de inventione duarum mediarum proportionalium linearum ex Herone, Menecmo, Archyta, Pappo et aliis].

20 Arithmeticorum Jordani, datorum, Maurolyci de formis numerorum, de terminis magnitudinum quae in 10° elementorum, et de eorum regulis et inventione et calculo.

Musicae ex Boetio, ex Graecis authoribus, ex Fabro. His additur compendium nostrum theoriam vocum et consonantiarum [modorum modulationum] omnem paucis comprehendens.

4^v 25 / Perspectivae totius, opticorum et catoptricorum Euclidis, Speculorum Ptolemaei, Joannis Petsan, Rogerii Bacchonis, Photismorum et Diaphanorum nostrorum cum speculatione de conspiciliis, deque iride.

Sphaerae, cum adnotationibus in Jo. Sacroboscum, In Theoricas planetarum, In motum 8^{ae} sphaerae, Sphaera Procli.

30 Magnae Constructionis Ptolemaei, quae Idea Ptolemaica appellari potest. Una cum traditionibus Alfraganii, Albategnii, Tebitii, Alfonsi, Jo. de monte regio, Georgii Peurbachii et aliorum.

Geographiae Ptolemaei, Strabonis ac recentiorum. Cum regulis inveniendi latitudines ac longitudines locorum, et inde distantias sive in globo 35 sive in plano, sive per calculum. De terrae ambitu inveniendo regulae.

Mechanicorum ex Vitruvio. Ex Herone de bellicis et hydraulicis machinis. Additur compendium problematum Aristotelis. Cum appendice super magnete et pyxide nautica, super iride [problemata notatu digna].

40 Instrumentorum ad geometriam et astronomiam spectantium, de quadrato, quadrante, Astrolabio, ex traditione Ptolemaei, Nicephori, Procli, 5^r Jordani et Stoeflorini. Item de instrumento armillarum, de / sphaera solida ex magna constructione Ptolemaei.

De lineis horariis regulae et theoria, in compendium tam a meridie quam ab occasu et ortu incipientium.

45 Calculi et computi in kalendarium et canonum in tabulas Alfonsi, Blanchini, Georgii eclipsium, Jo. de monte regio, diarii perpetui et alias.

Compendium Judiciariae ex priscis et recentioribus authoribus carptim ac summatim decerptum, quo ad observationes agricularum, medicorum et nautarum ac viatorum, et eorum quae citra superstitionem conceduntur.

50   Quaestiones arithmeticae, geometricae, astronomicae, cum regulis et exemplis.

[3]

5v   / Ordo servandus in legendis operibus

Euclidis Plana, in 4^{or} primis libris elementorum.

Eiusdem proportiones, in 5° et 6°.

Eiusdem Arithmetica, in tribus sequentibus libris.

5   Eiusdem Symmetria, in 10° tradita et per numeros in arithmetica.

Eiusdem Solida communia, in 11° et 12°.

Eiusdem Solida regularia, in tribus ultimis libris, cum additionibus.

Theodosii sphaerica elementa, 3. lib.

Menelai Sphaerica, totidem libris.

10   Sphaerica Maurolyci, cum tabellis sinus recti, faecunda, benefica.

Jordani et nostra arithmetica, de figuris planis et solidis et aliis ab Euclide omissis.

Archimedis de circuli dimensione.

Eiusdem Isoperimetra.

15   Eiusdem de Sphaera et Cylindro.

Apollonii Pergaei Conica lib. 8. et eius compendium.

Sereni Cylindrica, et eius brevissima demonstratio.

Archimedis Aequalia momenta, cum additione de centro solidorum.

Eiusdem Quadratura parabolae.

20   Eiusdem de spiralibus lineis.

Eiusdem Conoides et sphaeroides figurae.

Eiusdem sive Ptolemaei de speculo comburente.

Boetii Arithmetica, et Musica, cum compendio Jacobi Fabri et tractatu nostro brevissimo.

25   Joannis Petsan et Rogerii Bacchonis Perspectiva.

Euclidis Optica, Catroptica (!). Item Photismi nostri et Diaphanorum ac de iride tractatus.

Sphaera Procli, Autolyci, Sacrobosci, Campani et nostra. Iginii et Arati Asterismi.

30 Euclidis Phaenomena.

Georgii Peurbachii Theoricae planetarum.

Tabularum Alfonsi, Blanchini, Jo. Regiomontii, Peurbachii Canones cum appendicibus. Computus.

Ptolemaei Magna Constructio, cum compendio Jo. de monte regio. Et
35 Idea nostra brevissima.

Eiusdem Geographia cum Pappi figurationibus. Et cum navigationum recentium traditionibus.

Instrumenta geometrica et astronomica, ex Ptolemaeo, Mescealla, Proclo, Niceforo, Stoflerino, et ceteris.

40 Alfragani, Albategnii, Tebitii, Gebri traditiones.

Pappi, Heronis, M. Vitruvii mechanica et machinae.

Arithmetica, geometrica, astronomica problemata.

De iudiciis astrologicis, de temporum signis, ex Ptolemaeo, Alcabitio, Hermete, Dorotheo, aliisque latinis, graecis, ac peregrinis authoribus, Isa-
45 gogicae regulae.

17. sep. 1570.

[4]

[Sequuntur metrica quaedam opuscula in tres (!) tomos.]

6r / In primo tomo

Disticha Catonis.

Aurea Carmina Pythagorae.

5 Phocylidis poëma admonitorium. Monosticha Menandri.

Dicta Sapientium carptim Collecta.

Mimi publiani.

Isocratis ad Demonicum Paraenesis.

Argumenta Iliadis.

10 Argumenta Odysseae.

Philosophicae ac theologicae sententiae versu heroico.

In 2º Tomo

Elegum Carmen pro secessu Maronis montis.

Ad Simeonem Ventimillium Strategum epistola heroica.

15 Petri Lunae et Elisabetae Vegae nuptiae.

Simeonis Ventimillii et Mariae Epithalamium.

De Contemptu mundi Elegia.

Joannis Ventimillii Epicedium.

In fratris Jacobi obitum Egloga.

20   Pro fratrum ac nepotum obitu Sylva.

Ad Jeronimum Balsamum epistola.

In lectione Sphaerae Carmen hexametrum.

Ad Alfonsum Ruisium hexasticha 25 de diis gentilium.

Disticha, Epigrammata et Inscriptiones in ingressu Caroli V imperatoris.

25   In ingressu Mercurii Card. Carmina.

In coronatione Philippi, vivente patre praedicto Carolo, Epigrammata quaedam.

In adventu praesidum Carmina.

In adventu ductorum regiae classis et ducum Carmina.

30   Epigrammata et superscriptiones portarum, pontium, propugnaculorum et fontium.

Ad principes, primates et patritios epigrammata.

Epitaphia principum, reginarum, praesidum et patritiorum.

6v   / In 3° tomo

35   Hymni elegiaci, sapphici, pindarici, iambici, nocturni, matutini, vespertini, et per horas diurnas.

Hymni pro diversis festis Salvatoris, beatae Virginis, apostolorum, martyrum, confessorum, virginum, et per horas officii beatae Virginis.

        Principia, appendices et conclusiones hymnorum, pro diversis tempo-
40   ribus et pro horis canonicis.

Disticha et Epigrammata pro diversis festis per anni circulum.

Benedictiones metricae super lectiones nocturnas.

In 4° tomo

        Ordo Canonicarum horarum bipartitus in ferialem et festivum, in quo
45   satisfit tempori, per sua invitatoria, hymnos, antiphonas, versiculos. Collectas, atque etiam officio beatae Virginis, et memoriae festorum simplicium quotidie occurrentium et sanctorum omnium per singulas horas. Tres lectiones sumi possunt, prima ex veteri testamento, secunda ex novo. Ter-

50   tia de homelia, sermone, vel historia temporis instantis vel festi, seu ex
Martyrologio. Post quas, in festis recitari potest. Te deum laudamus. In
feriis autem unus, duo, tres, pluresve psalmi comissis his, quiqui quotidie
reputuntur per horas. Ita ut in spacio saltem trimestri percurratur, quidquid restat de psalteria una cum sex canticis. Nam canticum Benedicite,
in officio festivo ponitur ad laudes.

20 Apr. 1569.

Variant Readings to the *Index lucubrationum*

1   Index... Maurolyci *om. Ed1 et scr.* Aliena / Maurolyci *om. Ed2*

2   *ante* Euclidis *add. Vi* In primis aliena quaedam emendatiora per ipsius
operam et laborem facta / Euclydis *Mv hic et ubique*

2-3   tam... ignorantis *om. Ed1*

4   solida : quinque solida *Vi*

6   astronomiae *Ed2*

7-8   ad... triangulorum *om. Ed1*

8   sphaeralium triangulorum : triangulorum sphaeralium opportuna *Vi* / triangulorum : trium angulorum *Mv* triangulorum pertinentia *Ed2*

9   elementa *om. Ed1*

10   oportunis *om. Ed1* opportunis *ViEd2*

11   celindrica *Mv*

12   Archymedis *Mv hic et ubique*

14-15   Cum... facilitatis : In quibus multae demonstrationes additae et facilitas
Ed1

15   et... facilitatis : facilius demonstrata *Ed2* et artificio facilitata *Vi*

16   elementa arithmetica *Vi* / Et... Data[2] *om. Ed1* / Theonis Data geometrica *Ed2*

17   Bacconis *MvEd2* / *post* Perspectivae *add. Ed2* breviatae *et Vi* singulae
libris tribus / annotationibus *Mv*

18   errorum : quibusdam *Ed1*

19   Specula... comburente : De speculis Comburentibus *Ed1* / comburente :
ustorio *Ed2*

20   De : libellus de *Vi*

23   Aristotelis *tr. Ed1 post* mechanica / et iis : notatu dignis et quaedam *Vi*

23-24   additionibus... spectant : oportunis et notatu dignis additionibus *Ed1*

25 Propria *MpMv* Nostra vera sunt *Ed1* Propria ipsius authoris *Ed2* Authoris proprii labores *Vi*

26 quiidam *Mv*

27 Mathematicae *Ed1*

29 De centris *om. Ed2*

30 prospectiva *MvVi*

31 Arithmetrica *Mv* / libri duo *Vi*

32 tam... solidis *om. Ed1*

34-35 cum... conclusionibus *om. Ed1*

35 conclusionibus: demonstrationibus *Ed2*

36 quattuor *Ed2 hic et ubique* / demonstrata *om. Ed1*

37 *post* rei *add. Vi* atque cosae / sive capita *om. Ed1* / sive: vel *Ed2*

38 *post* redactae *add. Vi* quae algebra vocantur

39 *post* neglecta *add. Ed2* vel omissa

40 pro: super *Vi* / pro... scientia *om. Ed2* / triangulorum scientia *tr. Ed2*

41 in... motus: octo capitibus *Ed1* / circulis... motus: circuli primi motibus *Vi*

42 calorum *Ed2* / elementorum et coelorum *Ed1*

43 Bembro *Mv* / dedicata *Ed1*

44 elementorum: Apollonii *Ed1*

44-45 post... locandi *om. Ed1*

47 tam... solidae *om. Ed1*

47-48 Averroes... indicatur: Averrois ignorantia manifestissime constat *Ed1*

48 indicatur: demonstratur *Mv*

49 libri: libelli *Ed1* / quattuor *Ed2*

50 Archymede *Mv* / Et... paraboles *om. Ed1* / parabolici *Ed2*

52 Astrolabii *Mv*

54 libelli *Ed1*

56 coeptae *Vi*

57 et: vel *Ed2* / exorsae: exensae *Ed2*

59 hyperboles *Vi*

60-61 ad... facientes *om. Ed1*

61 facientis *Mv* / ad... calorem *Mp om. MvEd1Ed2* illuminationem et calorem *Vi* et *tr. ante* facientes

63-64 autem... formis: de oculi structura et conspiciliorum *Ed1*

65 libelli [1, 2] *Ed2*

66 Astronomicorum problematum: astronomicatum (*!*) libri *Ed1* / tres: libri tres *Vi* / In... traduntur *om. Ed2* / et exempla: cum exemplis *Ed2*

68 Mathematicae: geometriae *Vi*

68-70 in... mobilis: Et in diversas tabulas canones *Ed1*

69 Alfonsii *Mv* / Blankini *Vi*

71 brevissimum: ex praecipuis authoribus *Ed1, et hic desinit Ed1*

72 Euclydis Epythomae *Mv*

72-73 praesertim circa proportiones: in quintium (*!*), in arithmetica, in decimum, et in solidorum libros *Ed2*

73 et diffinitionibus *MpVi om. MvEd2*

74 libri tres *Mv*

75 quo... necessarias *MpVi om. MvEd2*

76 sive: sive circuli *Ed2*

79-81 Compendium... Regimontii *om. Mv*

79 Ptolomaicae *Ed2*

80 *ante* seriem *scr. Vi* et calculi

82 *post* Musicae *del. Mp* Ptolemaei / Musicae Boetii: Boetianae Musicae *Ed2*

82-83 quibusdam... facientibus: optimis speculationibus et calculo ac modulatuum (*!*) ratione, et systematum proportione *Ed2*

84 complectens: comprehendens, cum motuum secundorum Theoria *Ed2*

85 brevis et exactus *om. Mv*

86 Jo. Sacroboscum *Mp* Joannem Sacrobustum *Mv* sphaeram Io. Sacrobusti *Ed2* in sphaeram Sacrobosci *Vi* / Theorias *Mp* Theoriam *Mv* Theoricas *ViEd2*

87 Astrolabii *Mv*

90 orizonte *Mv* horologio *Vi*

91 et... horologio *Mp om. MvEd2* in quocumque horizonte *Vi* / *post lineam 91 add. Vi lineas 116-125*

92 Historiae... sicanicarum: compendium Sicanicae historiae *Ed2* compendium rerum sicanicarum in sex libris *Vi; et ante* Compendium *habet Vi rubricum* Sequuntur tractatus quidam circa historias

93 correctum *Vi* / correctum et instauratum *om. Mv*

94 urbium *om. Ed2*

95-97 Hymnorum... duo *om. Vi*

95-96 Poemata... latina *om. Mv* / Pythagoriae latina: Pythagorae moralia Latino moetro *Ed2 et tr. Ed2 titulum post* duo *in lin. seq.*

98 Boccacii *Ed2* / Cum multis: Item *Vi*

99 collectae *Vi* / genealogiis: prosapiis *Ed2* genealogiae *Vi*

100 necessariae *Vi*

101 Rhythmi *MvEd2* / Rhytmi... crucis *om. Vi*

102 orbe condito *Mp* urbe condita *Mv* Adamo protoplasto, Christi *Ed2*

104 hystoriis *Mv*

105 aethaneo *Vi*

106 Tridentini *Ed2*

107 *pro* Breviarium *hic et in sequentibus lineis scr. Ed2* Brevaria *ut singulum rubricum et Vi* Breviaria duodecim scilicet / Plautinae *Mv*

109 *ante* Laertii *scr. Ed2* Decem librorum

111 *ante* Polydori *scr. Ed2* Octo librorum / Polidori *Mv*

112 conciliorum: consiliorum Synodalium *Ed2* / de here *Mp om. MvEd2* De haeresibus *Vi; et tr. Vi* Conciliorum, De haeresibus *post* Siculi *in lin. 114*

113 L. flori... neopolitani *MpVi om. MvEd2* / L.: Lucii *Vi* / Neapolitani *Vi*

114 fabularum Diodori: Sex librorum Diodori *Ed2* / Diodorii *Mv*

116 cum demonstratione *MpVi om. MvEd2*

117 scilicet *om. Ed2*

117-118 beneficae et faecundae *Ed2*

118 sphaeralium triangulorum *tr. Ed2* / utiles *Ed2*

119 quae... est *MpVi om. MvEd2*

120 quibusquam *Vi*

121 *post* naturis *add. Vi* et proprietatibus / domiciliorum, planetarum: domorum 12 septemque planetarum *Ed2*

122 de regulis: Regulae *Ed2 et tr. post* quaestionum / perfectionum *Vi* / revolutionum, nativitatum: honoscoporium (!) *Ed2* / nativitatum: et quaestionum nativitatum *Mv*

123 nautas, medicos *tr. Ed2Vi*

XIII

196

124   spectantibus... agitur: et exclusis superstitionibus directae *Ed2*

124-125   Item... concessis *MpVi om. MvEd2*

125   *post* concessis *add. Vi* Item de meteorologicis signis temporum ex variis authoribus per Augustinum Sessam

126-143   De... Mess. *addidi ex Vita, pp. 40-41*

144   Notandum... operibus: Ex supra scriptis quae fuerint impressa *Vi*

145   Autolici *Ed2* / Sphaera: sphaerica *Mv*

147   foecundae *Ed2*

148   unum volumen *Ed2*

149   Spira: Spina *Ed2* / filio... germani *om. Mv* / Spirae: Spinae *Ed2*

150   Item *om. Vi* hic et ubique

151   *post* dicata *add. Ed2* 3. lib. *et Vi* cum distantiis coelorum et magnitudinibus stellarum

152   Oporinum: Oponimum *Ed2*

152-153   Item... 1558 *MpVi om. MvEd2*

153   Guillelmum *Vi*

155   Jo.: d. Jo. *Ed2Vi* / Impressa fuit *om. Ed2*

156   Bessaninum *Ed2* / sal. *om. Vi*

158   Grammatica... rudimenta: Grammaticae institutiones impressae sunt *Vi* / quaedam *om. Mv* / eundem: dictum *Vi* / Spinae *Ed2*

159   et nepotes *om. Ed2*

160   Rhythmi... 1552 *om. Vi* / quoque *om. Mv* / vernaculo sermone: materni de laude S. C. *Ed2*

162   20 Apr. 1568 *Mp om. MvEd2Vi*

164   conrectum *Vi* / Reverendissimo domino *om. Vi*

165   Card.: card. Veneto *Vi* / dicatum *Ed2Vi* / opographia *Ed2* / et: cum *Ed2* / multis *om. Vi*

166   anno... Septembri: mense septembri 1567 *Vi et tr. post* Iunctas / praeterito: salutis *Ed2* / 1567 *om. Ed2* / in *om. Ed2* / Septembris *Ed2*

167   fuit *om. Ed2* / *post* fuit *add. Vi* in 4° folio / parva: minima ibidem *Vi* / mense Julio: apud eosdem *Mv*

167-168   Et... 1568 *Mp om. MvEd2* Et tertio ibidem similiter, anno Domini 1570. Item quarto impressum fuit in 8° folio Neapoli, per Horatium Salvianum 1572 *Vi*

170   compendium *tr. Vi ante* Historiae

171   impressum... Messanae: Messanae impressum *Ed2* / Spinae *Ed2*

171-172   Georgii... nepotes: Petrum Spiram *Mv*

172   sal.: Domini *Vi* / 22 apr. *Mp om. MvEd2Vi*

173-178   Item... impressi *om. Mv*

173-175   Item... 1552 *om. Ed2*

175 salutis *om. Vi*

178 libellis *Vi* / libelli 8 *tr. Ed2 ante* senariis / Bindonum *Ed2* / impressi
     om. Ed2 et tr. Vi ante per / anno Domini 1556 *Vi*

Variant Readings for [2]

1   Ordo... compendiorum: Decrevi tandem, omissis perplexitatibus, omnia
     in compendia redigere, hoc scilicet ordine *Vi*

2   *ante* Elementorum *scr. Vi* Primum / novis *om. Vi*

3   ac brevibus *om. Vi* / exclusis: amputatis *Vi*

6   *ante* Sphaericorum *scr. Vi* Secundum

7   Phaenomenon *Vi*

9   *ante* Conicorum *scr. Vi* Tertium / magis: praecipuis magisque *Vi*

12-13 quae... omissis: unde sumitur theoria pro lineis horariis. Cylindricorum
       Sereni, omissis superfluitatibus *Vi*

14   *ante* Operum *scr. Vi* Quartum

15   *post* aequalibus *add. Vi* libellus

16   inventione... omissa: centris solidorum *Vi*

17   de inventione: Inventio *Vi*

17-18 mediarum... linearum: linearum proportionalium *Vi*

18   ex... aliis: per Heronem, Menecmum, Archytam et caeteros *Vi*

19   *ante* Arithmeticorum *scr. Vi* Quintum / Maurolyci *tr. Vi post* numerorum

20-21 de... inventione: cum regulis *Vi*

22   *ante* Musicae *scr. Vi* Sextum / ex ²·³ *om. Vi* / His additur *om. Vi*

23   *post* nostrum *add. Vi* circa / et *om. Vi*

23-24 modulationum... comprehendens *om. Vi*

25   *ante* Perspectivae *scr. Vi* Septimum / totius *om. Vi* / et *om. Vi*

26-27 Photismorum... iride: Photismi, Diaphana *Vi*

28   *ante* Sphaerae *scr. Vi* Octavum

29   Sphaera Procli *om. Vi*

30   *ante* Magnae *scr. Vi* Nonum / quae... potest *om. Vi*

31   Una... traditionibus: Traditionum *Vi*

31-32 Alfonsi... aliorum: Alphonsi, Regiomontani, Peurbachii *Vi*

33   *ante* Geographiae *scr. Vi* Decimum

33-35 Cum... regulae: Inventio latitudinum et longitudinum per eclipsim; di-
       stantiarum in globo, in plano per calculum. Modus inveniendi am-
       bitum terrae *Vi*

36   *ante* Mechanicorum *scr. Vi* Undecimum / Ex Herone: Herone, Vegetio *Vi*

37   Additur compendium *om. Vi* / problematum: Problematum mechanico-
      rum *Vi* / Cum appendice super: de *Vi*

38 et *om. Vi* / super *om. Vi* / problemata... digna *om. Vi*

39 *ante* Instrumentorum *scr. Vi* Duodecimum / *post* spectantium *add. Vi* hoc est

40 Astrolabio: astrolabo armillari *Vi*

40-42 traditione... Ptolemaei: Ptolemaeo, Nicephoro, Proclo, Jordano, Stoeflerino, De sphaera solida *Vi*

43-44 regulae... incipientium: a meridie et occasu in omni horizonte, seu verticali seu meridiano, describendis *Vi*

45 *ante* Calculi *scr. Vi* Decimumtertium / et [2] *om. Vi* / Alphonsi *Vi*

46 Georgii... alias: Peurbachii, Regimontii *Vi*

47 *ante* Compendium *scr. Vi* Decimumquartum

47-49 Judiciariae... conceduntur: denique judiciariae, et superius expositum est *Vi*

50 *Add. Vi* rubricum Decimumquintum

50-51 Quaestiones... exemplis: Quaestionum arithmeticarum, geometricarum, astronomicarum ut supra *Vi*

Variant Readings for [4]

1 Sequuntur... tomos *Vi om. Mp* / tres *in Vi corrigend. ad* quatuor *pro contextu Mp*

2 Primus tomus *Vi*

6 sapientum *Vi*

7 Mimi: Munii *Vi*

12 Secundus tomus *Vi*

13 Elegiacum *Vi*

15 *ante* Petri *scr. Vi* Epithalamion in / nuptias *Vi*

16 Ventimilliorum *Vi et tr. post* Mariae / epithalamion *Vi*

18 *post* Ventimillii *add. Vi* in fluvio Letoanni demersi / epicedion *Vi*

19 Jacobi fratris obitu ecloga *Vi*

21 Hieronymum *Vi*

23 Alphonsum *Vi* / hexastica *Vi*

24 et *om. Vi*

25 *ante* Mercurii *scr. Vi* Jo. Andreae / Carmina: messanensis epigrammata *Vi*

26-27 Epigrammata quaedam *om. Vi*

34 Tertius tomus *Vi*

43-55 In... 1569 *om. Vi*

INDEX

(The bibliographical preface is not included in this index)

Bassolis,Joannes de: III 137
al-Battānī: II 282;XIII 168,
176 n.,188,190
Baudoux,Mlle: VIII 27 n.26
Baur,L.: I 29;XII 254 n.11
Bellantius: XIII 174 n.
Bellifemine,G.: XIII 171,175 n.
6
Bembo,Pietro: XIII 153,155,172,
173 n.,181,184,186
Benedetti,Giovanni Battista:
I 43-44;XII 250,253,257 n.31
Bessarion,Cardinal: XII 252
Besthorn,R.O.: VIII 18 n.7,
19 n.13,20 n.13,24 n.20
Bianchini (or Blanchini),
Giovanni: XIII 150 n.,151 n.1,
155,168,174 n.,183,188,190
Billingsley,Henry: XII 255 n.
11,258 n.31
Birkenmajer,A.: VIII 26
al-Bīrūnī: IX 79;XII 243
Björnbo,A.A.: VIII 27
Blasius of Parma: I 38;II 284;
IV 23 n.44;VII 215-16,226
Blume,F.:IX 79 n.5
Boccaccio,Giovanni: XIII 174 n.,
184
Boethius: I 30-31;VIII 17;XII
243;XIII 151 n.1,159,162,
171 n.,172 n.2,175 n.9,176 n.,
180 n.31,182 n.34,183,188-89
Bonatus: XIII 174 n.
Boncompagni,B.: VIII 29 n.28;
IX 80 n.11;XIII 167,175 n.7
Bonucci,A.: XII 258 n.32
Bordone,Benedetto: XIII 150 n.
Borgia,Francesco: XIII 167
Boscovich,R. J.: V 215,221
Bradwardine,Thomas: II 283-85;
III 138,140 n.24,145 n.34,
156-57,158 n.51;IV 22 n.41,
24 n.46;V 215;VI 3-5;XI 40;
XII 244
Pseudo-Bradwardine version of
De Mensura Circuli: XI 51-54,
59;XII 245
Brahmagupta: IX 79
Branca,V.: III 139 n.20
Bredon,Simon: II 282
Brewer,J. S.: I 38 n.71
Bricot,Thomas: I 41-42
Brown,J. E.:VII 233 n.2;
X 409 ns.1-2,411 n.5
Bruni,Leonardo: III 139 n.20;
XIII 185
Bruno,Giordano: I 43;II 277
Bryson: XI 55

Brytte,Walter: II 282
Bubnov,N.: VIII 16,17 n.1
Buridan,Jean: I 38-44;II 278-
79,302;III 137 n.16;IV 4;
V 216,217 n.;VII 228
Burley,Walter: III 135
Busard,H. L. L.: II 292 n.30,
293 ns.32-33,294 n.,295 n.
35,299 n.40,300 n.43;IV 2 n.
Buttimer,C. H.: I 29

Caldo,Matteo: XIII 155,163
Calepino,Ambrogio: XIII 150 n.
Callisthenes: XIII 174 n.
Campanus of Novara: II 282;
IV 6 n.19;VIII 19,21,28-30;
IX 80,83-84;XI 56-57;XII
258 n.25;XIII 150 n.,153,
174 n.,175 n.9,176 n.,178,
189
Cantor,G.: VIII 17 n.1
Cantor,M.: IX 79 n.5
Cardano,Girolamo: III 139;
XIII 174
Carton,R.: I 36
Casali,Giovanni: III 140 n.
24;IV 4-5,8 n.21,12 n.,
13 n.26,23 n.43;VI 4;
VII 232
Cassiodorus: I 31;VIII 17;
XII 243
Castaldo,Giacomo: XIII 186,
and see Gastaldo
Cato: XIII 190
Cavalieri,Bonaventura: XII 253
Celaya,Jean de: II 302
Censorinus: VIII 17 n.1
Cerda,Juan de la: XIII 165
Charles I of Anjou: XII 248
Charles V: XIII 163,191
Cicero: XIII 155,185
Clagett,M.: I 29;II 276 n.3,
278 ns.4-5,283 n.20,284-85,
286 n.,287 n.,289 n.26,
292 ns.29-30,295 n.36,
298 n.,300 n.42,301 n.,
302 ns.46-50; III 137 n.16,
138 ns.17-19,157 n.49;IV
1 ns.1-2,2 n.,3 n.9,4 n.10,
7 n.19,9 n.22,11 n.,12 n.,
13 ns.26-27,15 n.29,16 n.,
17 n.,19 ns.34-35,21 ns.
37-38,22 ns.39-41 and 43,
23 n.44,24 n.47;V 216 n.3,
218 n.5;VI 4,5 n.,6 n.8;
VIII 233 ns.2-6 and 9 and
11 and 14-16,234 ns.19-21
and 23-25 and 27-29;VIII